HEALTHCARE TECHNOLOGIES SERIES 37

Technology-Enabled Motion Sensing and Activity Tracking for Rehabilitation

IET Book Series on e-Health Technologies

Book Series Editor: Professor Joel J.P.C. Rodrigues, College of Computer Science and Technology, China University of Petroleum (East China), Qingdao, China; Senac Faculty of Ceará, Fortaleza-CE, Brazil and Instituto de Telecomunicações, Portugal

Book Series Advisor: Professor Pranjal Chandra, School of Biochemical Engineering, Indian Institute of Technology (BHU), Varanasi, India

While the demographic shifts in populations display significant socio-economic challenges, they trigger opportunities for innovations in e-Health, m-Health, precision and personalized medicine, robotics, sensing, the Internet of things, cloud computing, big data, software defined networks, and network function virtualization. Their integration is however associated with many technological, ethical, legal, social, and security issues. This book series aims to disseminate recent advances for e-health technologies to improve healthcare and people's wellbeing.

Could you be our next author?

Topics considered include intelligent e-Health systems, electronic health records, ICT-enabled personal health systems, mobile and cloud computing for e-Health, health monitoring, precision and personalized health, robotics for e-Health, security and privacy in e-Health, ambient assisted living, telemedicine, big data and IoT for e-Health, and more.

Proposals for coherently integrated international multi-authored edited or co-authored handbooks and research monographs will be considered for this book series. Each proposal will be reviewed by the book Series Editor with additional external reviews from independent reviewers.

To download our proposal form or find out more information about publishing with us, please visit https://www.theiet.org/publishing/publishing-with-iet-books/.

Please email your completed book proposal for the IET Book Series on e-Health Technologies to: Amber Thomas at athomas@theiet.org or author_support@theiet.org.

IET The Institution of Engineering and Technology

Technology-Enabled Motion Sensing and Activity Tracking for Rehabilitation

Wenbing Zhao

The Institution of Engineering and Technology

Published by The Institution of Engineering and Technology, London, United Kingdom

The Institution of Engineering and Technology is registered as a Charity in England & Wales (no. 211014) and Scotland (no. SC038698).

The Institution of Engineering and Technology
Futures Place
Kings Way, Stevenage
Herts, SG1 2UA, United Kingdom

www.theiet.org

British Library Cataloguing in Publication Data
A catalogue record for this product is available from the British Library

ISBN 978-1-83953-410-2 (hardback)
ISBN 978-1-83953-411-9 (PDF)

Typeset in India by MPS Limited
Printed in the UK by CPI Group (UK) Ltd, Croydon

Cover Image: Andersen Ross Photography Inc / DigitalVision via Getty Images

Contents

About the author

Wenbing Zhao is a professor at the Department of Electrical Engineering and Computer Science, Cleveland State University, USA. He has over 200 peer-reviewed publications and has delivered 10 keynotes at international conferences. He is an associate editor for IEEE Access, MDPI Computers, and PeerJ Computer Science, and a member of the editorial board of several international journals, including *Applied System Innovation*, *Internal Journal of Parallel, Emergent and Distributed Systems*, and *International Journal of Distributed Systems and Technologies*. He is an IEEE senior member.

List of figures

List of tables

Introduction

The aim of this book is to document the state-of-the-art technology that has been developed to facilitate rehabilitation both for physical and for mental health. As the technology advances, we anticipate that technology will play an even greater role in rehabilitation. The scope of the book is limited to rehabilitation via non-intrusive physical activities. The topics on the use of exoskeleton and more intrusive methods for rehabilitation are intentionally excluded from the book, not because they are not important, but to make the book more coherent. Those topics deserve books on their own. The book includes three parts.

Part I introduces various motion sensing modalities, including inertial measurement units, pressure sensing, eTextile, muscle activity sensing, vision-based motion sensing, and instrumented gloves. Part II presents common human motion recognition algorithms, ranging from simple single-parameter determination, such as the determination of range of motion in terms of angles, to sophisticated rule-based and machine-learning-based activity recognition algorithms. Part II also includes a chapter on exergames (also referred to as serious games). Such interactive games are built on top of the motion recognition technology and are integrated with clinical considerations in game design for more effective rehabilitation. The first two parts lay the technology foundation for adopting and understanding technology in rehabilitation. Part III introduces various types of rehabilitation from the clinical perspective, and discuss the technology use for each type of rehabilitation. Clinical professionals could gain understanding on the various technologies and algorithms available to facilitate rehabilitation, reduce cost, and improve the effectiveness of rehabilitation. Technology developers would find Part III useful in learning the clinical theories, frameworks, and best practices on various types of rehabilitation, and may find new opportunities to improve the effectiveness of rehabilitation via technology.

Part I includes six chapters. Chapter 1 introduces sensors that measure the inertial movements, namely the accelerometer and gyroscope. Sometimes, magnetometer is also used to improve the measurement accuracy. These sensors are referred to as inertial measurement units (IMUs). With the technology advancement, IMUs are becoming much smaller and they are packaged with microprocessor, local storage, and wireless transmission hardware. Hence, IMUs nowadays are also called wearable sensors because they can be conveniently placed on desirable locations of the patient. Because the sensors have been incorporated in smartphone, smartwatches, wrist-band devices, and even game controllers such as Wiimote, such devices have also been used for rehabilitation study. This chapter additionally summarizes IMU devices (and vendors when applicable) that have been used in rehabilitation studies.

Chapter 2 introduces force and pressure sensing technology. Pressure sensing is usually used in gait-related studies. Common force and pressure sensing devices include force resistive sensors, force-plates, and smart insoles. Some studies also used the Wii balance board. Force and pressure sensing complements the measurement from IMUs.

Chapter 3 presents electronic textile (eTextile), which is a new wearable technology that has great potential in rehabilitation. Studies have shown that eTextile can be made into many forms of garment, such as shirts or gloves. The eTextile has built-in electronics to measure movements. Some eTextile could even measure pressure and temperature. The eTextile may have several major advantages over IMUs. It could be more convenient to use and be less error-prone to wear.

Chapter 4 is about muscle activity sensing. Motion sensing can be augmented by measuring the muscular contraction and relaxation using various myography techniques. An important benefit for measuring muscular movements is that one can infer the intention of a patient, i.e., what activity the patient is attempting to do with the information provided by the muscles that are not damaged, even if the patient could not actually perform the intended activity due to the presence of paralyzed muscles (e.g., due to stroke). Muscular contraction can be measured in several ways and this has led to several types of myography techniques being developed, including electromyography, mechanomyography, force myography, and optical myography.

Chapter 5 introduces vision-based motion sensing. Camera-based systems have long been used to measure human motion with computer vision algorithms. For precise measurement of 3D human motion, multiple cameras would have to be used to capture the images of the human subject from different angle to build a 3D model of the person. The traditional method is to have the human subject wear many special markers, while newer systems are becoming markerless. However, such systems are very expensive and require periodical maintenance, and hence, are limited to the lab setting. The recent development of depth sensing with the companion real-time skeletonization software development kit (such as Microsoft Kinect) made it possible to perform fairly accurate human motion tracking with a single depth camera at the comfort of home. Very recently, techniques for 2D skeletonization using a regular webcam to collect vision data have been developed (such as OpenPose). This chapter focuses on these newer marklerless technologies where a single depth camera or webcam is used for vision-based human motion tracking.

Chapter 6 provides an overview of various instrumented gloves for hand and figure motion tracking. Hand rehabilitation requires accurate tracking of fingers and hand orientations. Many different types of instrumented gloves have been developed for this purpose. Even though eTextile has been used to make instrumented gloves, many other sensing modalities, such as bend sensors, IMUs, fiber optics, have been incorporated into instrumented gloves.

Part II includes four chapters. Chapter 7 discusses how to measure basic parameters of human movements in rehabilitation exercises, such as range of motion, step count, and gait parameters. In addition, this chapter introduces some common measurement theories and how to evaluate a new measurement instrument.

Chapter 8 introduces machine-learning-based activity recognition for more sophisticated activities. This requires automated segmentation of different physical activities, the segmentation of each repetition, and the assessment of the quality of each repetition. A typical approach is to extract features from the collected motion data, and classify the data using machine learning algorithms such as support vector machine, based on some labeled training dataset. With the development of deep learning, it is possible to omit the manual feature extraction step.

Chapter 9 covers rule-based activity recognition methods. Rehabilitation exercises and many related physical activities are carefully defined by healthcare professionals and are highly repetitive. Hence, rule-based approach is a good fit. While machine-learning-based approach has the advantage of being flexible and not depending on a clinician to manually define the specification of the exercise, it is weak in providing specific feedback to the patient regarding exactly was done wrong. The rule-based approach has the advantage of providing precise feedback regarding the specified movements in a rehabilitation exercise.

Chapter 10 introduces exergames, which are also referred to as serious games. Traditionally prescribed rehabilitation exercises can be dry and boring. Many studies have shown that putting the exercises in the context of game playing would engage patients better and hence promote adherence. This is where technology could play a big role. This chapter provides an overview of the exergames that have been used in various rehabilitation programs that promote physical health, the study design, the technology used to render the exergames and that to facilitate data collection, the feedback provided to the user, and the clinical outcomes of the studies.

Part III includes six chapters. Chapter 11 provides an overview of technology-facilitated physical rehabilitation from the clinical perspective. Physical rehabilitation works to improve movement dysfunction, such as hemiplegia due to stroke. Technology may be used to help patients to restore movement, strength, stability and/or functional ability and reduce pain via targeted exercises and a range of other treatment methods.

Chapter 12 provides an overview of technology-facilitated occupational rehabilitation from the clinical perspective. Occupational rehabilitation technologies focus on restoring an individual's ability to perform necessary daily activities, such as working to improve fine motor skills, restoring balance, or assisting patients in learning how to increase their functional ability. This chapter also discusses studies on the tracking of activities of daily living, which is becoming increasingly popular.

Chapter 13 reviews technology-facilitated speech rehabilitation from the clinical perspective. Speech rehabilitation technologies is used in patients with difficulties in speech, communication and/or swallowing. Computer-based speech therapy is becoming increasingly popular because it lowers the cost of therapy and facilitates a patient to practice individually at the comfort of home.

Chapter 14 provides an overview of technology-facilitated pulmonary rehabilitation. Pulmonary rehabilitation technologies aid patients who have breathing disorders or difficulties. Pulmonary rehabilitation helps them decrease respiratory distress and maintains open airways. Exercise training is a core component of pulmonary rehabilitation.

Chapter 15 reviews technology-facilitated cognitive rehabilitation from the clinical perspective. Cognitive rehabilitation technologies work with patients to improve memory, thinking, and reasoning skills. One important study in cognitive rehabilitation is the detection of cognitive impairment levels by observing activities of daily living. Exercise training is also a core component in cognitive rehabilitation.

Chapter 16 provides an overview of technology-facilitated mental health rehabilitation. In mental health rehabilitation, exergames are pervasively used. For this type of target population, the games often go beyond encouraging physical activities, such as to teach the patients how to improve communication and other social skills in a virtual reality setting. Exercise training is also a core component in mental health rehabilitation.

Part I

Motion sensing technologies

Chapter 1
Inertial measurement units

An inertial measurement unit (IMU) is an electronic device that consists of a set of inertial sensors. The purpose of an IMU is to measure inertial parameters of a moving object, such as the position (including altitude), orientation, velocity, rotation (i.e., angular speed), and sometimes the heading of the object. Inertia is a physical property of a mass to remain at its current state (such as moving along a straight line with a fixed velocity) as long as there is no external force acted upon it. An IMU is equipped with at least an accelerometer sensor and a gyroscope sensor. If the moving direction needs to be determined, one can use an IMU that is equipped with the magnetometer sensor in addition to the accelerometer and the gyroscope sensor. IMU was initially used almost entirely for navigation. That is why the device is also referred to as inertial navigation unit (INU). INU was first used in World War II as part of the inertial guidance system for rockets. It was later also used for spacecraft and aircraft. With the rapid technology development in microelectromechanical systems (MEMS), IMU has significantly shrunk its size and it has been embedded in many wearable devices (such as smartwatches and wristbands) and mobile devices (such as smartphones and tablets). That is why IMU is also referred to as a kind of wearable device.

1.1 Accelerometer

An accelerometer is a sensor that measures acceleration in one, two, or three dimensions. A one-dimensional accelerometer measures the acceleration along a single direction. It can be thought of as a proof mass on a spring, as shown in Figure 1.1. A proof mass is one with known mass m so that the acceleration can be measured via the compression or extension of the spring. In practice, the proof mass is damped so that it does not oscillate when a force is acting on it. In MEMS accelerometer, the spring is replaced with a cantilever. The way an accelerometer works is that the acceleration of the proof mass m caused by a force F can be measured by the extension/compression of the spring. Given the spring coefficient k, the force will cause acceleration to the mass by Newton's law $F = ma$, and the force will also cause the spring to displace by the amount of x, which will cause a tension of kx. Hence $ma = kx$, which will lead to $a = (k/m)x$. There are several ways of measuring x, such as resistive, capacitive, and inductive techniques. Most MEMS accelerometers are capacitive

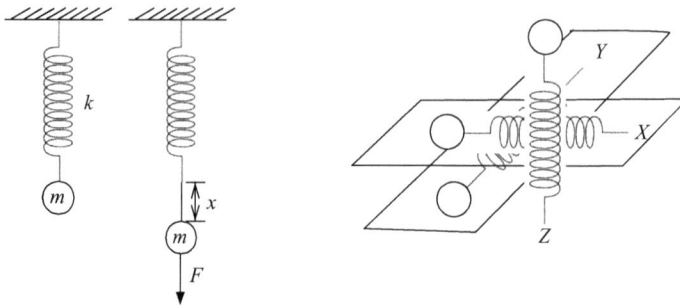

Figure 1.1 Working principle of accelerometer

accelerometers. Two other common types of accelerometers are piezoelectric and piezoresistive accelerometers.

A 2-dimensional accelerometer combines two 1-dimensional accelerometers perpendicular in the same plane. Adding a third accelerometer that is perpendicular to the plane would create a 3-dimensional accelerometer, as shown on the right side of Figure 1.1. On the planet Earth, a 3-dimensional accelerometer can be used to measure not only acceleration but also orientation due to the existence of the gravity. The standard gravity on Earth g is defined to be 9.80665 m/s^2. It is roughly the gravity measured at 45° latitude. If an accelerometer is placed on a flat surface with its x–y plane being in parallel to the surface, its z-axis reading would be close to the standard gravity g. When the sensor is dropping, its z-axis accelerometer reading would be 0 m/s^2. We should note that always a 3-dimensional accelerometer is convenient for measuring acceleration in all 3-dimensional space, high-end accelerometers are typically single dimension.

The specification of an accelerometer consists of several parameters. The most obvious parameter is the range of the accelerometer. The range is typically described in terms of the standard gravity, which is the maximum acceleration the accelerometer can measure accurately in each dimension. The range varies from ±1g to ±250g. To measure subtle movements, an accelerometer with small range will give a much more accurate reading that that with larger ranges. The high g range accelerometers can be used for rockets. It is recommended to use the lower 20% of the accelerometer's range for measurement to accommodate unexpected accelerations. For example, if the estimated maximum range is 0.5g, then an accelerometer with 2.5g or larger range should be used.

Another important parameter is the sensitivity of the accelerometer. It defines the amount of measured signal in terms of millivolts mV or picocoulombs pC per standard gravity g. For digital output accelerometers, the sensitivity is defined as least significant bit per g, or LSB/g. This can be confusing. In the blog posted by Henley [14], he explained how to interpret the sensitivity for digital output accelerometers. For example, given a digital output accelerometer's sensitivity of 256 LSB/g, this means that the resolution is really $1/256g = 0.0039g$ (i.e., how much acceleration can be represented by a single bit). The sensitivity of the same sensor can also be defined as having a resolution of 10-bit in $2g$ mode. The total range would be $4g$ and the decimal number of 10-bit is $2^{10} = 1,014$. Then each bit would represent $4g/1,024 = 0.0039g$.

Another important parameter is frequency response [14]. It is also referred to as bandwidth of the accelerometer. It is defined as the maximum deviation from its sensitivity over the frequency range. The deviation can be specified in percentage or dBs. The data sheet of an accelerometer often provides a frequency response curve. In practice, the motions measured by an accelerometer typically have some frequency range that one can estimate. The range of operation of the accelerometer must covers the desirable frequency range of the motion to be measured. Otherwise, the measurement would not be accurate. For example, if the 0 frequency shows a large deviation, the sensor cannot be used to measure static acceleration (such as to determine orientation) or slow vibrations, and it cannot be used to measure distance traveled for dead reckoning [15] because the acceleration will be integrated twice.

Noise is another parameter. It is commonly defined as a broadband root-mean-square (RMS) value for residual noise measured by the sensor without any mechanical excitation. Some accelerometer vendors choose to provide spectral noise density for different ranges because it varies with the frequency.

Often the temperature sensitivity of the sensor is given as well. It indicates the shift in sensitivity when the temperature changes in terms of percentage per Celsius (%/°C).

1.2   Gyroscope

Gyroscope is a device that measures orientation and angular velocity. Its working principle is conservation of angular momentum, where the total angular momentum of a closed system would not change without external forces. Any spinning mass or a mass that rotates around a center would have angular momentum. In contrast to angular momentum, linear or translational momentum is a measure for a mass that move translationally (i.e., without rotation). A force applied on a mass would alter its translational momentum. Similarly, a torque applied on a mass would change its angular momentum. The total of internal torques of any system is always zero.

Traditionally, gyroscope consists of a spinning disk or wheel along one axle. It can be mounted into two or three gimbals. When a torque is applied to the spinning disk along the input axis (which is perpendicular to the spin axis), as shown in Figure 1.2, the conservation of angular momentum will lead the spinning disk to rotate along the output axis, which is perpendicular to both the spin axis and the input axis. This phenomena is called gyroscopic precession. This principle constitutes the basis for the measurement of orientation and angular velocity.

To fabricate gyroscope in MEMS, a different type of movement, i.e., vibration, is used instead of spinning. MEMS gyroscope is called vibrating structure gyroscope (VSG). VSG is also referred to as a Coriolis vibratory gyroscope. The underlying principle is that a vibrating mass would continue vibrating in the same plane even when its supporting structure rotates. However, when the supporting structure rotates,

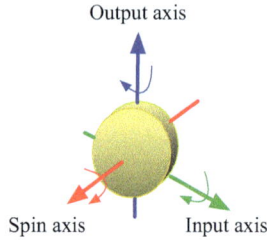

Figure 1.2   Working principle of the gyroscope

the Coriolis effect would cause the mass to exert a force on its support, which can be measured to derive the rate of rotation.

Like accelerometers, gyroscope can also be one-dimensional, two-dimensional, or three-dimensional, which are often referred to as uniaxial, biaxial, and triaxial gyroscopes. There are also many types of gyroscopes based on its underlying construction methods. For convenience and cost, MEMS gyroscopes are the most common type and they are virtually all VSG and they typically have digital output (with 16-bit the most often). We should note that research on MEMS gyroscope is still very active. It is likely that newer generations of MEMS gyroscopes will have much better performance [16].

The specification of a gyroscope also includes several parameters, including angular rate range, sensitivity, bandwidth (or frequency response), temperature sensitivity, zero-rate level, angular transverse sensitivity, bias stability, and random walk. The primary output parameter for a gyroscope is the angular rate. The angular rate range specifies the range of the rates that the sensor can measure accurately. Consumer-grade MEMS gyroscopes typically offer a sequence of ranges for the users to choose, with the highest range of 2,000 degrees per second (dps), for example, 250/500/2,000 dps.

The sensitivity provides the smallest angular rate that can be accurately measured, and it is typically given in terms of mdps/LSB or mdps/digit, where mdps stands for millidegrees per second, and LSB is the lease significant bit, which is the same as that for accelerometers (with digital output). The notation of mdps/LSB can be confusing. It is really given the equivalent of LSB in terms of mdps. Using the data sheet of ASM330LHH sensor as an example (https://www.st.com/resource/en/datasheet/asm330lhh.pdf), the angular rate sensitivity given for the ±125 dps range is 4.37 mdps/LSB. It means that the sensitivity is 4.37 mdps represented by the least significant bit. Similar to accelerometers, a gyroscope with smaller ranges would have better sensitivity. For example, for ASM330LHH, the sensitivities for the ranges of ±250 dps, ±500 dps, ±1,000 dps, ±2,000 dps, ±4,000 dps are 8.75, 17.5, 35.0, 70.0, and 140.0 mdps/LSB, respectively. The consideration on bandwidth is rather similar to that of accelerometers. Unfortunately, not all gyroscope vendors provide the bandwidth information. The temperature sensitivity is given as the percentage of changes per Celsius.

Zero-rate level, which is also referred to as offset calibration error, is the angular rate output from the sensor when it is stationary without any movement or rotation. MEMS gyroscopes typically have fairly large zero-rate levels. For example, the ASM330LHH sensor has a zero-rate level of ± 10 dps (https://www.st.com/resource/en/datasheet/asm330lhh.pdf). Sometimes, the zero-rate level is given in terms of LSB, in which case, one would need to check with the sensitivity given to do the conversion. One factor that causes the zero-rate level output is the extensive mechanical stress to the sensor after being mounted to the circuit board.

The angular transverse sensitivity, sometimes also referred to as angular rate cross-axis sensitivity, defines the maximum output regarding the rotation rate about an axis that is orthogonal to the rotation axis. Apparently, the lower the value the better the sensor. It is typically expressed in terms of percentage of the orthogonal input angular velocity. For example, the ASM330LHH sensor has $\pm 1\%$ of angular transverse sensitivity (at 25°C).

Bias stability (also referred to as bias instability) defines the maximum deviation of the reported angular rate from the mean value over time at constant temperature. It shows how stable the measurement output is over time. This is another parameter that indicates the level of imperfection of the sensor. Hence, the lower value, the better. For example, the ASM330LHH sensor has a 3 degrees per hour bias stability.

Another parameter that indicates the level of imperfectness of the sensor is random walk. Random walk is used to describe a random process in many different fields. For example, in physics, random walk is used to model Brownian motion and diffusion. Random walk can be expressed in terms of angular random walk and rate random walk. For gyroscope, random walk is calculated using the Allan Variance [17] of the bias data [18]. The rate random walk is said to be "a random process of uncertain original, possibly a limiting case of an exponentially correlated noise with a very long correlation time" [18]. The rate random walk is rarely reported in gyroscope data sheets. Instead, the angle random walk if often reported. It is derived from the Allan Variance of the integrated bias data [18]. It is impacted by high-frequency noise terms with a correlation time significantly shorter than sampling time. The angle random walk is expressed in terms of degrees per square root of hour. For example, the ASM330LHH sensor has a 0.21 deg/$\sqrt{h}$ angle random walk. We can convert between the Allan Variance and the angle random walk by dividing or multiplying 60. For example, the Allan Variance for ASM330LHH is $60 \times 0.21 = 12.6$ deg/h.

1.3 Magnetometer

A magnetometer is a device that measures the magnetic field. In this book, we will only discuss MEMS-based magnetometers. Also, the magnetometer included as part of an IMU is used to help determine the heading of the movement by sensing the magnetic field of the Earth. Therefore, the magnetometer must be sensitive enough to measure the rather weak magnetic field of the Earth, which has a range of 25–65 μT [19], where T is the unit for the magnetic field, Tesla.

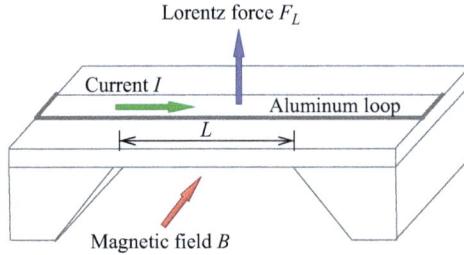

Figure 1.3   The working principle of the Lorentz force-based resonance method for making magnetometers

There are many different ways of measuring magnetic field. Even for MEMS-based magnetometers, there are several different manufacturing technologies, and this is still a very active research field [20]. It appears that the state-of-the-art method of making MEMS magnetometers is the Lorentz force-based resonance method. Sensors fabricated this way are often referred to as resonant magnetic sensors. The design principle is shown in Figure 1.3. In a resonant magnetic sensor, an aluminum loop is placed on top of a beam structure, current (I) is flowing throw the loop. In the presence of magnetic field (B) that is perpendicular to the current, the beam will experience a force upward F_L, which is called Lorentz force. The force is proportional to the strength of the magnetic field B, the current I, and the length of the beam L, i.e., $F_L = IBL$. Because the beam is clamped at both ends, it is often referred to as clamped–clamped beam.

Because the magnetic field is very weak, the Lorentz force is very small. To make accurate measurement, another physics phenomena, resonance, is exploited to magnify the displacement caused by the force. Every physical structure has a resonant frequency. When an oscillating force is applied to the structure at the resonant frequency, the structure would oscillate with a much higher amplitude than when the same force is applied at non-resonant frequencies, essentially magnifying the force by many times. The clamped–clamped beam is designed exactly for the purpose of creating resonance.

The displacement of the beam can be measured via piezoresistive sensing, capacitive sensing, and optical sensing [21]. In piezoresistive sensing, active piezoresistive materials are placed on the beam and their resistance will change proportional to the longitudinal strain experienced by the beam when there is a displacement of the beam caused by the Lorentz force in the presence of magnetic field. Capacitive sensing involves more complex structures and electronic circuit to measure the capacitance changes in the presence of the displacement of the beam. Optical sensing is perhaps the easiest to understand. In optical sensing, the displacement of the beam is directly measured by shining light on the beam surface with a laser, and using a detector to collect the reflected light.

MEMS magnetometers are characterized by range, sensitivity (or resolution), temperature sensitivity, and RMS noise. The data sheets typically use Gauss (G)

instead of Tesla as the unit for the magnetic field strength, where $1\,T - 10,000\,G$. Hence, Gauss is a much smaller unit than Tesla. The range is typically in tens of Gauss. For example, the LIS2MDL magnetometer has a dynamic range of $\pm 49.152\,G$ (https://www.st.com/resource/en/datasheet/lis2mdl.pdf). The sensitivity is typically expressed in milligauss (mG). For example, LIS2MDL has a sensitivity of 1.5 mG/LSB with 7% deviation. Here the interpretation for LSB is identical to that for gyroscopes. The temperature sensitivity is given in terms of percentage per Celsius, similar to that for accelerometers and gyroscopes. For LIS2MDL, the temperature sensitivity is $\pm 0.03\%/^{\circ}C$. Some sensors provide zero Gauss drift data. For example, IST8303 sensor's data sheet (http://isentek.com/data/IST8303%20Datasheet.pdf) provides a zero Gauss drift of 3 mG. The noise level may also be provided. For example, the LIS2MDL provides a RMS (room mean square) noise of 3 mG in high-resolution mode.

1.4 Rehabilitation studies using IMUs

We divide the studies according to types of IMUs that have been used, from systems or devices constructed using low-level IMU chips, to high-level prepackaged sensing devices designed for professional use, to consumer devices. For each type, we summarize the findings in a table, and then we provide an outline of the studies. The motion recognition and classification perspective and the clinical perspective will be discussed in later chapters.

1.4.1 Studies using low-level IMUs

The studies using low-level IMUs are summarized in Table 1.1. To use these low-level IMUs, one would have to add other essential components to build a complete system for motion sensing, such as microprocessor, storage, communication model (e.g., Bluetooth or WiFi), a PCB board, and a case appropriate for attaching to the human body.

Abbasi-Kesbi *et al.* [22] developed a system that can be used to measure joint angles accurately using an MPU-9150 IMU from InvenSense, which consists of a 3-axial accelerometer, a 3-axial gyroscope, and a 3-axial magnetometer. The output of the three sensors are fused together using a complementary filter to derive the joint angle.

Abdelhady *et al.* [24] reported a system they developed to measure gait parameters using MPU-6050 IMU from InvenSense, which consists of a 3-axial accelerometer and a 3-axial gyroscope. The raw data collected are augmented using a Kalman filter based on the knee kinematics. The goal of the system is to enable data collection for patients at the home setting.

Ai *et al.* [25] developed a system to measure the range of motion of the ankle and the trajectory of the leg using two low-level IMUs. One is MPU-6050 from InvenSense. The other is a magnetometer HMC5883L from Honeywell. They fused the outputs of both sensors to achieve more accurate and smoother trajectory of the

Table 1.1   Studies using low-level IMUs

IMU brand	IMU	Use case	Refs
MPU-9150, InvenSense	acc, gyro, mag	Joint angle with complementary filter	[22]
MPU-9150, InvenSense	acc, gyro, mag	Detecting wrist motion using dart thrower's movement and differentiating healthy subjects and those who have injured wrists using DTW and clustering	[23]
MPU-6050, InvenSense	acc, gyro	Joint angle with Kalman filter	[24]
MPU-6050, InvenSense	acc, gyro	Angle ROM with simple fusion of three sensors	[25]
HMC5883L, Honeywell MPU-6050	mag acc, gyro	Knee angle for TKR patients	[26]
MPU-6000 InvenSense	acc, gyro as part of system	Position & movements of lumbar spine	[27]
MPU-9250	acc, gyro, mag	Assessment of upper limb function using ML for post-stroke hemiplegic patients	[28]
MPU-9250	acc, gyro, mag	For measuring angle in gait training for patients with knee osteoarthritis	[29]
MPU-9250 IMU and Omron D6T thermal sensor	acc, gyro, mag	Detect 24 predefined motions using a wrist-mounted device using ML. Showed thermal sensor improves classification accuracy	[30]
ADXL202 (4 used)	1 and 2-axial acc	Rule-based activity detection	[31]
ADXL345 acc, Analog Devices and L3G4200D gyro, STMicroelectronics		FTSST and gait parameter tracking	[32]
LIS3LV02DQ, ST Microelectronics	3-axial acc	Activity recognition using ML & rule-based methods	[33]
LSM9S0, Adafruit, used with a magnet	acc, gyro, mag	Knee angle measurement based on magnetic field	[34]
H34C 3-axial acc, from Hitachi metals; ENC-03R, 1-axial gyro, from Murata		Measure trunk, thigh, and shank angles during gait	[35]
KXM52, Kionix	acc	Three IMUs are used to measure the trunk posture changes	[36]
Epson	gyro	Vicon 370 is used as reference system to validate the IMU result	
Unbranded IMU	acc, gyro, mag	Monitoring the amount of hand use using support vector regression for stroke survivors	[37]
Unbranded IMU	acc, gyro, mag	IMU is used to collect motion data for detection of gait abnormality in patients with neurological disorders	[38]
Unbranded IMU	acc, gyro, mag	IMU is used to collect motion data to determine gait characteristics	[39]
Unbranded IMU	acc, gyro	Three IMUs are used to collect motion data to determine compensatory movement patterns in upper extremity	[40]

movements. The sensors were placed on the frontal plane of the shank for tracking of leg trajectory, and they were placed on the middle of the instep plane to track the ankle range of motion.

Bussmann *et al.* [31] proposed a system for daily activity monitoring based on up to four accelerometer sensors. Two 1-axial sensors are attached to the lateral side of each thigh, one 2-axial sensor is attached to the lower part of the sternum, and an additional sensor can be applied to the sternum. The activity detection is done via an activity knowledge base using the extracted features from the collected accelerometer data. This is one of the early works attempting to determine activities of daily living.

Cortell-Tormo *et al.* [27] developed a system to monitor the lumbar spine motion. The system consists of 7 MPU-6000 sensors from InvenSense. The sensors are carefully placed together to a subject's spine area to measure the curvature and spinal motion. The sensors are placed on top of the foot.

Gonzalez-Villanueva *et al.* [33] reported a system aimed to monitor human motion for physical rehabilitation. The human motion data are collected via a number of 3-axial accelerometers (LIS3LV02DQ from ST microelectronics) placed on different parts of the body depending on the rehabilitation exercises to be recognized. They experimented with various machine learning algorithms to classify daily activities such as standing, sitting, and walking. They found that a rule-based method called fuzzy finite-state machines (FFSMs) is suitable to recognize complex exercises such as the Sun Salutation exercise. The sensor data collected would help determine the key poses as defined in the FFSMs.

He *et al.* [28] developed a wearable sensing device with one MPU-9250 IMU to collect data from patients who have had stroke under various movements (hand to lumbar spine, shoulder flexion, and forearm pronation). A feature vector consisting of the mean, standard deviation, and root square mean values is extracted from the raw measurement data. The data are then labeled and trained with four machine-learning algorithms, KNN, Random Forest, Bayesian, Support Vector Machine (SVM) to classify each patient based on the Fugl–Meyer scale, which was designed to evaluate the seriousness of stroke patients (from 0 to 6). They found that SVM gave the best accuracy.

He and Yi reported a wearable system for gait rehabilitation for senior patients with knee osteoarthritis [29]. The system uses a MPU-9250 IMU to measure the *z*-axis angle. The angle is measured using the gyroscope based on the angular velocity. The system contains an additional type of sensors called pressure-sensitive electric conductive rubber sensors, which is used to measure the knee adduction moment curve.

Huang *et al.* [26] presented a system they developed to monitor the progress of the patients and to encourage the participation of rehabilitation at home after the total knee replacement surgery. A key parameter being monitored for recovery is the range of motion of the knee, which requires the measurement of the knee flexion angle. Other important parameters include the repetition count and the practice number. The knee angle measurement is done using two self-made sensors containing the MPU-6050 IMU. The inputs from the accelerometer and the gyroscope are fused together using a Kalman filter to derive the knee angle.

Hullfish *et al.* [34] reported their work on using wearable sensor to measure knee flexion. They followed a rather unconventional method. Instead of relying on accelerometer and gyroscope to measure the motion directly, they used a rare-earth magnet in conjunction with a LSM9S0 IMU from Adafruit, which consists of a 3-axial accelerometer, a 3-axial gyroscope, and a 3-axial magnetometer. The knee angle is derived by measuring the magnetic field using the magnetometer in the IMU.

Liu *et al.* developed a system to monitor the amount of hand use for stroke survivors [37]. The system consists of one device attached to the finger and another on the wrist. The device incorporates a unspecified IMU with a 3-axial accelerometer, a 3-axial gyroscope, and a 3-axial magnetometer. They identified particular features that are relevant to the amount of hand use using a correlation-based feature selection algorithm, then employed support vector regression to estimate the hand use. The performance was evaluated using the normalized root mean square error.

Lui and Menon reported their work on using a wrist-mounted device to detect 24 predefined motions using machine-learning algorithms. They showed that the addition of a thermal sensor (Omron D6T) to the MPU-9250 IMU significantly improved the detection accuracy [30].

Motoi *et al.* proposed a wearable gait monitoring system [35]. The system incorporates a 3-axial accelerometer (H34C from Hitachi metals) and 1-axial gyroscope (ENC-03R from Murata). They used the custom IMU to measure trunk, thigh, and shank angles during walking training.

Nguyen *et al.* [23] reported their study on characterizing the motion of the wrist in the dart thrower's movement using two IMU devices, one placed on the back of the throwing hand, and the other on the forearm. Each device consists of an MPU-9150 IMU. From the raw data, they derived the quaternion trajectory of the wrist and compared the distances between different throws using dynamic time warping. Then they applied clustering to differentiate the movements of the healthy participants and those with injured wrists.

Papi *et al.* [32] reported their work on validating three different sensors (2 IMUs and one none-IMU-based) for measuring parameters related to the five-time sit-to-stand test (FTSST) and treadmill walking. One of the IMU is custom-made with an ADXL345 accelerometer from analog devices and an L3G4200D gyroscope from STMicroelectronics. The other IMU is from Opal. A 10-camera motion tracking system from Vicon is used as the reference system for validation of the measurement using the three sensors.

Qiu *et al.* proposed an IMU-based system for collecting data in patients with neurological disorders to detect gait abnormality [38]. Their system consists of a custom-made IMU with accelerometer, gyroscope, and magnetometer. They developed their own algorithm to fuse the input from the three sensors for better accuracy. In a closely related paper, Qiu *et al.* reported their work on IMU design and how to compute the gait characteristics using the motion data collected by the custom IMU [39]. With their collected data and the proposed method, they compared the gait parameters for healthy subjects, stroke patients, and convalescence patients.

Ranganathan *et al.* [40] presented their work on using three IMUs (specifics on the accelerometer and gyroscope in IMU were not given) to collect motion data to determine compensatory movement patterns in upper extremity while doing a set of predefined reaching and manipulation tasks. The classification of whether or not the compensatory movement is present is done via the Native Bayes classifier.

Wong and Wong reported a study of using a custom-made system with three IMUs, each consisting of a IMUKXM52 3-axial accelerometer from Kionix and 3 uni-axis gyroscope from Epson, to measure the trunk posture changes [36]. They used the Vicon 370 motion capture system to validate the result using their IMU-based system.

1.4.2 Studies using prepackaged professional sensors containing IMUs

These types of studies are the most common because the IMU devices are designed for researchers with both low-level and high-level application programming interfaces. These IMU manufactures often perform scientific studies themselves and report the accuracy of their devices with great details. The studies using this type of IMUs are summarized in Table 1.2.

Allseits *et al.* [42] studied how to perform continuous measurement of the knee angle during gait using leg-mounted IMU sensors. Their primary goal is to introduce a new algorithm called GO algorithm for knee angle calculation based on gyroscope only. They used a multi-camera-based system to establish the ground truth, and compared with a well-known algorithm, which is referred to as the CF algorithm [69]. In their experiments, they compared with results obtained via an Opal IMU from APDM in conjunction with the CF algorithm, and those obtained using an in-house IMU in conjunction with their GO algorithm. They showed that the GO algorithm performs as well as the CF algorithm and aligns well with the ground truth.

Annegarn *et al.* [41] reported their validation study on an IMU device called CAM from Maastricht Instruments. The goal is to collect objective physical objective data for patients with chronic organ failure (such as chronic heart failure and COPD) during a 1-h period with a set of predefined activities. The ground truth is established by manually analyzing the recorded video during the experiment. They showed that the IMU device could accurately measure the time spent in several different postures, including the weight-bearing postures, nonweight-bearing postures, and dynamic activity.

Argent *et al.* [47] proposed to use machine learning-based regressions to model the joint angle movements. They used the Shimmer IMU and collected the accelerometer and gyroscope data. To obtain the ground truth, they used a 3-camera motion capture system called CODA. They experimented with four regression machine learning methods, linear regression, polynomial regression, decision tree regression, and random forest regression. While doing regression, they also incorporated the anthropometric data of the participants, such as height, weight, and limb segment length. The performance is evaluated in terms of the root-mean squared error and the coefficient of determination.

Bai *et al.* [53] reported a study on position tracking using Xsens MTx IMUs. They provided some technical details on the composition of the MTx IMU, where it consists

Table 1.2   Studies using prepackaged professional sensors containing IMUs

IMU brand	IMU	Use case	Refs
CAM, Maastricht Instruments	acc	Activity monitoring for selected set of exercises	[41]
Opal, APDM In-house IMU	acc, gyro gyro	Knee angle in gait using CF algorithm Knee angle in gait using gyro-only algorithm	[42]
Opal, APDM	acc, gyro, mag	Knee ROM with 2 IMUs	[43]
Opal, APDM	acc, gyro, mag	Hip angle while doing walking, jumping jack, squat, jump turn	[44]
Opal, APDM		Measure head and trunk ROM and peak rotational velocity while doing a set of standing and walking tasks for patients with mild traumatic brain injury.	[45]
Opal, APDM		FTSST and gait parameter tracking	[32]
Opal, APDM		Measure segment angular displacement while performing functional tasks	[46]
Shimmer IMU	acc and gyro	Various joint angles using various regressions	[47]
Shimmer IMU	acc, gyro, mag	Detect Parkinson's disease using machine learning	[48]
Shimmer IMU	acc, gyro, mag	Rule-based detection of upper limb activity (making a cup of tea)	[49]
Shimmer IMU	acc, gyro, mag	Gait parameters for hemiparetic patients	[50]
Shimmer IMU	acc, gyro, mag	6 IMUs are used to collect motion data for classification activities of daily living in hemiparetic rehabilitation patients	[51]
Xsens MTx IMU (used a reference)	acc, gyro, mag	Measuring various joint angles	[52]
Xsens MTx IMU (config 1) (config 2)	acc, gyro, mag 4 MTx 1 MTx	Position tracking (PT) PT with kinematic model PT with dead reckoning	[53]
Xsens MTx IMU	acc, gyro, mag	Spine angular patterns and ROM during gait	[54]
Xsens MTx IMU		Joint angle ROM and kinematic model for activity detection	[55]
Xsens	acc, gyro	Activity classification with KNN	[56]
Xsens MTw IMU		A method to measure elbow angle using two IMUs without the need of specific IMU-to-segment alignment	[57]
Xsens MTw IMU		Capturing the motion of a random motor task and estimate the upper-limb impairment level in stroke patients using a clustering algorithm and a regression model	[58]

(*continued*)

Table 1.2   Continued

IMU brand	IMU	Use case	Refs
Xsens MTi IMU		Proposed a kinematic model to estimate upper limb motion based on the data collected via 4 IMUs	[59]
Xsens MT9-B IMU	acc, gyro	Measurement upper limb movements with kinematic models and a simulated annealing optimization to reduce errors	[60]
Xsens MTx IMU	acc, gyro, mag	Measure upper limb movements with 2 IMUs	[61]
MetaMotionR, MbientLab	acc, gyro, mag	Classification of rehab exercises with ML	[62]
Trivisio IMUs (10 IMUs)	acc, gyro, mag	Upper extremity movement monitoring and feedback with HMM	[63]
Inertia-Link IMU, MicroStrain	acc, gyro, mag	Pelvis orientation in gait, STS, step-up	[64]
Inertia-Link IMU, MicroStrain	acc, gyro, mag	Used to measure functional capacity of patients who complain about lower back pain (LBP)	[65]
acc from GC Dataconcepts	acc	Used to measure gait parameters in daily life of the patients with LBP	
BNO055, Bosch Sensortec	acc, gyro, mag	Measuring knee angle during walking	[12]
IGS-180 motion capture suit, Synertial UK	17 IMUs	Rule-based segmentation and detection of 7 activities	[66]
Vitaport 3, Tempec BV	acc	To objectively estimate stroke patients' functional ability scale score based on data collected with an IMU using random forests	[67]
Physilog	acc, gyro	Identify postural positions and several types of movements in daily living	[68]

of two ADXL202E biaxial accelerometers, three ENC-03J uni-axial gyroscopes, and three KMZ51 uni-axial magnetometers. They experimented with two configurations. In one configuration, four MTx IMUs were used to perform position tracking based on a kinematic model. In the second configuration, a single IMU was used to perform position tracking based on dead reckoning.

Baraka *et al.* [48] reported their work on using a Shimmer sensor that consists of both IMU and surface electromyography (sEMG) to collect motion data on Parkinson's tremor in upper limb. They then experimented with several machine learning algorithms to classify the tremor, including decision tree, linear discriminant analysis, KNN, SVM, boosted tree, and bagged tree. Among these classifiers, the bagged tree gave the best result.

Bavan *et al.* [62] reported their study using several machine learning algorithms to classify rehabilitation exercises based on motion data collected by an IMU called

MetaMotionR (which consists of accelerometer, gyroscope, and magnetometer) from MbientLab, San Francisco, CA. The sensor was placed on the upper arm, and the participants were patients who have shoulder pains. The participants were asked to perform five rehabilitation exercises designed for the study, which involve movement of a single plane, including shoulder abduction, should flexion, wall slide, wall press, and shoulder rotation. The experimented with four classifiers, including decision tree, SVM, KNN, and random forests. They performed two different types of training and validation schemes. One is the standard ten-fold-validation, and the other is less-known method called leave-one-subject-validation, where the test is done on the subject that is not used for training the classifier. The latter is more challenging. The SVM gives the best accuracy on the ten-fold-validation, where the random forests achieves the highest accuracy for the leave-one-subject-out-validation.

Biswas *et al.* [49] proposed a method to detect arm movements using a single wrist-worn Shimmer sensor. The authors developed a set of rules for activity detection based on the characteristics of the particular arm movements while making a cup of tea. They validated their algorithm with four healthy subjects and four stroke patients.

Bleser *et al.* [63] proposed a system to function as a personalized rehabilitation exercise trainer. The system used up to 10 IMUs from Trivisio to collect motion data when a participant performed the predefined eight upper extremity strength training exercises. These exercises are designed for rehabilitation of cardiac patients. Each IMU consists of accelerometer, gyroscope, and magnetometer. The system uses HMM to automatically determine whether or not an exercise is done correctly. They confirmed that the algorithm is more strict than a human trainer.

Bolink *et al.* [64] conducted a validation study on using a single IMU called Inertia-Link from MicroStrain attached onto the lower back to measure pelvis orientation during gait, sit to stand, and step up activities. An optical motion capture system was used as the gold standard for comparison. They concluded that the IMU can be used for the intended purpose.

Chiang *et al.* [43] used two APDM Opal sensors to collect the knee range of motion data with 18 patients who have done the total knee arthroplasty surgery. One sensor is placed on the thigh, and the other is placed on the shank with predefined alignment. They used the accelerometer and the gyroscope data to determine the range of motion.

Derungs, Schuster-Amft, and Amft [50] used four Shimmer sensors to collect movement data while the patients were walking to extract gait parameters for a longitudinal study. The parameters include stride duration, cadence, and stride count. The goal of the study is to see if the movement patterns of the affected side and the non-affected side would converge.

Digo *et al.* [54] used five Xens MTx sensors placed on different segments of the spine to measure the spine angular patterns and the range of motion of the spine angular movement during walking.

In [55], Gil-Agudo *et al.* developed a sensor system using five Xsens sensors placed on the trunk, the back of the head, the right arm, the forearm, and the hand, respectively. The authors developed a kinematic model based on the sensor input to analyze functional complex movements in activities of daily living, such as drinking from a glass. More specifically, the model captures the movements of the head and the upper limbs in terms of the shoulder flexion and extension, abduction and adduction, external and internal rotation; elbow flexion and extension, and pronation and supination; wrist flexion and extension, and radial and ulnar deviation; head flexion, extension, and inclination. The first tested the accuracy of the measurement of single joint angles in terms of range of motion using the Codamotion system as the gold standard reference. They tested the robustness of the system in the detection of the drinking task. Both configurations achieved desirable results.

Horenstein *et al.* [44] reported a study of measuring the hip joint angle using an Opal sensor while the participants doing several tasks, including walking, jumping jack, bilateral squat, and jump turn. They utilized the magnetometer included in the Opal sensor to establish a sacrum coordinate system to map the thigh orientation.

Kayaalp *et al.* [12] presented a validation study on using an IMU-based device to monitor the knee flexion and extension angle during a walking task for patients who have received the total knee arthroplasty. The authors performed a thorough statistical analysis on the results obtained using the IMU and those obtained using a multi-camera motion capture system as the gold standard.

Laudanski *et al.* [56] proposed to classify gait activities of patients with stroke using two Xsens IMUs (one on each leg) and the KNN algorithm. The study investigated what time and frequency domain features are most pertinent to each gait activity.

Muller *et al.* [57] proposed a method to measure the elbow angle using two IMUs (Xsens MTw) without the need of specific IMU-to-segment alignment. The method enables self-calibration within 10 s and was validated using a multi-camera system as the ground truth.

Nguyen *et al.* [66] reported their study on automatically segment the raw data and detect seven activities in the context of a cleaning task as part of daily living. The motion was captured using an IGS-180 motion capture suit from Synertial UK, which contains 17 IMUs (OS3D from Inertial Labs). They developed a set of rules to automatically segment the raw data and to detect the activities including finding the maximum or minimum in the raw data, angular velocity of relevant body segments, the magnitude of acceleration, normalized angle of relevant body segments, etc.

Oubre *et al.* [58] reported proposed a new way of estimating the upper-limb impairment level in stroke patients using a random motor task using a single Xsens MTw IMU. They aimed to shorten the task that the patients must be performed to reduce their burden. They experimented with the algorithms for the estimation, one unsupervised clustering algorithm, and the other supervised regression model.

Parrington *et al.* [45] reported a validation study on using two Opal IMUs (from APDM) to measure head and trunk range of motion and peak rotational velocity while doing a set of standing and walking tasks for patients with mild traumatic brain injury. During these tasks, the participants were asked to move their head with and without moving the trunk. These tasks are referred to as vestibular rehabilitation tasks. The motion tracking using IMUs were validated with respect to a 12-camera motional analysis system called Raptor-E from motional analysis.

Patel *et al.* [67] reported their work on using an IMU (Vitaport 3 from Tempec BV) to collect motion data while a stroke patient is doing a set of designated motor tasks.

They subsequently extracted features from the raw data and used random forests to estimate the score for each individual task. They then computed the overall functional ability scale score based on the scores for individual tasks.

Pérez *et al.* [59] presented a kinematic model to estimate upper limb motion based on the data collected via four Xsens MTi IMUs. They validated their model using a BTS SMART-D multi-camera motion analysis system. Although the work was intended for neurorehabilitation, the model could be used for other purposes.

In [65], the authors studied how the perceived lower back pain (LBP) would impact the patient's functional capacity and actual physical activity (in terms of walking). They used an Inertia-Link IMU from MicroStrain to capture motions related to the functional capacity using three tasks, gait, sit-stand-sit, and block stepping. They used an accelerometer from GC Dataconcepts to measure gait parameters in the patient's daily life. They found that LBP would impact the patients' functional capacity, but not their actual physical activity.

Schwenk *et al.* [68] reported their study on predicting future falls in people with dementia using motion data collected via an IMU called physilog, which contains an accelerometer and a gyroscope. The device can output a person's postural positions and several types of movements in daily living such as walking, standing, sitting, or lying. They then calculated the physical activity parameters from the raw data in terms of the percentage of different types of activities daily, and used that to predict falls.

Seiter *et al.* [51] proposed a method to automatically classify activities of daily living in hemiparetic rehabilitation patients using motion collected by six Shimmer IMUs. Only accelerometer data are used in their study. After extracted the basic features in the time and frequency domains, they developed a set of activity vocabulary using two alternative ways, one based on clustering, and the other based on a set of rules. The activity words were then used to perform activity topic discovery using Latent Dirichlet Allocation (i.e., to detect recurring activity routine patterns). The final step was to use KNN to classify the activity topics into activity routines.

Tulipani *et al.* reported their work to validate the use of Opal IMU to measure segment angular displacement while performing functional tasks [46]. They used a 9-camera motion analysis system from Qualisys AB as the reference system to compare with the result obtained from the IMUs.

In an earlier work, Zhou *et al.* proposed a method to estimate upper limb movements with kinematic models and a simulated annealing optimization to reduce errors [60] using two Xsens MT9-B IMUs (one attached to the elbow and the other attached to the wrist). In a follow-up paper, they reported a similar work with two newer generation of Xsens MTx IMUs to estimate arm movements [61].

1.4.3   Studies using consumer-grade devices containing IMUs

It is also common for consumer-grade devices being used in research and clinical studies due to their low-cost, availability, and perhaps better usability. These devices contain various IMUs and are often not designed for professional-level motion capture. Nevertheless, they have been shown to be accurate enough for collecting motion

Table 1.3   Studies using consumer-grade devices containing IMUs

IMU brand	IMU	Use case	Refs
Wii Remote	acc	Measure roll and pitch angle	[52]
Wii MotionPlus	acc and gyro	Measure 2D movement on vertical plane w.r.t. ground	
Sony Move	acc, gyro, mag	Measure various joint angles and position tracking	
Samsung Galaxy SII	acc, gyro, mag	Measure various joint angles and position tracking	
iPhone 4	LIS331DLH (acc)	Measure angular movement	[70]
Samsung Galaxy S5 mini		Activity level classification using SVM for COPD patients with motion data	[71]
iPod	2 units used at scapulae and L5 level	Measure angular excursion of thoracic and pelvic trunk motion	[72]
iPod touch 4	LIS331DLH (acc) and L3G4200D (gyro)	Estimate gait parameters from the phone IMUs	[73]
Android smartwatch	acc, gyro	Gait patterns classification using ML for patients who have gone through lower body orthopedic surgery	[74]

data objectively for various physical activities. The studies using this type of devices are summarized in Table 1.3.

Bai *et al.* [52] examined the accuracy of several consumer devices with embedded IMUs using the Xsens MTx IMU as a reference for comparison. The consumer devices included the Wii Remote, Wii MotionPlus, and Sony Move game controllers, and the Samsung Galaxy SII smartphone. The Wii Remote is only equipped with an accelerometer sensor. The Wii MotionPlus incorporated a gyroscope in addition to an accelerometer. The Sony Move controller and the Samsung smartphone both consist of accelerometer, gyroscope, and magnetometer. More specifically, the authors provided technical information regarding the IMU components used in Sony Move, the accelerometer is Kionix KXSC4, the gyroscope is a triaxial, and the magnetometer is AKM AK8974. Hence, they are capable of measuring joint angles and positions of the upper limbs fairly accurately.

Bittel *et al.* [70] validated the accuracy and the precision of a mobile app running on iPhone 4 for angular movement measurement using a isokinetic dynamometer system.

Cheng *et al.* [71] used smart phones (Samsung Galaxy S5 mini and LG Optimus Zone2) to collect activity data to automatically classify the type of patients with chronic obstructive pulmonary disease (COPD). More specifically, they aim to determine the pulmonary function levels that are termed as global initiative for chronic obstructive lung disease levels 1, 2, and 3 using a machine-learning method called support vector machine based on a set of features extracted from the activity data collected. The feature vector consists of eight spatiotemporal gait parameters.

Kegelmeyer *et al.* [72] reported a study on using two iPod to collect data on trunk control for patients with Huntington's disease. One iPod is placed at the lower border of scapulae. The other is placed at the L5 level. The iPod is equipped with professional software for data collection and for providing biofeedback when the pelvis moved greater than eight degrees. The participants were asked to perform several activities, including sitting, standing, and walking.

Kyristsis *et al.* [74] proposed to use a smartwatch worn on the wrist to recognize gait patterns for patients who have done lower body orthopedic surgery. They identified nine gait patterns and used machine learning algorithms to classify the data collected via the smartwatch. Prior to training and testing using the machine learning algorithms, the raw data were segmented (with a 5-s window) and features were extracted in the time and the frequency domain.

Steins *et al.* [73] reported their work on extracting gait parameters using IMUs embedded in an iPod touch 4, which consists of a LIS331DLH accelerometer and a L3G4200D gyroscope, both from STMicroelectronics. They compared the result with both that of an Xsens IMU and that of an optical motion capture system called Oqus 300 from Qualisys. More specifically, they used the device to first estimate the center of mass of the participant. They then derived gait parameters such as the individual and average step data based on the peak-to-trough differences in position and acceleration.

1.4.4   Studies using wearable trackers

This is a special type of consumer-grade devices that are designed to collect activity information to promote healthy living styles for the general population. Typically, these devices are inexpensive and easy to wear on the wrist or other designated places (such as waist) and have very limited functionality and user interfaces. However, in recent years, these devices are providing more and more sophisticated interfaces and functionalities, some even resemble a smartwatch. The studies using this type of devices are summarized in Table 1.4.

Appelboom *et al.* [75] used Fitbit Zip wearable trackers to measure the gait parameters, such as gait velocity and step length, for both postoperative patients and healthy participants. To see which sensor placement gives the most accurate measurement, they used four Zip devices and placed them in different positions: left and right hips, and left and right ankles, respectively. They found that the ankle placement of the Zip device provided better accuracy than the hip placement for the postoperative patients. The placement does not impact the accuracy for healthy participants (as control).

Braun *et al.* [76] reported a study using a StepWatch Activity Monitor from Orthocare Innovations to monitor the walking activity of youth with lower limb salvage. The device was used to collect daily step counts. The purpose of the study was to see how long they have to monitor to collect enough data to determine the outcome of rehabilitation of these individuals.

Capio, Sit, and Abernethy [78] experimented with using a tracker called MTI from Actigraph to objectively measure the activity levels of children with Cerebral

Table 1.4   Studies using wearable trackers

Brand	IMU	Use case	Refs
Fitbit Zip	acc	Gait parameters	[75]
StepWatch Activity Monitor, Orthocare Innovations		Walking activity monitoring for youth with lower limb salvage	[76]
StepWatch Activity Monitor, Orthocare Innovations; Piezo Step MX; ActivPAL3		Step counting accuracy	[77]
MTI, Actigraph	acc	Physical activity counts tracking for children with cerebral palsy	[78]
GT3X+, Actigraph	3-axial acc	Step counting during sub-acute rehab post stroke	[79]
SenseWear Armband & Digiwalker SW701		Estimating step count and energy expenditure	[80]
SenseWear Armband		Study on side effects and usability	[81]
Active Style Pro HJA-750C, Omron	3-axial acc	Duration of different levels of physical activities	[82]
DynaPort Activity Monitor and SenseWare		To objectively measure the physical activities of patients with COPD in daily life	[83]

Palsy. MTI consists of a one-dimensional accelerometer and outputs an activity count. The device was placed in a pouch and positioned over the right hip of the participant using an elastic belt to attach to the participant's waistband. They made sure that the device is aligned with the person's midaxilla. In addition, each participant also wore a heart rate monitor where the heart rate was expected to be consistent with the activity levels. The authors found strong correlations between the MTI output and the heart rate with a gold standard.

Furlanetto *et al.* [80] reported their study on the accuracy of two consumer-grade devices for step counting and the measurement of energy expenditure in patients with COPD and healthy seniors. One device is SenseWear armband with multisensors, and the other is a Digiwalk SW701 pedometer. In their study, the participants were video tapped, and the actual step counting was manually determined as the ground truth. They found that at high speed, the pedometer performed accurately when the participants were walking at high speed, but had underreported significantly in low speed. The SenseWear armband failed to register step count, but somehow reported fairly accurate energy expenditure.

Joseph *et al.* [79] proposed to use a Actigraph GT3X+ to monitor the physical activity level during the sub-acute rehabilitation phase of stroke patients. The participants each wore an Actigraph device around the hip using an elastic band to record the number of steps taken.

McCullagh *et al.* [77] reported their study on the step count accuracy with three IMU-based devices, including a StepWatch Activity Monitor (SAM), a Piezo Step MX pedometer, and an ActivPAL3 3-axial accelerometer. They used the video tape of the sessions to establish the ground truth. They showed that the ActivPAL3 provides the highest accuracy in step counting when worn at the ankle.

McNamara *et al.* [81] reported their findings on the side effects and usability of the SenseWear Armband. They surveyed 314 people, among them, 252 had COPD, 36 had dust-related respiratory disease, and 26 were healthy. Many of them reported discomfort or even adverse side effects such as skin irritation, and the device was hard to take-off and put-on. These findings are important because indeed many such devices failed to adequately consider usability and comfortness to their users.

Miyamoto *et al.* [82] presented a study to validate a consumer-grade motion monitor called Active Style Pro HJA-750C from Omron, which consists of a 3-axial acc, to keep track of the physical activities of COPD patients. The device can report the durations of different levels of physical activities. They showed strong correlation of the data acquired by this device with respect to other more well-studied devices such as DynaPort Move Monitor or Actimarker.

Probst *et al.* reported their study on the outcomes of two exercise training programs on the physical activity level in daily life in patients with COPD [83]. They used two wearable sensors. One is a DynaPort Activity Monitor, which measures the time spent per day on walking, standing, sitting, and lying. The other is a SenseWare armband, which estimates the energy expenditures (total per day and each significant activity).

Chapter 2
Force and pressure sensing

In addition to IMU, human motion can also be measured by the force exerted on the entity that a person is in contact with, such as the floor when walking. Furthermore, while a person's joint moves, such as joint extension or flexion, the movement of the body segment around the joint could also be estimated by the force applied to the wearable materials if they are used. Similarly, the trunk posture can be estimated by the force applied to the sensors attached to the body segment.

When some force is applied to an object, the force can be measured through the pressure experienced by the object per unit of 2-dimensional area. Given a force F, and the area A that the force is in contact with an object, the pressure P can be defined as $P = F/A$. The smaller the area, the higher the pressure. The unit of pressure is Pascal, or Pa for short, which is defined as one Newton per square meter.

The fundamental idea of force sensing is to transform the pressure into some measurable quantity, such as the change in voltage (in piezoelectric pressure sensors), resistance (in resistive pressure sensors), capacitance (in capacitive pressure sensors), and wavelength (in optical pressure sensors).

2.1 Types of pressure sensors

2.1.1 Piezoelectric pressure sensors

This type of sensors utilizes a peculiar characteristic of the piezoelectric materials to measure the pressure applied to them. More specifically, when the pressure is applied to the piezoelectric material, the material is compressed, which in turn will generate a voltage proportional to the applied pressure, as shown in Figure 2.1. Piezoelectric sensor has many advantages over other types of pressure sensors. The most notable advantage for motion tracking is that the sensor does not require any external energy source for pressure sensing. Furthermore, piezoelectric materials have very little deflection, i.e., there is little structural displacement under pressure, which gives the sensor high degree of ruggedness. Hence, piezoelectric sensors can be used in smart shoes or insoles for pressure sensing. Other advantages might not be directly relevant for human motion sensing. For example, piezoelectric materials have very high natural frequency, provide high linearity over a wide amplitude range, and they

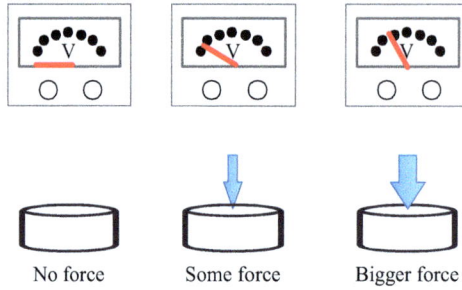

Figure 2.1   Working principle of piezoelectric pressure sensors

are insensitive to radiation and electromagnetic fields. Some piezoelectric materials can operate in very high temperatures.

However, piezoelectric pressure sensors are not suitable to measure static forces. This is due to the fact that a static force would cause the piezoelectric material to generate a fixed amount of electric charge. Furthermore, common electronics that are used to read the output would cause a constant loss of electrons and the signal would decrease over time.

Piezoelectric pressure sensors typically use one of the three materials: (1) piezoelectric ceramics, (2) single crystal materials, and (3) piezoelectric thin films. Piezoelectric ceramics can be fabricated using a sintering process and usually have fairly high piezoelectric sensitivity. However, the sensitivity would degrade overtime, especially in high temperature. Single crystal piezoelectric materials have smaller sensitivity than piezoelectric ceramics, but they have much better long-term stability. Piezoelectric thin films are used in MEMS devices, which offer much smaller footprint than the previous two types [84]. According to a comprehensive review on piezoelectric MEMS [84], piezoelectric thin films typically have 1–3 μm thickness.

2.1.2   Resistive pressure sensors

In some metals and semiconductors, the strain applied could trigger significant but linear resistance changes. This effect is referred to as piezoresistive effect. In metals, the piezoresistive effect can be observed simply due to the stretch of the metal where the resistance would be increased if the cross-sectional area becomes smaller using the Ohm's law, as shown in Figure 2.2. When the across-sectional area shrinks from A to A' and the length increases from l to l', the resistance R' will become larger than the original resistance R. We should note that in some semiconductors, the piezoresistive effect can be much larger than the resistance changes caused by geometrical changes.

The piezoresistive effect has been used to make strain gauge, which is a form of pressure sensor. In strain gauges, typically a metallic foil pattern, which consists of zig-zag lines, is used. Strain gauges often provide a gauge factor, GF, as a key specification. GF is defined as $GF = \frac{\delta R / R_G}{\varepsilon}$, where δR is the change in resistance,

Figure 2.2 Working principle of piezoresistive pressure sensors

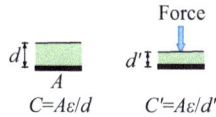

Figure 2.3 Working principle of capacitive pressure sensors

R_G is the resistance without any load on the gauge, and ε is the strain applied to the gauge. The gauge factor is typically slightly over 2 for metallic strain gauges.

2.1.3 Capacitive pressure sensors

The capacitive pressure sensors measure the pressure based on the changes of the capacitance. The sensor consists of a capacitor with two capacitive plates . One of the plates is a diaphragm that will move under pressure, and the other is an electrode and it is fixed. When the diaphragm moves under pressure, the gap between the two plates shrinks, which increases the capacitance, as shown in Figure 2.3. The capacitance C is determined by the area of the plates A, the dielectric permittivity of the insulating medium between the two plates ε, and the distance between the two plates d, i.e., $C = A\varepsilon/d$. In the presence of a force, the distance will be reduced (to d' as shown in the figure), then the capacitance will be increased to C'.

2.1.4 Optical pressure sensors

Optical pressure sensors utilize various ways to capture the effects exhibited due to the deformation of the structure through the measurement of light. The simplest way is to mechanically block part of the light in the presence of pressure. Newer sensors depends on interferometry. A big difference between this type of optical pressure sensing and other types is that it requires a laser light source and a detector outside the sensor itself. One way to implement the interferometry is via the Fabry–Perot interferometer (FPI). The other way is Fiber Bragg Grating (FBG). The FPI can operate on cavity length changes or cavity refractive index changes in the presence of pressure [85]. The FBG method operates on the wavelength change due to the deformation caused by the pressured applied either longitudinally or laterally [86], as shown in Figure 2.4.

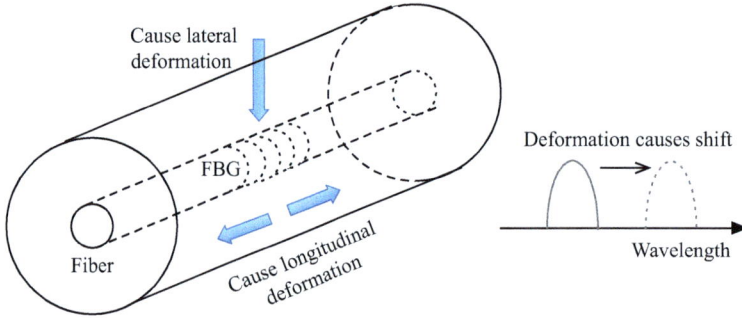

Figure 2.4 Working principle of capacitive pressure sensor

2.2 Applications in motion tracking for rehabilitation

2.2.1 Epionics SPINE system

The Epionics SPINE system is developed by Epionics Medical GmbH, Germany. The system consists of two stripes of strain gauge sensor arrays. These sensors are used to measure the bending angles on the back surface of a participant, one stripe is placed on each side of the spine. Each stripe has 12 sensors so that the angle at each spine location can be measured accurately. This system has been used by several studies to measure lumbar back motion during functional activities [87–89].

In [87], Suter *et al.* asked participants to perform a list of activities, including standing, sitting, rising from chair, box lifting, walking, and running while collecting data on the lumbar lordosis angles using both the Epionics SPINE system and a multi-camera Vicon motion analysis system. The latter was used as a reference to validate the Epionics SPINE system. The authors performed Bland–Altman analyses on the results obtained by the two systems, and they found that the two systems generally have high agreement.

In [88], Consmuller *et al.* studied the velocity of lordosis angle during maximum flexion and extension activities with 429 healthy participants. The objective of their research is to study the range of functional kinematics of the spine in terms of the velocity of lumbar shape changes. The Epionics SPINE system allows the rapid assessment of functional kinematics in the lumbar spine. One direct benefit for their study is that their results could serve as normative data to compare patients with spinal pathology or after surgeries.

In [89], Dideriksen *et al.* studied spinal movements in functional tasks for both patients with low back pain (LBP) and healthy participants using the Epionics SPINE system. While they found no difference in task-related spine angular movement between the two groups, they did find that the movement variability not direct related to task execution was much higher for patients with LBP than that for healthy participants. They concluded that the structure of the variability of spinal movement was different in patients with LBP.

In [90], Vaisy *et al.* measured the lumbar spine functional movement using the Epionics SPINE system in patients with low back pain. They used the system to measure the maximum range of motion and the maximum time of motion of the lumbar and pelvis tilt. They also studied the association between kinematic characteristics obtained from the sensor measurement and the clinical characteristics.

2.2.2 Force plates

Force plate is a mechanical device that measures the group of reaction forces of a human while doing activities such as walking, running, or jumping. In addition to vertical force, multi-axis force plates can also measure lateral/horizontal forces, which is referred to as shear forces. Typically, a force plate uses load cells to measure the forces. The underlying pressure sensing technologies could be one of those introduced in the previous section.

It is worth mentioning that the Wii balance board can be considered a consumer-grade force plate with a single purpose of measuring the center of pressure. The Wii balance board has been proven accurate and reliable enough in the clinical setting to study stand balance [91–93]. The technical details of the Wii balance board are given in the patent application filed by Nintendo [94]. The board has four load cells, one at each of the board's four corners.

Due to the age of the technology (the first force plate was introduced in 1975 [95]), and the high cost of traditional force plates, they are mostly used as the gold standard to validate newer technologies such as IMUs [96–98]. Occasionally, force plates are used as the primary means for motion tracking in sports training and rehabilitation [99,100].

Portable systems that resemble force plates have been developed and embraced by researchers [101–104]. One such system is the GAITRite walkways system. The GAITRite walkways system can be rolled up and deployed on flat surfaces, and it has many varieties (https://www.gaitrite.com/gait-analysis-walkways). One version of the GAITRite walkways is reported to consist of eight sensor pads in the form of roll-up carpets, and once deployed, it is 61-cm wide and 488-cm long, containing a grid of 18,432 sensors in the grid pattern 48×384 with 1.27 spacing between adjacent sensors [105]. In another version, it has an active sensor area of 61-cm wide and 427-cm long, with 16,128 pressure sensors [106].

2.2.3 Smart insoles and smart shoes

Smart insoles are insoles that have been instrumented with various sensors. Some researchers prefer to use the term smart shoes. In most cases, smart shoes would contain smart insoles. Some so-called smart shoes do not use instrumented insoles. They have been used to collect data on gait characteristics via pressure sensing and possibly other sensing modalities such as IMU. Researchers have used various pressure sensing technologies to measure the pressure applied to the insole, including piezoelectric, piezoelectric, piezocapacitive, optical pressure sensing, and even barometers. Typically, an array of sensors is placed on specific locations to detect pressure points of the foot. Some smart insoles/shoes are also equipped with IMUs to complement pressure sensing.

Most recently published papers on smart insole are reporting custom-made smart insoles that aim to improve the insoles' motion detection accuracy (such as by fusing pressure sensors with IMUs), and improve the insoles' usability (such as by adding the wireless transmission feature). There are also commercial-off-the-shelf in-shoe pressure sensing systems. The Pedar system seems to be a highly established product that became available as early as 2000 [107]. Another one is the F-Scan system from Tekscan (https://www.tekscan.com/products-solutions/systems/f-scan-system).

The Pedar product page (https://www.novelusa.com/pedar) claims that Pedar is "the gold standard for measuring pressure distribution inside footwear." The currently (as of writing in April 2021) posted specification indicates that the system has a maximum of 256 sensors. The product now offers several ways of storing and fetching data, i.e., through tethered method using a fiber optic cable or USB cable, wirelessly through Bluetooth, or store locally on a SD card. The system has an internal 32MB flash storage. The type of pressure sensor used, however, is not disclosed. In a paper published in 2005, the Pedar system was said to have 99 sensors and the foot was divided into 10 regions (heel, mid foot, 1st–5th metatarsal, hallux, second toe, and toes 3–5) [108].

The F-Scan system has also a long history. The first paper on F-Scan is at least as early as 1993 [109], and it has been actively used since [110–115]. The F-Scan product page provided virtually no technical details regarding the product (but provided a long list of papers that have used their product). The papers that used the F-Scan system contain no technical details regarding the product either [109–115].

The definition of foot regions differs. Previously we mentioned that the foot is divided into 10 regions in [108]. In [116], the foot is defined to have nine regions, as shown in Figure 2.5, which include (1) the great tole, (2) other toes (those labeled by s, e, o, and T), (3) the medial metatarsals (i.e., Med Met), (4) the middle metatarsals (i.e., Mid Met), (5) the lateral metatarsals, (6) the medial arch (i.e., Med Arch), (7) the middle arch (i.e., Mid Arch), (8) the lateral arch (i.e., Lat Arch), and (9) the heel. In Figure 2.5, we superimposed the 16 sensors used in [117] in the foot regions. As can be seen, the 16 pressure sensors are predominately placed on 8 of the 9 regions because normally the medial arch does not apply much pressure on the insole.

Prior to putting the smart insoles in practical use, they must be calibrated carefully. According to [117], there are primarily two ways of performing calibration. The first method is to use a combination of force plates and multi-camera-based motion analysis system (such a Vicon) to corroborate the data collected by the smart insoles. This method essentially is a system-level validation to access the consistency of the gait characteristics obtained by the smart insoles and the reference systems. The other method is to use controlled load on the sensors to make sure the measurement is accurate. This is analogous to microbenchmarking a software system. For the latter approach, the CETR University Micro-Tribometer was used to calibrate smart insoles by applying weights in all three X, Y, and Z directions [118]. A robotic platform was used for this type of calibration in [119]. A bar-type load cell was used for calibration in [117]. The performance of some pressure sensors with respect to static and dynamic loads have also been studied [120], which confirmed that piezoresistive sensors have lower accuracy for static pressures.

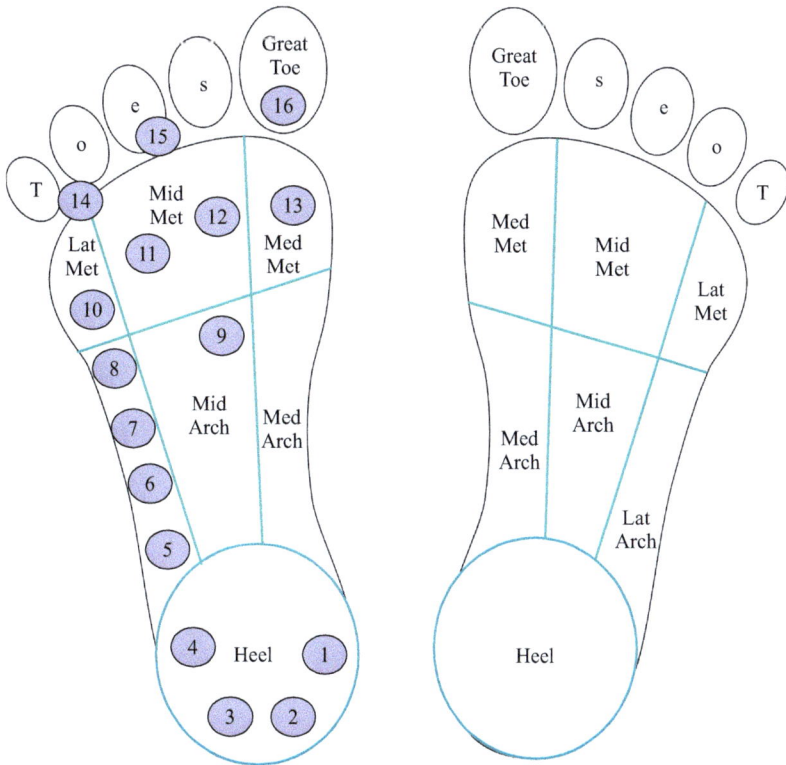

Figure 2.5   Foot regions and an example placement of pressure sensors in the sole

Tahir *et al.* [117] reported the design, calibration, and performance of smart insoles for gait measurement in a comprehensive manner. As mentioned previously, each insole is equipped with 16 pressure sensors, as shown. They experimented with three different types of pressure sensors, including a sensor based on piezoresistive pressure sensing from Interlink Electronics, which they refer to as force-sensitive resistors (FSRs), a sensor based on ceramic piezoelectric materials from Murata Manufacturing, and a MEMS pressure also based on piezoelectric materials (LDT0-028K) from Measurement Specialties. The authors reported extensive calibration and human subject test performance of their smart insoles with these three types of sensors in the paper.

Mohammed *et al.* [113] presented a study on using a commercial-off-the-shelf product that contains smart insoles to study gait characteristics. The product is called F-Scan wireless in-shoe pressure mapping system from Tekscan. Unfortunately, the company website does not disclose the details on the design of the system. The only known fact is that the system uses pressure sensors in the in-shoe components. In [113], Mohammed *et al.* aimed to use machine learning algorithms to automatically segment the gait cycle phases. Ultimately they hope to be able to detect abnormal gait behaviors.

Vilarinho *et al.* [121] proposed a smart insole for foot plantar pressure monitoring using optical pressure sensing. The cork insole is instrumented with polymer optical fiber with five FBGs for pressure sensing. The paper reported the calibration results using a testing machine. The authors also conducted a human subject test on the usability of the proposed insole. However, considering that the light source and actual analysis of the signal are outside the insoles connected by a optical fiber, the actual usability and potential cost could be a concern.

Tao *et al.* [122] designed a smart insole system for real-time pressure mapping. Each insole consists of 24 capacitive pressure sensors. They characterized both static plantar pressure mapping and dynamic pressure mapping using the developed smart insole system. For static plantar pressure mapping, they experimented with three different standing postures, including normal posture, underpronation, and overpronation. For dynamic plantar pressure mapping, they experimented with walking straight, turning around, and going upstairs. The smart insole was connected to an acquisition chip placed at the back of the shoe. Then, the acquired data were transmitted to a nearby computer wirelessly.

Lin *et al.* [123] reported their work on developing a smart insole and the characterization of the smart insole. Unlike other smart insoles, they incorporated IMUs into the insole. Each insole consists of 48 piezoresistive polymer pressure sensors and an IMU (BMX055 from Bosch Sensortec, which has a 3-axial accelerometer, a 3-axial gyroscope, and a 3-axial magnetometer). They tested the smart insole with hallway walking, ascending stairs, descending stairs, and slope walking. They could measure the plantar pressure from both heel and toe groups in these scenes. In addition, they also measured the number of steps taken and the center of pressure.

Howell *et al.* [124] presented their work on using custom-made smart insoles to perform kinetic gait analysis of both normal and stroke patients with walking disabilities. Each insole is equipped with 12 FSR sensors (model 402 from Interlink Electronics). The sensors were calibrated with an iLoad Mini 50-pound load cell from Loadstar Sensors. Initially, they built a 32-sensor insole to test which parts would experience pressure during walking. They found that placing the sensors under the heel, metatarsal–phalangeal joints, great toe, and the lateral arch would be the most effective for gait analysis. That is why they eventually decided to use 12 sensors to cover those areas. The sensors were connected to a ribbon cable, which was then connected to a box with a microcontroller, wireless transmitter, and batteries. They used a 10-camera motion analysis system (Vicon) to validate the results obtained from the custom-built smart insole.

Lee *et al.* [125] studied how to objectively quantifying the walking ability in patients with degenerative spinal disorder using smart shoes. Each shoe is instrumented with 5 pressure sensors, which are placed inside the shoe at different locations to detect heel strike, mid-lateral plantar pressure, toe pressure, and other spatiotemporal information of the gait. In their experiment, the patients were asked to walk on a 10-meter trail at a self-paced speed, then make a turn and walk back along the trail. From the collected pressure sensor data, they obtained 76 spatiotemporal features. Based on these features, they performed higher level classification and lower level regression. Finally, they derived clinical scores using support vector regression.

Kong and Tomizuka [126] reported the development of smart shoes using air pressure sensors (referred to as air bladders) to measure ground reaction forces for gait monitoring. The four pressure sensors are placed at the following areas of the foot under the insole: great toe, the first and second metatarsophalangeal joints (Meta12), the fourth and fifth metatarsophalangeal joints (Meta45), and the heel. Similar smart shoes have used by others [127,128].

Zhang, Tomizuka, and Byl [128] proposed a motion tracking system for gait physical therapy. The system consists of several IMUs to be placed on key joints, and a pair of smart shoes. The smart shoe design followed that of [126]. In [128], the authors referred the sensors as barometric sensors. The battery and the signal processing unit were placed outside the shoe. Their system was validated by using a 10-camera PhaseSpace IMPULSE motion capture system.

2.3   Energy harvesting in smart shoes

Before we close this section and this chapter, we want to highlight an exciting new research direction on energy harvesting in smart shoes. The research is motivated by the observation that the energy could have been harvested when the foot strike the ground due the pressure applied on the insoles and the shoes. Jung *et al.* [129] presented their work on the design and experimentation with an energy-harvesting device in smart shoes. The device incorporates several polymer piezoresistive tension-type sensors and piezoelectric ribbon harvesters. The ribbons are inserted in the insole of the shoe. The authors have experimentally shown that the incorporating of the energy harvesting ribbon could extend the service time of the primary battery, and improve the primary power efficiency. When one is running with the smart shoes, the power generated by the ribbon could approach to that of the primary battery. Besides exploiting the piezoelectric materials, the triboelectric effect has been exploited to harvest energy. The triboelectric effect refers to the phenomenon where certain materials would become electrically charged when they are separated from some other materials with which they were in contact. Triboelectric nanogenerators (TENGs) have been developed to harvest the energy generated as a result of the triboelectric effect. Zou *et al.* provided an excellent review on the use of TENGs in smart shoes for energy harvesting [130]. TENGS can be placed on the insole, embedded into the insole, engineered into the insole, integrated into the sole of the shoe, or placed under the sole.

Chapter 3
E-Textile-based sensing

If textile can be instrumented with stretchable thin electronics, human motion could be monitored much more conveniently and comfortably than using IMU devices by wearing smart clothes made from such textile. With the advancement of stretchable electronics [131], researchers are indeed making inroad in this type of textile, which is often referred to as electronic-textile, or e-textile for short [132].

The different types of e-textiles proposed so far all use some form of strain sensors to detect the strains caused by human motion. The predominant approach is to apply a small DC current and measure the resistance changes in the strain sensors attached to the clothing. Some work used AC current and measure the impedance changes in the strain sensor instead [133]. Due to the particular requirement of applying the strain sensors on clothing (or sometimes directly on the skin), the strain sensors must be thin (typically under a 1 mm) and highly stretchable. Hence, traditional strain sensors are not a good fit. For example, according to [131], the skin on the related body segments (feet, waist, knees) can stretch and contract by up to 55% while traditional strain sensors can only allow 5% stretch/contraction. While several different types of thin wearable strain sensors have been proposed, it appears that the conductive elastomer has been the dominating approach since the early days because they can be applied to the fabric either using a mask or attached to the fabric easily [134–141]. Newer proposed sensors relied on various materials, from commercial stretch sensors [142] and commercial sliver-coated yarn [143], to nanotube-based sensors [131], to copper wire stitched to the shirt [133].

That said, compared with IMU, e-textile is still in its infancy. Although various e-textile-like sensors have been proposed, they need to add electrodes and wires connected to a processing unit with a microprocessor, storage, and wireless transmitter. The processing unit and the associated wiring would make the actual e-textile far less-convenient from the dream of monitoring human motion by only wearing clothes made from e-textile alone. Furthermore, e-textile sensors do not come in the form of textile that can be used to make clothing. They need to be attached to the clothes or directly to the human skin at particular locations. Nevertheless, this is an exciting new research area for human motion tracking.

3.1 Conductive elastomer

Elastomer is a rubber or rubber-like material. More formally, elastomer is a polymer with viscoelasticity, or elastic polymer. Conductive elastomer is a type of elastomer that has been mixed with additives to make the material electrically conductive. Common additives include carbon and silica. The latter has lower cost but poorer conductivity. When conductive elastomer is stretched, its conductivity changes. This characteristic has been used to make wearable strain gauges that can be printed to fabrics to form a type of e-Textile [134,135]. They can also be fabricated as elastic strain gauge sensors that can be attached to the fabric or directly to the skin.

3.1.1 Working principle

The working principle of conductive elastomer is that when it is stretched, its resistance increases (or conductivity decreases). It is basically the same as that for the piezoelectric pressure sensors as we have introduced in the previous chapter and illustrated in Figure 2.2. The sensitivity of this type of materials is characterized by a gauge factor (sometimes referred to as strain factor). Given the length of the material at rest (i.e., non-stretched) is L_0 with resistance R_0, and when stretched at the length L with resistance R, the gauge factor GF is defined as:

$$GF = \frac{(R - R_0)/R_0}{(L - L_0)/L_0} = \frac{(\delta R)/R_0}{\varepsilon} \tag{3.1}$$

where $\varepsilon = (L - L_0)/L_0$ is referred to as the strain.

3.1.2 Attaching conductive elastomer to fabric

To apply the conductive elastomer materials to the fabric, Tognetti *et al.* [134] designed an adhesive mask to cover the fabric. After covering the fabric with the mask, they first applied the lycra substrate, then they smeared the conductive elastomer materials (a solution of elastosil and trichloroethylene). The next step is to remove the mask and treat the instrumented fabric in an oven set to 130°C for about 10 min. The mask was designed to reflect the desired topology of the strain sensor network.

Giorgino *et al.* [135] used a long-sleeved shirt printed with conductive elastomer sensors in a way similar to that of [134] to measure a set of upper-limb rehabilitation exercises, namely, shoulder sagittal abduction (30° flexion, 60° flexion, and 90° flexion), shoulder elevation, trunk lateral flexion, and humeral-upper limb external rotation. The authors deployed 19 conductive elastomer segments over the arm, forearm, and shoulder on one-side of the shirt depending on the patient's side of hemiplegy.

Tormene *et al.* [136] also used conductive elastomer printed on the back of a corset. They used a total of 13 sensors. Their primary objective is to measure the trunk movements, including flexion, extension, rightward and leftward bending. In addition to the classification of different kinds of trunk movements, they also estimated (1) qualitatively the range of motion in terms of small, average or large flexion; (2) the speed qualitatively in terms of slow, average, and fast; and (3) the counting of the number of repetitions. They compared the results with the measurements obtained via an inertial sensor.

Conductive elastomer can be attached to clothing in other ways too. For example, Mattmann *et al.* [137] attached a set of commercial thermoplastic elastomer to a shirt at specific locations using silicone film. To be precise, they placed 21 strain sensors on the back of the shirt to detect shoulder movements, arm movements, bending forward, torso rotations, bending sidewards, lifting shoulders.

3.1.3 Motion tracking with conductive elastomer

To infer the movements from the resistance measurement using conductive elastomer is a highly complex problem. There are predominately two approaches: (1) develop a model on how the body would move for particular movements and use the model to predict the strain placed on the sensor; (2) use machine learning to classify specific set of postures in an exercise using the sensor measurements as input.

Tognetti *et al.* [134] developed a kinematic model of the upper limb. The model considers two factors: (1) the time it takes for each movement (such as shoulder abduction), and (2) the deformation of the body, which would cause the increase or reduction of the length of certain part of the body that corresponds to where the conductive elastomer will be placed on the clothes. As one can image, this calculation is quite complex. There is also a concern of uncertainty due to the imperfect fit of the clothes with the body (which would lead to the uncertainty of the model). Nevertheless, the authors demonstrated highly accurate measurement of the flexion and abduction angles of the three upper limb joints, i.e., wrist, elbow, and shoulder. In earlier work, the same research group explored using symbolic linear reduction algorithm to derive the upper limb joint angles [144].

Instead of developing a sophisticated kinematic model, Giorgino *et al.* [135] used machine-learning algorithms to classify the types of exercises that the subject is performing as well as the neutral position where the limb is parallel to the body with supine forearm. The input is the information gain from each sensor. Their intuition is that those that have the strongest information gain are the sensors that have been stretched the most during a specific exercise. Hence, the information gain from the set of sensors can be used as the input to differentiate different postures of the upper limb. They experimented with the nearest neighbor, logistic regression, and C4.5 algorithms with fairly good result. They also analyzed their results and acknowledged the weakness of using information gain as input, i.e., redundant sensors could have identical information gain values.

Mattmann *et al.* [137] also used the machine learning based approach to identify 27 different postures using the input from the 21 strain sensors with the Naive Bayes classifier. These 27 postures include a set of postures while the subject is sitting, and a set of postures while the subject is standing. The sitting postures include (1) rotation of trunk to the right, (2) rotation of trunk to the left, (3) bending trunk sidewards to the right, (4) bending trunk sidewards to the left, (5) lifting right shoulder, (6) lifting left shoulder, (7) lifting both shoulders, (8) slumped with the shoulders over hip, (9) bending trunk forward with bent back, (10) bending trunk forward with straight back, (11) bending trunk forward maximally with hands beside the feet, (12) forced upright, (13) arms to the front, (14) arms to the sides, and (15) arms over head. The standing postures include (1) rotation of trunk to the right, (2) rotation of trunk to the left, (3) bending trunk sidewards to the right, (4) bending trunk sidewards to the left, (5) lifting both shoulders, (6) slumped, (7) bending trunk maximally forward with hands approaching toes, (8) forced upright, (9) extending arms to the front, (10) squatted while maintaining straight back, (11) flexing torso sidewards to the right, and (12) flexing torso sidewards to the left. The results are quite good. The authors also did a thorough analysis on the results.

To process the data obtained using conductive elastomer sensors, Tormene *et al.* [136] first applied the principal component analysis to reduce the dimensions, they then used the dynamic time warping (DTW) to compute the similarity between the movements. Based on the DTW distance, they performed classification of different movements.

3.1.4   New development

Bergmann *et al.* [138] reported the fabrication and evaluation of a new type of elastic strain gauge sensor that belongs to the family of conductive elastomer. They used a combination of graphitized carbon black nanopower with polyether resin Texin 985 from Bayer AG to make the sensor. They found that adding 20% carbon nanopower produced the best result. Then, they made individual sensors with electrodes that are ready to be connected for the data acquisition electronics. To use the sensors for motion tracking (i.e., knee joint angle measurement), they sewed the sensors to clothing (legging). Instead of developing a formal kinematic model or using machine learning, they explored which part of the raw data could be used to best correlate to the target knee joint angle and then used that portion of data to predict the knee joint angle. The ground truth was established using a Cybex dynamometer from CSMI Inc. to simultaneously measure the knee joint angle. In a follow-up study [139] using the same sensor, Papi *et al.* adopted a different approach in the classification of a set of activity while placing a sensor to the knee area in the leggings. First, the raw data acquired from the sensor were converted into the frequency domain. The frequency domain characteristics were computed, including the median frequency, power of the spectrum, peak frequency, and maximum spectral amplitude. Then, the frequency domain features, the output range of the time domain signal, and the anthropometric information of the participants including the gender, age, height, and leg length were used to classify the a set of activities, including walking, running, and ascending and descending stairs, using random forest machine learning algorithm.

Recently, Esfahani *et al.* [140] presented the design, fabrication process, the validation of a new type of stretchable sensor. It is referred to as a conductive flexible nylon/lycra fabric strain gauge, but broadly speaking, it is also a form of conductive elastomer-based sensor. The sensor is developed using polypyrrole (PPy) 1,5-naphthalenedisulfonic acid with screen printing and chemical vapor deposition. It has high linearity and low hysteresis error. Hysteresis is a phenomenon that the sensor reading would be dependent on the history, which is not desirable. Hence, the lower the hysteresis error the better. The sensor was later used to measure upper body motions [141]. The back of a shirt is attached with 11 of the sensors to measure thoracic flexion and extension, lateral bending, axial rotation, and bilateral shoulder motions. Using the sensor measurements, the authors experimented with doing both posture classification and angle estimation using a two-layer feed-forward artificial neural networks.

3.2   Commercial elastic sensors

One commercial elastic sensor used in rehabilitation research is the sensor from StretchSense, Auckland, New Zealand (https://stretchsense.com/). The company appears to be specializing on instrumented gloves now, but the company website provides no technical details on their products. In [142], Poomsalood *et al.* used the sensor from StretchSense to measure knee range of motion. According to [142], the sensor from Stretch Sensor can stretch to up to three times of the rest length. This is important so that the sensor does not restrict the movement of the subject. The authors used a Cybex dynamometer to establish the ground truth for validating the sensor in knee range of motion measurement. Their study showed that seven out of nine healthy participants had high coefficient of determination and low root mean square error of less than 5 degrees.

In [143], Hu *et al.* used the commercial product from Mitsufuji Corporation called Agposs T1, which is a silver-coated yarn, to fabricate a stretchable sensor for human motion capture. They used the sensor to measure the knee movement both indoors and outdoors. They used the multi-camera system VICON to establish the ground truth to validate the fabricated sensor.

3.3   Other approaches

In [133], Sardini *et al.* instrumented a shirt using enameled thin copper wire of 1-mm diameter on both the front and the back side. More specifically, they sewed the copper wire in a zigzag pattern allowing the stretching of the shirt (therefore the wire as well). The wires serve as a sensor to detect the person's posture. They applied AC current to the wires and measured the inductance using a commercial impedance analyzer HP4194A. Similar to a strain gauge sensor where its resistance changes when strain is applied to the sensor, the impedance would change when the sensor is stretched and resorted to its rest position. Hence, by measuring the impedance, the posture can be inferred. They validated their design using a marker-based optical measurement system Codamotion, which consists of three Coda CX1 cameras.

In [131], Yamada *et al.* reported the fabrication and testing of a stretchable carbon nanotube film strain sensor that can be attached to fabric or the skin directly for human motion tracking. They demonstrated a variety of applications of their sensor, including placing the sensor on a stocking (for knee joint angle measurement), adhering to the throat (to monitor phonation via detecting movement of the laryngeal prominence), and attaching to the glove (for finger movement tracking).

Chapter 4
Muscle activity sensing with myography

Motion sensing can be augmented by measuring the muscular contraction and relaxation using various myography techniques. An important benefit for measuring muscular movements is that one can infer the *intention* of a patient, i.e., what activity the patient is *attempting* to do [145] with the information provided by the muscles that are not damaged, even if the patient could not actually perform the intended activity due to the presence of paralyzed muscles (e.g., due to stroke). Conventional motion sensing modalities can only measure what activity the patient is doing. Understanding what the patient is attempting to do is especially import for home-based self-monitoring system that is designed to help the patient perform rehabilitation exercises at home. Such a system could provide more specific and constructive feedback to the patient towards regaining muscle function [145].

Muscular contraction can be measured in several ways and this has led to several types of myography techniques being developed. The most well-known technique is electromyography, where the muscle activities are detected via the measurement of electric potential generated by the muscle cells. One could argue that electromyography is the most direct way of measuring muscular contractions. Another one is mechanomyography, which is a way of studying muscular contraction via measuring the low-frequency vibrations caused by the contraction movements. Force myography is another way of studying muscular contraction by measuring the force exerted by the muscular expansion and contraction at the surface using force sensors. Muscle contraction and relaxation can also be measured optically by placing markers on the muscles. This method is referred to as optical myography.

4.1 Electromyography

Electromyography, or EMG for short, is the oldest type of myography. The history of the EMG technology was documented in [146]. Some of the history, and particularly on who coined the term "electromyograph" was attributed to an earlier book on EMG [147]. Unfortunately, that book does not seem to have mentioned this fact. Nevertheless, the development of modern EMG technology started around 1800s. In that era, an important pioneering researcher in this field is Emil du Bois-Reymound. His research contributions on mechanical neuroscience is documented in an article [148].

There are two types of EMG. One is nonintrusive called surface EMG where the EMG electrodes are attached to the skin of the patient to collect signals. The other type is intramuscular EMG where the electrodes are inserted into the muscle, which is intrusive. Due to its intrusiveness, intramuscular EMG is typically only used in clinical settings. Surface EMG is predominately used in the rehabilitation and other cases.

EMG signal processing is quite sophisticated due to the need to remove various artifacts and noises, and the need to amplify the signal. The skin, electrode, leads, and the amplifier all could introduce noises. For the signal to be useful, it will be amplified, filtered, and quantified [147]. The final result is often expressed in terms of the root mean square (rms) of the amplitude in microvolts. In general, the stronger the muscular contraction, the stronger the EMG signal. For classification purposes, the EMG results would be further analyzed in the time and/or frequency-domain to extract distinctive features. Although there are vast literature on EMG, most of them are used in the context of prosthesis control, exoskeleton control, robotics [149]. Some used EMG for the purpose of gesture recognition [150] or clinical diagnosis [151].

In the context of rehabilitation, EMG is typically used in conjunction with inertial sensors to achieve higher accuracy in recognizing the exercise tasks performed during different stages of rehabilitation. In the two studies that focused on rehabilitation [145, 152], the same sensing equipment, KinetiSense (apparently different versions because one study was published in 2008 and the other was published in 2015), from Cleveland Medical Devices Inc. KinetiSense consists of a command module and multiple sensor units. Each sensor unit has an EMG sensor and IMU with a 3-axial accelerometer and a 3-axial gyroscope [152]. By combining EMG with IMU, KinetiSense can collect information on both the forces associated with the action (i.e., kinetic sensing via EMG), and the motion of the action (i.e., kinematic sensing via IMU) [149]. The two studies are summarized in Table 4.1.

4.1.1   EMG in upper-extremity stroke therapy

In this subsection, we introduce the work by Giuffrida *et al.* on using an early version of the KinetiSense device to perform task discrimination for the purpose of upper-extremity stroke therapy [145]. Six sensor units are used together to collect a total of 12 EMG channels (i.e., each sensor unit has two channels) with each channel collecting information from a particular muscle, which has a corresponding target action. For example, the target action for the biceps muscle is the flexion and supination of the forearm. For the upper-extremity stroke therapy, five tasks were carefully selected

Table 4.1   Studies based on EMG

Setup	Human activities	Method	Refs
KinetiSense	Upper extremity stroke therapy	K-means clustering	[145]
KinetiSense	Walking and balance of ACL-R patients	Fuzzy clustering	[152]

for the study: (1) finger and wrist extension, (2) wrist extension and finger flexion, (3) arm sawing motion, (4) forearm supination and wrist deviation, and (5) forearm pronation and wrist deviation. For each task, every subject was asked to perform 10 trials (i.e., basically each repetition) if possible. Specifically, practice sessions were intentionally omitted to simulate a worst-case scenario for the performance of the proposed task recognition algorithm. The study has five subjects with stroke and eight normal subjects without neurological impairments.

One of the primary research objectives of the study by Giuffrida *et al.* is to develop an adaptive algorithm that could identify accurately the task the patient is practicing while the patient is regaining the muscle functions over time [145]. The authors chose to use the K-means clustering algorithm because this algorithm can be made adaptive by gradually updating the cluster center as new data come in. The input to the K-means clustering algorithm is based on time-domain features extracted from the EMG data and IMU data. To minimize the sensors involved, the authors determined that it is best to combine EMG and gyroscope data, i.e., the accelerometer data are not used for task identification. The rationale is that the gyroscope data reflect the angular movement of the tasks better than the accelerometer data.

The features used for task recognition was created in several steps. First, each channel of EMG was averaged using a moving window (for every 30 data points), which was followed by normalization step that produced a zero-mean distribution. Second, every pair of the 12 normalized EMG channels was multiplied to create a new vector, and the elements in this new vector were summed together to produce a single value. This will create the final feature vector of 66 dimension (there 66 different combinations from the 12 channels). From the explanation of the paper, it appears that each trial is represented by this single 66-dimension vector for the EMG modality. One would have expected a matrix instead. How exactly this final vector is computed, therefore, is unclear. Nevertheless, the accelerometer and gyroscope data were processed exactly the same way to produce a feature vector for classification.

Five clusters were used so that each cluster corresponds to a different exercise task. The initial center for each cluster is determined by using a training set, which is taken randomly from 80% of the collected data. For testing, the Euclidean distance between the feature vector of a trial and each of the five cluster centers is computed, and the trial is assigned to the cluster that has the smallest distance. One place of the paper mentioned that the cluster center would then be updated after the trial is assigned to a cluster. First, we suppose this is done only when the 20% of the generalization data were tested because the cluster center already reflected the 80% training data. Second, because the classification accuracy is not 100%, adjusting a cluster center this way could exacerbate the accuracy of the algorithm, even although it is adaptive.

The authors compared the task recognition accuracy of seven different combination of sensing modalities, including using only EMG. Additionally, the authors performed an analysis of variance (ANOVA) using patient type (stroke or normal), therapy task, and sensor-type combination as independent variable, respectively, and the task recognition accuracy as the dependent variable. The overall task recognition accuracy is 89.61% for normal subjects, and 82.27% for stroke subjects. The sensor type combination is confirmed to be a significant variable only when EMG is included. The authors decided to use the EMG and gyroscope combination to minimize the sensors involved (which also simplifies the processing).

4.1.2   EMG in recovery progress evaluation of anterior cruciate ligament reconstructed subjects

In this section, we introduce the work by Malik, Senanayake, and Zaheer [152]. The authors used a later version of KinetiSense as the instrument for data collection, where each sensor unit has a 12-bit EMG sensor (i.e., 960 Hz sampling rate) and an IMU with an accelerometer and a gyroscope. The study aims to validate an algorithm to assess the recovery progress of patients who had their anterior cruciate ligament reconstructed (ACL-R).

Due to the nature of the rehabilitation needed, five tasks are centered around ambulatory activities and balance: (1) normal walking on flat surface without speed requirement; (2) high speed walking on a treadmill at 7 km/h; (3) high speed walking on a treadmill at 8 km/h; (4) single leg balance testing while eyes open on the balance training platform; (5) single leg balance testing while eyes closed on the balance training platform. For each task, a subject was required to perform three trials with about 1 min rest in between two trials. Each trial of the ambulatory activity lasts about 1 minute and each trial of the balance task lasts about 20–30 s.

To collection information regarding the gait tasks, four IMUs were placed at the thighs and shanks, and four EMG sensors were placed on the four different knee extensor and flexor muscles. For the balancing tasks, all IMUs and EMG sensors were placed on the weight bearing leg. The collected IMU data were filtered and segmented. The collected EMG data were normalized to have zero-mean and then band-filtered. For ambulatory activities, the segmentation was done with respect to the gait cycle. Then, for each trial, ten gait cycles were identified and averaged. For balancing activities, three different segmentation sizes (10, 15, and 20 s) were used.

The preprocessed IMU (accelerometer and gyroscope) data were fused together to derive kinematics features, which include flexion and extension, internal and external rotations, abduction and adduction of the knee joint along the sagittal, frontal, and transversal planes. This is done for seven phases of a gait cycle. Hence, the total number of kinematics features for the ambulatory activities is $3 \times 3 \times 3 \times 7 = 63$. For balance activities, the features include the mean, max, and min values for the 3D rotations were computed for each segmentation size (i.e., 9 features). The preprocessed EMG data were transformed via the multilevel discrete wavelet decomposition analysis. This resulted 840 EMG features for each trial of the ambulatory activities, and 150 EMG features for each trial for the balance activities.

Next, the authors performed a multivariate analysis of variance (MANOVA) to determine if the normal group and the ACL-R group are statistically different when represented by the means of the set of features. The final result is indeed that the two groups are significantly different. Because the number of features for each activity is quite large, the authors performed a principal component analysis to reduce the dimension by retaining only those principle components.

After the feature reduction step, the principal components in the features were then used to construct a set of clusters using the fuzzy clustering and classification techniques. For each activity, four different clusters were created, which represented the healthy subjects, and the ACL-R subjects at three different stages. The later the stage, the closer to the healthy subjects. For classification, 75% of the data were used as the training set, and the remaining 25% were used as the testing set. The accuracy for ambulatory activities (over 90%) are significantly higher than that for the balance activities (slightly over 80%). The inclusion of EMG features increased the accuracy compared with using only the kinematics features extracted from IMU data, except the walking at normal speed, which actually lowered the accuracy by about 1%. For the balance activities, the 15-s segmentation gave the highest accuracy. The ground truth for the classification was established by relying the experts' subjective evaluation of the stroke subjects.

4.2 Machanomyography

Machanomyography (MMG) is to study the muscular contraction by detecting the vibration resulted from the lateral movement of muscle fibers. It was first proposed by Orizio in 1993 [153] as a way to study human muscle contractions, although muscular sound has been known since 1800 according to Orizio. Posatskiy and Chau proposed a new MMG sensor in 2012 [154], which becomes the de facto device for MMG study in recent years. The device is based on microphone with cylindrical and conical acoustic chambers [154]. MMG sensor is relatively new and the technology is not as mature as EMG. However, it does have several advantages over EMG. The MMG sensor is much easier to use than EMG sensors. The MMG sensor has higher signal-to-noise ratio. Perhaps more importantly, MMG sensors are immunity to changes in skin impedance, which eliminates the use of shaving and conductive gel [155]

Like EMG, MMG has also been predominately used in prosthesis control. We are aware of only a single study that used MMG in the context of rehabilitation [155]. In [155], Woodward, Shefelbine, and Vaidyanathan reported a new MMG sensor that they adapted based on the design introduced in [154]. The device measures low frequency vibrations caused by muscles during contraction in the range of 2–200 Hz. To facilitate the monitoring of both motion and muscle contraction together, the MMG device incorporates an Aktiv IMU as well as a barometer. In addition to carefully validate the performance of the proposed MMG sensor, two case studies were performed to demonstrate the utility of the MMG sensor.

In the first case study, six healthy subjects were recruited. They were asked to perform a set of activity, including standing, lying, walking, running, ascending and descending stairs, taking an elevator up and down. During this case study, both the MMG sensor and the IMU were placed on the right leg, where the MMG sensor was placed on the lateral head of the gastrocnemius, and the IMU was placed on the lateral side of the leg. While the subjects were performing the activities, each subject was provided with a push button, and on changing to a different activity, the subject was asked to push the button. In addition, an investigator followed each subject the entire duration to record the activity performed. This is to ensure that the collected data can be reliably labeled correctly.

The collected data were first preprocessed prior to classification. This includes filtering and segmentation. The activity classification was done using a custom algorithm based on K-means clustering in two steps. In the first step, an activity was first assigned to one of three clusters: (1) stationary activities (including standing, lying, and taking elevator); (2) dynamic activity (walking and running); and (3) dynamic altitude activities (ascending and descending stairs). The clustering steps at this stage follow the traditional clustering algorithm, which starts with a random cluster centroids and iterates through all the data until results converge. At the second stage, the activities within each cluster were then further classified into respective activity by considering the specific attributes for each activity. For example, in the stationary cluster, if a window shows some barometer change, then it must correspond to the taking-elevator activity. The lying and standing activities will show no barometer change, but the two would have different gravity plane readings. The value of MMG in the feature set is used to different walking and running where running would have larger MMG value. The average classification accuracy reached 98.4%.

In the second case study, three subjects were recruited. Among the three subjects, one is a healthy subject as the control. Another subject has hemiplegia and cerebral palsy. The third subject is a unilateral below-the-knee transtibial amputee. The IMU and MMG sensor were placed on exactly the same location on the right leg. Unlike the first study, the sensors were used to collect the activities of daily living of the three subjects. Typically a subject would put on the sensors at the beginning of the day in the morning and collect about 5 h of data. The walking data were identified and extracted in the study. The identification of walking was achieved by exploiting the repeatable gyroscopic pattern for each stride. From the collected data, the following features were extracted: MMG rms, cadence, and gyroscope magnitude. The temporal characteristics of the three subjects were identified in terms of these three features.

4.3   Force myography

Muscular contraction can also be measured directly using force sensing resistors [156]. This type of myography is called force myography (FMG). The studies used MFG are summarized in Table 4.2.

The paper by Wininger, Kim, and Craelius appears to be the first one that adopted the term force myography [156]. The authors constructed a myokinetic FMG cuff to study the grip force by measuring the pressure signature of the forearm. The device is equipped with 14 force-sensitive sensors (FSRs) from Interlink Electronics (model 402). An FSR is a form of resistive pressure sensor, as we have described in Chapter 2.1.2. These sensors were placed at the seven anterior and seven posterior areas of the forearm. In addition to the force myography cuff, the study used a custom grip force dynamometer, which is also constructed using FSRs and it is used to directly measure the grip force of the hand. Nine healthy subjects were recruited to test whether or not there is a strong correlation between the grip force and the forearm pressure

Table 4.2 Studies based on FMG

Setup	Human activities	Method	Refs
FMG cuff on forearm	Hand grasp and release cycles	Intraclass correlation coefficient	[156]
Wearable FSR sensors	Hand grasp and arm lift motion	Visual comparison	[156]
Three-layered device with FSR sensors	16 upper extremity gestures	KNN/C4.5 at frame level, and majority voting for final prediction	[157]
A strap with 8 FSR sensors	Upper extremity activities	Extreme learning machine	[158]

signature. These subjects were asked wear the cuff and hold the force dynamometer, and perform repeated grasp and release cycles according to the prompt displayed on a computer screen. The correlation between the grip force and the forearm pressure signature was established via the intraclass correlation coefficient (ICC).

Prior to [156], several related studies were published, even though the authors did not formally use the term FMG. The earliest result was published in 2006 by Amft *et al.* [159], where two different sensors were used to measure muscular activities, one is FSR and the other is fabric stretch sensor, which is essentially a form of conductive elastomer used in e-Textile that we have discussed in Chapter 3.1. Ten subjects were recruited to evaluate the feasibility of using these two forms of force sensors to study muscle activities in comparison with using EMG sensors. The subjects were asked to perform four actions: (1) move the lower arm upward against resistance; (2) bend the hand outward; (3) making the grasping motion by first open the hand and then close the hand; and (4) lift a heavy object with the right arm. The trials demonstrated indeed that there are very distinct features visible in the measured obtained by the force sensors for the four different actions.

In [157], Ogrs, Kreil, and Lukowicz reported their study on using FSR to recognize 16 different upper extremity gestures performed by 2 subjects. The authors designed a three-layered device for each subject to wear in order to accurately measure the force of the muscles of the forearm. Two rings of FSR were embedded in the device as the second layer. The inner layer is a thin stocking, and the outer layer is a bicycler's sleeve, which is "tight but stretchable" [157]. With this device, one FSR ring was placed on the lower part of the forearm, and the other ring was placed on the upper part of the forearm. The 16 gestures are those that one would do while giving a talk, which includes four for whiteboard operations (i.e., open the marker, write on whiteboard with the marker, close the marker, and erase what on the whiteboard), four for beamer screen (i.e., scroll the screen down, point on screen with hand, scroll the screen up, turn off the projector using a remote), four for computer operations (i.e., open the laptop, type some words on the laptop, make some mouse movement, and close the laptop), and the four for drinking a bottle of water (i.e., open the lid, pour water to a glass, close the lid of the bottle of water, and take a sip of water from the glass).

Each subject was required to perform these 16 gestures 10 times. The FSR signals were sampled at 20 Hz. A sliding window of 30 with a step size of 15 was used to process the raw data. For each window, the mean and the variance were extracted as the features. Two different types of classification frame-based classification and hidden Markov models (HMM) were carried out. In frame-based classification, the k-nearest neighbor (KNN) and the C4.5 tree were used. The overall classification for each gesture in the frame-based approach was done by doing major voting on the results based on each window. KNN appeared to be the best classifier among the three, with 76% accuracy. The authors also attached accelerometer and gyroscope together with FSR, which led to much better classification by fusing the accelerometer and the gyroscope data. One would have expected to see the result when all sensing modalities were combined together, but it was done reported.

In [160], Kreil, Ogrs, and Lukowicz reported their follow-up study on using the FSR sensors mounted on the upper leg to analyze muscle activity during bicycling.

In a relatively more recent paper [158], Xiao and Menon reported their work on using FSR sensors to track upper extremity activities. The FSR sensors were made by Interlink Electronics. Eight FSR sensors were incorporated into a strap with 3-cm spacing in between each pair of sensors. The strap is supposed to be wrapped around the proximal portion of the forearm. Six healthy subjects were recruited to validate the proposed system. The subjects were asked to perform a drinking task with a cup. This task was decomposed into six distinct postures: (1) arm completely relaxed; (2) raise the elbow to the horizontal level; (3) extend all fingers; (4) grasp an empty cup; (5) while holding the cup, flex the elbow; (6) pronate the wrist so that the cup is getting close to the mouth. These six postures were regarded as six classes.

Each subject would first perform a training phase so that labeled data could be collected for the classifier training. During the testing phase, the subject is prompted to perform the required postures according to the instructions displayed on a monitor. During the training phase, the subject was asked to maintain each posture for 7 s and repeat the entire sequence of six postures three times. This enables the extraction of 5 s of data for each posture. The FSR data were sampled at 1 kHz and the raw data were filtered with a low-pass filter with a 4 Hz cut-off frequency. The extreme learning model was used as the classifier. The classification of individual postures has fairly high accuracy for all subjects (over 90% for all except one subject). Because the ground truth was established based on the instruction sequence displayed on the monitor, the accuracy largely depends on how quickly and how well the subject follows the instructions. Hence, the experiment cannot be easily extended to the drinking activity in a natural setting.

4.4   Optical myography

Muscular activities can also be tracked by observing the visible muscle movements using a camera. The method is referred to as optical myography (OMG) and it was first proposed by Nissler *et al.* [161]. Typically, markers are placed on the surface of the skin to help track the muscular movements and the camera is positioned above the body segment. So far, OMG has only been used to study finger movements, where the markers are placed on the forearm. Also, recent publications have basically come from two research groups, as summarized in Table 4.3.

Nissler *et al.* adopted the AprilTag [162] as markers (placed on the forearm) for optical tracking of the muscle position changes [161,163]. AprilTag was introduced by Olson to enable the tracking of 3D position and the 3D orientation of the tag. While the use of AprilTag has the potential to enable highly accurate tracking, it requires that the entire setup to be fixed, i.e., the forearm would have to be fixed on a flat surface and the camera is placed above the forearm at a predefined fixed location [161,163]. Ten healthy subjects were recruited to validate the approach. During the experiments, five rows of markers with two markers per row were placed on the forearm. The subject then was required move the fingers as instructed via a computer monitor in front of the subject. The set of movements to be performed includes the flexion of the index finger, the rotation of the thumb, the flexion of the thumb, and a combined flexion of three fingers (i.e., little, ring, and middle) [161,163]. The validation of the approach is to compare the measured data with the instructed sequence of movements. The measured data are processed using ridge regression [164] to build regression models for the movements. The performance of the regression is evaluated in terms of the normalized root mean squared error.

Compared with other of the myography, this setup is much too restrictive. Hence, the authors proposed a different configuration for OMG where the camera can be worn on the arm using a simple sticker placed on the forearm [165]. The deformation of the sticker is tracked via a convolutional neural network from the video frames captured by the camera. The results are exceptional good. The classification accuracy for five different finger poses ranges between 96.21% and 99.3%. The ground truth is determined by the sequence of instructions on finger movement displayed on the monitor, where the subject is supposed to follow.

Wu *et al.* [166] adopted a very similar setup of [162] with a different set of markers. The set of markers consists of 17 solid circle disks of different colors that are arranged in three rows. Like other studies, the markers are placed on the forearm. The authors used an artificial neural network model to predict the finger postures. Each marker is identified by the 2D location and its color. Hence, there are 51 features used for the classification using the Keras API of TensorFlow framework. The authors

Table 4.3   Studies based on OMG

Setup	Human activities	Method	Refs
Camera + 10 AprilTag markers on forearm	Thumb/index/combo flexion + thumb rotation	Ridge regression	[161,163]
Camera + one big marker on forearm	Thumb/index/combo flexion + thumb rotation	Convolutional neural network	[165]
Camera + 17 circular color markers on forearm	8 hand postures	Artificial neural network with Keras API	[166]

Table 4.4   Different types of myography

Type	Technology maturity	Rehab applications	Restriction
EMG	Mature with commercial products	Upper and lower extremity	Subject can move freely
MMG	Research prototype	ADL	Subject can move freely
FMG	FSR sensors commercially available	Upper extremity and hand movement	Subject can move freely
OMG	Research prototype	Finger flexion/rotation and hand postures	Forearm fixed

compared the placement of the markers on the anterior side and the posterior side of the forearm. The placement of markers on the anterior side gave much better finger posture predication accuracy (i.e., average 92.2%).

4.5   Summary

In this section, we summarize the different types of myography in terms of their technical maturity and their applications in rehabilitation in Table 4.4. Among the four types of myography, EMG is undoubtedly the most mature and has the widest application rehabilitation studies. FMG appears to offer a viable alternative to EMG with inexpensive commercial FSR sensors available. MMG is promising. However, it would require the researchers to develop the sensor from scratch, which could limit its adoption. The setup of OMG is extremely simple and cost is very low because it only requires a conventional webcam and custom-made markers. Unfortunately, the current setup has strong limitations, e.g., the subject's forearm has to be fixed to a flat surface, and the camera has to be placed fairly close to the marker areas. So far, OMG has only been used to study finger movements and hand postures. How to create an untethered setup for OMG remains to be seen.

Chapter 5
Vision-based motion sensing

While vision-based human motion tracking systems have long been used for research, such as Vicon motion tracking systems. These systems use multiple cameras to obtain synchronized video frames from different angles around the subject and develop a 3D model of the subject for motion tracking and analysis. The subject typically would have to wear multiple markers at key positions depending on the activity to be tracked. Such systems are very expensive and requires periodically maintenance. Hence, they are only used in research laboratories and clinical facilities for rehabilitation studies. The release of the Kinect sensor by Microsoft in late 2010 completely transformed the paradigm of vision-based motion sensing. Although the Kinect sensor was released as an add-on gadget for the Microsoft's Xbox game console, soon it was used for motion tracking in many non-gaming applications, particularly for healthcare. A search using "Kinect" as the keyword at Web of Service Core Collection returned over 8,500 papers as of September 2021. A search using "Kinect rehabilitation" as the keywords returned over 1,135 papers. This revealed that a considerable portion of the Kinect-based research is for related to rehabilitation, as shown in Figure 5.1.

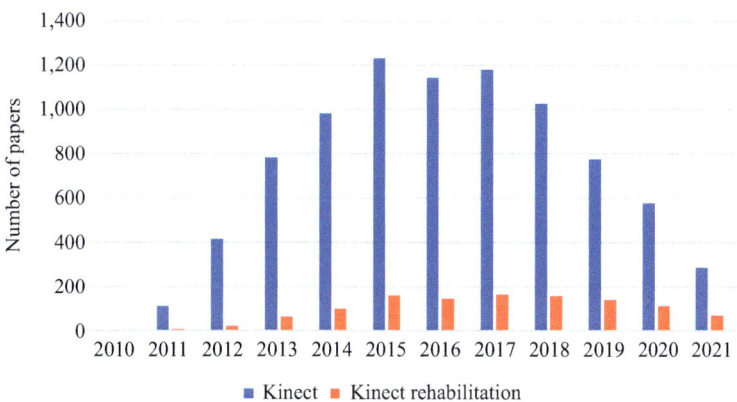

Figure 5.1   Number of publications per year on Kinect as indexed by Web of Science

5.1   Microsoft Kinect sensor

So far, the Kinect technology has gone through three generations, as shown in Figure 5.2. The most notable technical innovations include: (1) low-cost depth sensing; (2) real-time 3-dimensional human skeleton tracking. Depth sensing refers to the ability of perform 3-dimensional of sensing of the objects in the field of the view of the sensing system, i.e., specific dimension of these objects can be measured, including both along the frontal plane parallel to the sensing system, as well as how far they are from the sensing system. Traditionally, stereo-vision was the dominate approach to depth sensing, where two cameras are used to perform triangulation. However, this approach is bulky and expensive. PrimeSense patented a new way of performing depth sensing using structured light with a single camera and an infrared emitter [167]. While it was known for a long-time depth can be directly measured via time of flight calculation by emitting a laser to objectives in the field of the view and collecting the reflected light back, only in recent years such technique was made compact and low-cost [168].

PrimeSense's structured light was the foundation for depth sensing for Kinect's first generation products (typically referred to as Kinect v1), including the original Kinect for Xbox360 and Kinect for Windows, which was a special built for Kinect application development using a computer running the Microsoft Windows operating system. Kinect's second-generation products (typically referred to as Kinect v2) switched to using time-of-flight (ToF) for depth sensing. Kinect v2 products include the Kinect for Windows v2 and the Kinect for XboxOne game console. After the

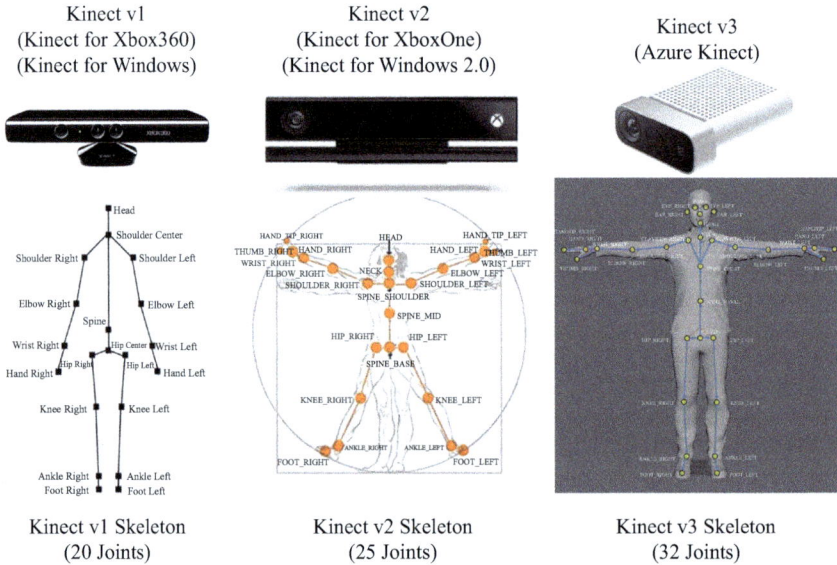

Figure 5.2   Three generations of the Kinect technology

announcement of the discontinue of the Kinect product lines, Microsoft resurrected its Kinect development, probably driven by its success in the HoloLens mixed-reality product line. In 2019, Microsoft released its third generation Kinect product called Azure Kinect. The third-generation Kinect (i.e., Kinect v3) also uses time-of-flight for depth sensing [168], and its form factor is significantly smaller than that of Kinect v1 and Kinect v2.

For in-depth technical details, we refer the readers to read a comprehensive survey that we authored previously on Kinect (Kinect v1 and Kinect v2) [169]. Here we summarize the key specifications of the three generations of the Kinect sensors in Table 5.1. Kinect v3 support two different modes: (1) narrow field of view depth mode (NFOV), and (2) wide field of view depth mode (WFOV). The two modes have different depth sensing resolutions and depth sensing ranges. For both modes, the depth sensing range can be extended at the expense of a reduced depth sensing resolution by doing some software manipulations. For color imaging, Kinect v3 also supports several native resolutions. In the 4:3 aspect ratio, it supports $2,048 \times 1,536$ and $4,096 \times 3,072$ resolutions. In the 16:9 aspect ratio, it supports $1,280 \times 720$, $1,920 \times 1,080$, $2,560 \times 1,440$, and $3,840 \times 2,160$ resolutions. The hardware specification for Kinect v3 is obtained from https://docs.microsoft.com/en-us/azure/Kinect-dk/hardware-specification. Compared with Kinect v1, Kinect v2 added the tracking of hand tip (left and right), thumb (left and right), and the neck joints. Kinect v3 made significant changes to the tracked skeleton joints. More torso and shoulder joints are now being tracked and some joints previously tracked appear to be relocated, such as the left and right shoulder joints. There are two spine joints (spine chest and spine naval joints) instead of a single spine joint, and there are two additional joints, left and right clavicle joints. The hip center joint is renamed to pelvis joint. There is only a single head joint in Kinect v1 and Kinect v2. Now, there are a total of

Table 5.1   Key specifications of three generations of the Kinect sensors

Feature	Kinect v1	Kinect v2	Kinect v3
Depth sensing	Structured light	ToF	ToF
Depth sensing resolution	640×480 30 fps	512×424 30 fps	640×576 (NFOV) $1,024 \times 1,024$ (WFOV)
Color image resolution	640×480 30 fps	$1,920 \times 1,080$ 30 fps	Several modes
Depth sensing range	0.4–3 m (near mode)	0.5–4.5 m	0.5–3.86 m (NFOV)
	0.8–4 m (default mode)	Up to 8 m without skeletonization	0.25–2.21 m (WFOV)
Depth sensing FoV	$43° \times 57°$	$> 43° \times 70°$	$75° \times 65°$ (NFOV) $120° \times 120°$ (WFOV)
Skeleton Tracking (with full skeleton)	Up to 2 subjects	Up to 6 subjects	No technical limit
	20 joints per skeleton	25 joints per skeleton	32 joints per skeleton

six joints in the head area, including nose, left and right eye, left and right ear, and head (appears to have moved down from its original location). Another interesting feature of Kinect v3 is that it does not impose an upper bound on the number of users it can track. The upper limit is imposed by practical space and by the computing power of the computer. The hardware requirements on the computer to run Kinect v3 applications are determined based on concurrent tracking of five users with full skeletonization.

5.2   Feasibility studies of using Kinect in rehabilitation

Earlier studies have focused on carefully characterizing Kinect sensor regarding its measurement accuracy. A common approach is to use a traditional marker-based multi-camera human motion tracking systems, such as Vicon, to concurrently measure the subject's movement, the results of which are regarded as the ground truth. Although some measurement errors have been reported, the results of these studies are largely positive. In addition to the measurement accuracy, the measurement consistency of the Kinect sensor has also been studied. In the following, we outline the feasibility studies roughly chronologically and the summary is provided in Table 5.2.

In [170], the accuracy of the distance between Kinect positions was evaluated using a marker-based reference system called iotracker in the context of chronic pain rehabilitation. More specifically, the vertical distance between the right shoulder joint and the right-hand joint is compared with the measurement obtained from iotracker. In addition, the foot position was also derived from the Kinect depth frames and compared with the measurement from iotracker. The authors showed that measurements from Kinect were quite close to those from iotracker. The discrepancy was believed to be due to the different definition of joints.

In [171], the 3D joint positions of the left and right hand/elbow/shoulder reported by the Kinect skeleton frame were validated against the measurements using a marker-based system called OptiTrack. The comparison was done when the subject was performing external rotation, scapular retraction, and shoulder abduction exercises. The results showed that the positions for hand and elbow Kinect joints are highly consistent with those from OptiTrack. However, the Kinect joint positions for the left and right shoulders differ from those from OptiTrack significantly.

In [172,173], Kinect joint displacements and Kinect joint angles were validated against a marker-based motion tracking system, Vicon, in the context of postural control and gait retraining. All movements of the subject are within the frontal plane that is facing to the Kinect sensor.

In [174], the knee, hip, shoulder joint angles calculated from Kinect skeleton frames were validated against a marker-based multi-camera system, Vicon. During the study, the subject was asked to move the knee, hip, and shoulder joints within anatomical planes. The mean errors for the knee joints range between $6.78°$ and $8.98°$. The mean errors for the hip joints range between $5.53°$ and $9.92°$. The mean errors for the shoulder joints are larger, which range between $7.19°$ and $13.19°$.

In [175], the Kinect upper body joints were validated against a model fabricated using plywood. The distance traveled for each joint along the x-, y-, and z-axes was measured when the subject is facing directly with the Kinect sensor while moving

Table 5.2   Feasibility/validation studies on Kinect sensor

Assessment	Context	Reference system	Ref
Right shoulder-hand distance; foot distance	Chronic pain rehabilitation	iotracker	[170]
3D joint positions of the left and right hand/elbow/shoulder	External rotation, scapular retraction, and shoulder abduction	OptiTrack	[171]
Kinect joint displacements and Kinect joint angles for movements within frontal plane	Postural control and gait retraining	Vicon	[172, 173]
Knee, hip, shoulder joint angles	Movements within anatomical planes	Vicon	[174]
Displacement of upper body joints	N.A.	A plywood model	[175]
Shoulder, elbow, hip, and knee angles	Movement within anatomical planes	Vicon	[176]
Displacement of head and knee joints, and maximum hip and shoulder joint angles	Exercises for patients with Parkinson's	Vicon	[177]
Joint positions and body geometry	6 exercises for elderly population	PhaseSpace	[178]
Upper extremity joint ROM	Shoulder flexion, shoulder abduction, shoulder adduction	Vicon	[179]
Displacement of joints and various joint angles	A set of five common exercises	Cortex	[180, 181]
Eight performance indices for evaluating upper body movements	Rehabilitation for stroke patients	N.A.	[182]
Chest wall motion	Adults with cystic fibrosis	Nikon laser scanner	[183]

the upper body joints. The results showed that the maximum joint displacement error was below 4 cm.

In [176], the Kinect shoulder, elbow, hip, and knee angles, calculated from Kinect joint data are compared with a marker-based system (Vicon) in the context of a set of exercises involving the movement within anatomical planes (i.e., frontal and sagittal planes). At the maximum range of motion (i.e., maximum joint angle during each exercise), the differences of the joint angles as measured by Kinect and Vicon are within 11 degrees. These discrepancies were likely caused by inaccurate estimation of joint centers by Kinect, as evidenced by the variation of bone lengths from frame to frame. The study also showed that Kinect has good reproducibility.

In [177], the displacement of head and knee joints, and the maximum hip and shoulder joint angles were assessed against the Vicon marker-based system where the subject was performing a set of exercises designed for people with Parkinson's disease, including standing still, reaching forward and sideways, stepping forward and sideways, and walking on the spot. The results showed some joint angles differ significantly, such as the trunk flexion.

In [178], Kinect skeletonization accuracy was compared against a marker based system called PhaseSpace in the context of six daily exercises for elderly population (knee lift, cops and robbers, deep breathing, pendulum legs, stand ups, and line tapping). The validation was done both on an individual joint level and on the body geometry (i.e., bone lengths) level. The results showed that Kinect's measurement on joint positions is comparable to that of the much more expensive PhaseSpace system for controlled poses such as standing. However, in other types of poses, Kinect measurement error can be 10 cm. The hip joints reported by Kinect are higher by about 20 cm than those determined by the PhaseSpace.

In [184], Kinect joint positions and joint angles were evaluated using PhaseSpace system as reference. Ten human subjects (5 male and 5 female) were recruited to perform upper extremity movements for reachable workspace evaluation. The results showed that joint positions were within 66.3 mm from the reference system, but the angle measurements have large deviations from 9° to 28°. However, the reachable workspace envelop has high agreement between the Kinect measurement and that of the PhaseSpace system.

In [179], Kinect joint angles were evaluated using the Vicon marker-based systems as the ground truth for shoulder range of motion movements. More specifically, the shoulder flexion angle, shoulder abduction angle, shoulder adduction angle, and the should angle while the subject was doing random reaching activities. The accuracy of the Kinect measurements is determined by two metrics, the average absolute error and the average absolute deviation. The range of motion (ROM) is defined to be the difference between the maximum and the minimum angles in the trajectory of the activity. Given the Kinect measurement of ROM, ROM_K, and the Vicon measurement of ROM, ROM_V, the absolute error is defined to be $|ROM_V - ROM_K|/180$ (ROM is measured in degree angles). The average absolute deviation is defined to be the square root of the average of the variances. Nineteen human subjects were recruited for the validation study. The absolute errors for different activities are mostly under 10% with the maximum being slightly over 20% for all subjects.

In our previous work [180], we showed that Kinect provides sufficiently accurate joint measurement to assess the correctness of five common physical therapy exercises including hip abduction, bowling, sit to stand, can turn, and toe touch. What matters most when assessing the quality of a rehabilitation exercise is the accuracy of the metrics important for each exercise. We note that typically there is a generous bound on the movement of the designated body segments for rehabilitation exercises. By clearly defining a set of correctness rules for each exercise with appropriate error bounds, one could judge the quality of an exercise performed using Kinect data as long as the measurement error does not exceed the tolerated bound. Furthermore, systematic deviation in Kinect measurement can be masked by adjusting the corresponding rule parameters [181].

There is also research on measuring parameters other than joint displacement and joint angle. In [183], Harte *et al.* reported their work on validating using four Kinect sensors to measure the chest wall motion in healthy volunteers and adults with cystic fibrosis. A Nikon laser scanner was used to capture complete detailed geometric scans to establish the ground truth. The four Kinect sensors are placed on the four sides around the subject. Instead of using skeleton frames, the Kinect depth frames were captures to build 3D point clouds, which enables the calculation of the chest wall volume and its changes. More specifically, the tidal volume, respiratory rate, and minute ventilation are extracted. The authors concluded that the Kinect-based system could provide fairly accurate measurement for chest wall motion with a much lower cost than spirometry.

Mobini, Behzadipour, and Saadat [182] reported their study on the reliability of measurements using the Kinect sensor instead of the accuracy. The reliability analysis provides information on how consistent of the measurements over different people and different sessions using the Kinect sensor, which is very important for clinical applications. In [182], the test–retest reliability of measurements using Kinect was studied in the context of the evaluation upper extremity recovery of stroke patients. Eight performance indices were used, including mean velocity, normalized mean speed, normalized speed peaks, logarithm of dimensionless jerk, curvature, spectral arc length, shoulder angle, and elbow angle. The reliability analyses include intra-class correlation coefficient, standard error of measurement, and coefficient of variation. The results of all reliability analyses are within the acceptable range.

5.3   Kinect-based systems in rehabilitation

In this section, we outline Kinect-based rehabilitation systems that have been reported in the literature. The availability of Kinect makes it possible to designed a home-based rehabilitation system where a patient could practice the prescribed exercises at the comfort of home. Unlike professional-grade devices, which often demand meticulous calibration before use, Kinect requires no calibration at all. Furthermore, Kinect can perform rather reliable motion tracking regardless of the types of clothes the patient wears, and it is also a rugged device that is suitable for use at the home environment. We roughly divide the work into three categories: (1) Kinect-based system with simple visual feedback only, (2) Kinect-based system with sophisticated performance quality feedback, and (3) integration of Kinect and other sensing modalities. Tracking human motion is inherently very challenging and it is far from a resolved problem. Hence, we intentionally included studies that focused on data collection and/or offline data analysis. Such systems have minimum graphical user interface and sometimes the participants are not shown any visual guidance at all.

Typically, a single Kinect sensor is used to track the movement of one participant. As shown in Figure 5.3, a Kinect-based system has four major components: (1) target population; (2) user interface; (3) exercises used for rehabilitation; and (4) feedback mechanisms, if present.

Figure 5.3   Major components of Kinect-based system for rehabilitation

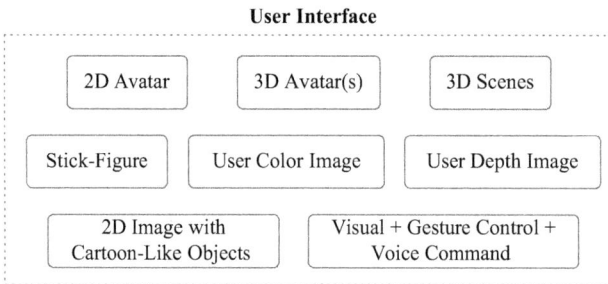

User Interface

2D Avatar	3D Avatar(s)	3D Scenes
Stick-Figure	User Color Image	User Depth Image
2D Image with Cartoon-Like Objects	Visual + Gesture Control + Voice Command	

Figure 5.4   Types of user interfaces used in Kinect-based system for rehabilitation

Some form of visual display is used in all Kinect-based systems, although in some systems, the visual display is used for data collection, data analysis, or showing the outcome to the healthcare professionals. Common types of user interfaces used in Kinect-based rehabilitation are summarized in Figure 5.4. One or two 3D avatars are typically used in the visual display. One avatar would mirror the movements of the participant doing an exercise. Sometimes an additional avatar is used demonstrate the ideal movement to guide the participant. Some systems incorporate 3D scenes, which aimed to create a virtual immersive environment for serious games designed for rehabilitation. Some systems use simpler 2D avatars. Unity is typically the development platform of choice. The Microsoft Kinect software development kit (SDK) offers application programming interface (API) for creating simple visual display with 2D stick-figure for each skeleton being tracked, color and depth images, which some systems chose to use. Some systems used a color image with superimposed 2D objects for the game-like exercise. The Kinect SDK also offers API for voice control. Based on the skeleton data, one can develop gesture recognition to enable natural user interface via gesture control.

The target population for Kinect-based rehabilitation is summarized in Figure 5.5. Studies that focused on stroke rehabilitation appear to be the most pervasive [185–188]. Studies on general rehabilitation are also popular [189–195]. Other studies target specific patient populations, include patients with Parkinson's disease [196], patients with traumatic brain injuries [197], patients with spinal cord injury [198],

Target Population

Stroke	Parkinson's Disease	Traumatic Brain Injuries	Spinal Cord Injury
Lower Back Pain	Lymphedema	Rheumatoid Arthritis	Cystic Fibrosis
Motor Disability	Chronic Pain	Elderly	General

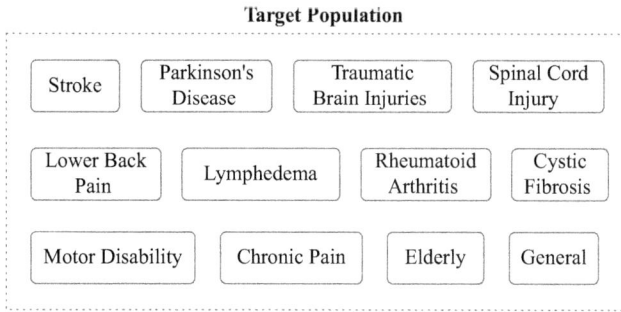

Figure 5.5   Types of user interfaces used in Kinect-based system for rehabilitation

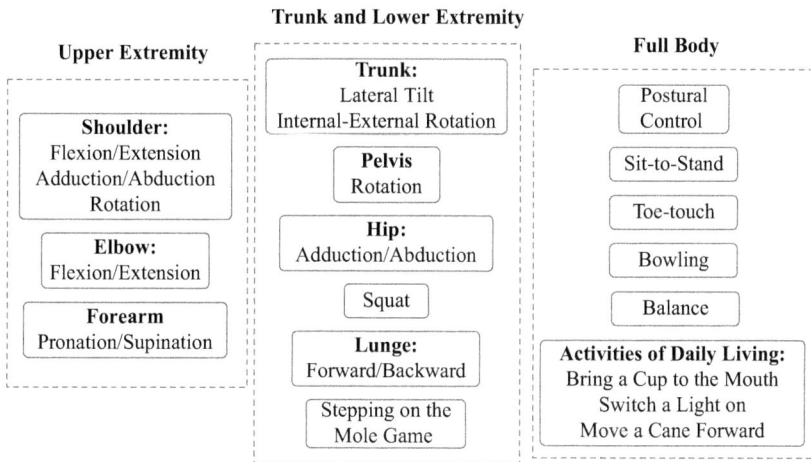

Upper Extremity

Shoulder:
Flexion/Extension
Adduction/Abduction
Rotation

Elbow:
Flexion/Extension

Forearm
Pronation/Supination

Trunk and Lower Extremity

Trunk:
Lateral Tilt
Internal-External Rotation

Pelvis
Rotation

Hip:
Adduction/Abduction

Squat

Lunge:
Forward/Backward

Stepping on the Mole Game

Full Body

Postural Control

Sit-to-Stand

Toe-touch

Bowling

Balance

Activities of Daily Living:
Bring a Cup to the Mouth
Switch a Light on
Move a Cane Forward

Figure 5.6   Common types of exercises used in Kinect-based system for rehabilitation

patients with lower back pain [199], patients with chronic pain [170], breast cancer patients with lymphedema [200–202], patients with rheumatoid arthritis [203], and patients with cystic fibrosis [183]. There are also studies on improving students with motor disability [204], and improving the health of the elderly [178].

A wide variety of exercises have been used in Kinect-based systems for rehabilitation. We roughly divide these exercises into three categories: (1) upper extremity; (2) trunk and lower extremity; and (3) full body, as shown in Figure 5.6. Upper extremity exercises involve the movements of the shoulder and elbow joints, and sometimes the rotation of forearm. The shoulder has three degree of freedom. Hence, shoulder exercises include shoulder flexion and extension, shoulder abduction and adduction, and shoulder rotation. The elbow has only a one degree of freedom, hence, the only exercises are elbow flexion and extension. The forearm could rotate (pronation and supination). Although there are many exercises involving with hand and fingers, Kinect is not capable of tracking such movements. Leap motion has been used to track hand and finger movements [205,206]. However, based on our own experiences, self-occlusion is too severe to have meaningful tracking for hand and fingers using leap motion.

Trunk and lower extremity exercises include trunk lateral tilt, trunk internal-external rotation, pelvis rotation, hip abduction and hip adduction, squat, forward and backward lunge. In one study, a serious game is used where the participant is asked to step on the mole appearing on the left or the right in the virtual scene [207].

Fully body exercises include balance and postural control, sit-to-stand, toe-touch, bowling. Some studies also considered activities of daily living, such as bring a cup to the mouth, switch a light on, move a cane forward [186].

Various methods and algorithms have been used to analyze the motion collected from the Kinect sensor either in real-time (sometimes referred to as online processing) or offline. Some studies included methods to perform repetition count provided what is performed is already known. Other studies focused on the recognition of the exercise performed, which could be achieved using supervised machine learning. The automated recognition of the exercise performed could also enable the repetition count without prior knowledge what exercise is going to be performed. It is much more challenging to assess the quality of the exercise performed. Some algorithms, such as dynamic time warping (DTW) [200] and Hidden Markov Model (HMM) [192], were used to determine a scalar score on the performance of the exercise by comparing with pre-recorded data as a reference (typically performed by a physical therapist to demonstrate the correct way of doing the exercise). Others aimed to provide specific information to the user exactly what is not performed correctly by formulating the exercise in terms of a set up of rules (i.e., a rule-based method) [193,194]. Fuzzy logic has been used to better capture the rules as defined by physical therapists [189,195,199]. Deep learning models have also been used to either classify the exercise performed or to assess the performance of the exercise performed, some of which in terms of the patient's capability or severity levels [206].

5.3.1   Kinect-based system with visual feedback only

Early research work focused on using Kinect-based rehabilitation games to engage patients. It is important for patients to adhere to the prescribed rehabilitation exercises at home for a proper and speedy recovery. Unfortunately, patients often fail to perform the required the rehabilitation exercises at home adequately for multitude of reasons, such as the difficulty in following the instructions and lack of motivation. Healthcare providers often do not have a good way of obtaining the progress (or the lack thereof) made by the patients. Kinect-based systems could be used to address these issues by providing a more intuitive visual interface and a more entertaining game-like environment for the patients to perform the prescribed rehabilitation exercises, and by collecting data regarding the quantity and quality of the rehabilitation exercised performed by the patients, which promote accountability of the patients towards recovery. The studies in this category are summarized in Table 5.3.

Chang *et al.* described a study [204] that uses a Kinect-based system to increase the motivation of two young adults with motor disability to engage in rehabilitating

Table 5.3   Kinect-based systems with visual feedback only

User interface	Exercises	Target population	Ref
No details provided	Arm movements	Students with motor disabilities	[204]
2D stick figure overlaps with the depth image	Elbow flexion/extension	Children with disabilities	[208]
Picture with superimposed avatars	Lower limb rehabilitation	Parkinson's disease patients	[207]
3D avatars	Posture tracking for occupational healthcare	General population	[209]
Stick-figure for skeletal joints and actual image of user	Upper extremity functional evaluation	Hemiplegic stroke patients	[210]

exercises in a public school setting. They confirmed that indeed, the Kinect-based system played a big role in increasing the motivation, interest, and perseverance for the two students to engage in physical rehabilitation. The technical details of the Kinect-based system were not provided. The system runs on a notebook computer and the output is projected to a screen. Audio feedback was mentioned in the paper but it is unclear what specific information was provided to the user. Rehabilitation activities included lifting both arms: (1) to the front, (2) to the side, and (3) upwards.

Rahman *et al.* described a preliminary interactive rehabilitation system designed for disabled children to practice at home without the direct supervision of a healthcare professional [208]. The child's motion is captured by a Kinect sensor and the system could determine the joint-based angular movements. The system presents a virtual reality interface using the online Second Life virtual world, where the child could view directly or download a prerecorded video using Kinect for visual guidance. The example exercise provided is the elbow flexion and extension, where the elbow joint angle can be measured.

Palacios-Navarro *et al.* reported a pilot study on using a Kinect-based system for lower limb rehabilitation in Parkinson's disease patients [207]. In the study, the Kinect-based system is used to provide a visual interface for the subject to carry out the prescribed exercise (i.e., stepping on the mole that appear from the left or the right). The visual interface consists of a real picture superimposed by cartoon-like mole avatars with two counters, presumably for the number repetitions of correct movements and that of incorrect movements. No details were provided on how the assessment was done. The study incorporated a 5-week training period where there are four sessions per week for a total of 10 h. Each session consisted of 30 min where the subject would exercise for 40 repetitions and take a 3-min or longer rest break. Seven patients participated in the study and the result showed that the exercise protocol is well received and has positive impact to the subjects' performance as measured using the 10-meters-walk test.

Valdivia *et al.* reported their preliminary work on posture tracking for occupational healthcare [209]. Example exercises include standing hamstring stretch and various flexion and extensions. The authors actually focused on the comparison of two different user interfaces and two sensing modalities. One user interface is implemented using Unity 3D game engine and can run as a mobile app. The other interface is a non-traditional wooden manikin instrumented with IMUs and the captured IMU data are transmitted to a computer running an Unity application for visual feedback. In the manikin interface, a user would manipulate the manikin to experiment with different postures. In the mobile app interface, a user would manipulate the avatar using the touch screen and can view the feedback directly. The authors also compared the accuracy of the two sensing modalities, but apparently the processing of the data are done offline. IMU-based data collection requires calibration while Kinect-based data collection does not. Kinect-based data collection suffers from occlusion issue for large flexion. Otherwise, the two methods provide similar measurement accuracies.

Kim *et al.* presented their clinical study results [210] using a Kinect-based system to collect motion data for upper extremity functional evaluation in hemiplegic stroke patients. The interface provided is rather simple with no direct feedback regarding the performance quality to the subject (a stick-figure showing skeletal joints on the left, and the actual image of the user). Each trial is rated by a therapist on the subject's performance with a Fugl–Meyer Assessment (FMA) score of 0, 1, and 2. The motion data collected using Kinect are then labeled accordingly and the features are extracted offline. An artificial neural network is developed to establish the pattern in the features. The upper extremity exercises include: (1) shoulder retraction, (2) shoulder elevation, (3) shoulder abduction, (4) shoulder external rotation, (5) elbow flexion, (6) forearm supination, (7) should adduction and internal rotation, (8) elbow extension, (9) elbow extension, (10) forearm pronation, (11) hand to lumbar spine, (12) shoulder flexion 0–90 degrees, (13) shoulder abduction 0–90 degrees, and (14) shoulder flexion 90–180 degrees.

5.3.2   Kinect-based system with performance quality feedback

Providing interactive feedbacks to patients is an important requirement for rehabilitation exercise monitoring applications. During rehabilitation training, patients are required to perform an exercise in a specific manner to meet the objectives of rehabilitation. The output of human motion recognition should be presented as feedbacks to patients in real time to inform them about any incorrect movement. Table 5.4 summarizes the key points of each study in this category. The mechanisms used in Kinect-based system for rehabilitation to access exercise performance are shown in Figure 5.7.

Lange *et al.* reported the development and assessment of an interactive Kinect-based rehabilitation system for balance training [211] for adults with neurological injuries using Kinect to monitor the subject posture. The system presents a game-like interface and it encourages the subject to go out of the base of the support. In a follow-up study, the same group authors reported further development and evaluation of a video game for balance rehabilitation training [212]. The graphical user interface is relatively simple with a rough avatar to mirror the user's upper-extremity movements. The system provides a numerous score as the feedback. It appears that the score was computed based on the measured range of motion.

Table 5.4   Kinect-based systems with performance feedback

User interface	Context	Feedback mechanism	Ref
Simple 3D avatars	Balance training	Numerical score based on range of motion measurement	[211,212]
2D with simple feedback	General rehab	Based on shoulder joint angle measurement	[213]
3D avatars	Telerehab	An algorithm to recognize the type and correctness of exercises	[191]
Polished 3D scenes	Rehab for TBI patients	No details are reported	[197]
3D avatars	General rehab	A kinematic model to identify static and dynamic axis in each prescribed exercise	[190]
Basic interface to record and replay activities	General rehab	Dynamic time warping on the movement trajectories and fuzzy logic	[189]
Depth image with performance feedback	Rehab for lower back pain patients	Fuzzy logic-based algorithm	[199]
3D avatar	In-home rehab for patients with lymphedema	Dynamic time warping	[200]
Stick figure and depth image	Functional assessment for patients with lymphedema	LDA, NB, SVM	[201,202]
2D cartoon-like	Rehab for stoke patients	Repetition count and scoring. No details reported	[185]
Display on mobile devices	Rehab for patients with Parkinson's diseases	Various machine learning algorithms and correctness rules defined by experts	[214]
No display for users	Rehab for stroke patients	Hybrid deep learning (CNN, CNN-LSTM, CNN-GRU)	[188]
Simple 3D avatar	Upper limb exercises for patients with spinal cord injury	Deep learning (CNN-LSTM)	[198]
Visual interface with gesture and voice controls	Four exercises for patients with rheumatoid arthritis	Joint angle measurements	[203]
Not applicable	Three exercises for stroke rehabilitation	A hybrid model combining machine learning and hand-crafted rules	[186]
3D scenes	Four exercises for general rehab	Similarity score determined by DTW and HMM	[192]
Not applicable	Identify abnormal postures in patients with Parkinson's disease	DT, SVM, and KNN were used to classifier 8 abnormal postures	[196]
2D stick-figure	8 shoulder and elbow exercises for stroke rehab	KNN to recognize exercise being performed	[187]
3D avatars	General rehab	Rule-based	[193,194]
3D avatars	General rehab	Rule-based with fuzzy logic	[195]

Range of Motion/ Joint Angle	Dynamic Time Warping (DTW)	Traditional Machine Learning Classifiers: LDA, NB, DT, SVM, KNN	Deep Learning: CNN, CNN-LSTM, RNN-LSTM
Rule-Based	Hidden Markov Model (HMM)		
Kinematic Model	Fuzzy Logic		

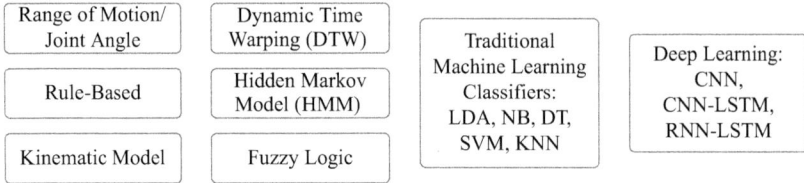

Figure 5.7   Common mechanisms used in Kinect-based system for rehabilitation to assess exercise performance

Da Gama *et al.* reported a Kinect-based system designed to facilitate rehabilitation [213]. The system is capable of measuring the shoulder abduction and flexion angles. Based on the angle measurement, the system could provide visual feedback on progress the subject is making and the correct completion of the prescribed exercises. Basic 2D user interface was used instead of the 3D avatar as used in later systems. The angle measurement based on the Kinect-system is reasonably good assessed using goniometry, with an error between 2 and 7 degrees. The authors also performed a usability survey with 10 subjects. The feedback was very favorable.

Anton *et al.* reported a Kinect-based system designed for doing telerehabilitation [191]. The system offers an avatar-based graphical user interface where the left avatar demonstrate the exercise to be performed by the user, and the right avatar mirrors the user movement. The authors stated that they developed an algorithm to recognize the type of exercises, determine the correctness of postures and exercises. However, no details were given on how the determination was done. The system could allow the recording of the movement for replaying to the user.

Ustinova *et al.* reported a pilot study with nine human subjects who have mild-to-moderate manifestations of traumatic brain injuries (TBI) to evaluate the efficacy of a Kinect-based virtual reality game system [197]. The system contains a number of carefully designed games. One type of games is referred to as virtual teacher where a personal instructional avatar explains and demonstrates each exercise (one 1-minute introduction and 16 2-min exercises). The other type of games is referred to as virtual challenger games, which consist of four games called octopus, courtyard, boat, and skateboard. The goal of the virtual teacher games is to train the patient's intra- and inter-limb coordination, manual dexterity, balance, and eye-head coordination. The octopus and courtyard games are designed to train dynamic balance and arm-postural coordination. The Boat game is designed to train the patient's postural stability under sensory conflict conditions. The skateboard is designed to train the patient's static and dynamic postural control, agility and stepping pattern. However, it is not clear if the system is capable of automatically scoring the patient's performance. The claim that "successful performance of each game is rewarded by a number of points that accumulate throughout the entire gaming session" [197] seems to indicate that the system has the mechanism, but no details given on how the scoring is done in the paper.

Velloso *et al.* developed a system aimed to facilitate at-home rehabilitation exercise monitoring [190]. The system employed a kinematic model to identify static and dynamic axis in a prescribed exercise. The model parameters are automatically fitted using an exemplar. This finalized model enables the system to continuously monitor violations of static axes in real time, and to count the repetitions for dynamic joints.

Su developed a similar system [189] that aimed to provide specific feedback to the patient. The primary evaluation method is dynamic time warping on the movement trajectories. Fuzzy logic is employed to capture the clinician's subjective requirements on performing the exercises. After the gesture recognition step on individual features, a single score was given to the patient after combining the fuzzy rules.

In [215], Dao *et al.* reported their study on the development and preliminary testing of a Kinect-based system for rehabilitation. For each participant, a 3D model was constructed, which requires a calibration step. The 3D model was then used to assess the flexion/extension movements of the subject with a graphical user interface with a 3D avatar. The details of the model for assessment are not provided.

In [199], Capecci *et al.* presented a Kinect-based system designed for telerehabilitation. The system provides real-time feedback to the patients using a fuzzy-logic based algorithm. The assessment criteria is defined in terms of primary targets, primary constraints, and secondary constraints. These targets are constraints are defined for each exercise and are derived from the Kinect skeleton data. For each target or constraint variant, a linguist description of good or bad is defined, which is followed by the formation of fuzzy sets used for doing fuzzy inference. Ultimately, a global score is computed based on the input variables and the fuzzy inference rules. To validate the proposed system and two groups of subjects were recruited, the experimental group has 20 patients with lower back pain, and the control group has 20 healthy subjects. Five exercises that target the rehabilitation of lower back pain patients were used in the study, including: (1) lifting the arms, (2) lateral tilt of the trunk, (3) trunk internal-external rotation, (4) pelvis rotation, and (5) squat. The user interface is relatively simple, consisting the depth image of the subject on the left side, the performance assessment result on the right side of the screen, and the overall score at the bottom. A reliability analysis was also performed on the experiments in terms of the Pearson coefficient on the correlation between the evaluation using the proposed algorithm with Kinect data, and that done by a trained physical therapist. Finally, the acceptability and the usability of the proposed system were evaluated using a questionnaire. In a follow-up study, the same group of authors collected more data from larger groups of human subjects with the set of five exercises [216].

In [200], Chiang *et al.* described a Kinect-based system designed for doing in-home rehabilitation for patients with lymphedema. A set of exercises that help improve lymph flow was experimented, including muscle-tightening deep breathing, muscle-tightening pumping, shoulder exercise, large muscle exercises such as walking and marching. Each exercise was broken into several sequences of movements. The performance evaluation is done via aligning the actual movements with ideal movements using dynamic time warping. The system provides a 3D avatar-based graphic user interface to the users.

Moreira *et al.* presented a Kinect-based system for assessing the upper-body function in breast cancer patients who suffer from lymphedema [201,202]. Due to the purpose of system, the user interface consists of stick-figure showing the skeleton data, and the depth image. The system uses both the skeleton joint data and the depth data as input to extract features needed for the assessment. The features include the upper-limb volume (which is computed using the depth data using computer vision technology), shoulder range of motion, hand height and hand width, elbow flexion, and hand acceleration. Several machine-learning algorithms were experimented to classify the upper-body function based on the features, including linear discriminant analysis, naive Bayes, and support vector machine. Support vector machine gave the best result compared with ground truth.

In [185], Foreman and Engsberg reported the design and evaluation of a Kinect-based system that they refer to as a virtual reality tool for measuring and shaping trunk compensation for stroke patients. The system provides a cartoon-like graphical user interface and is capable to doing repetition count and scoring for each exercise game (all games involve upper-extremity movements). Five patients with stroke were recruited to evaluate the system. In addition to motion data such as repetitions, trunk compensations, and some kinematic parameters, the authors performed usability evaluation of the system using the system usability scale, which appears to be widely used for evaluating technological systems [217].

In [214], Wei *et al.* reported the Kinect-based system for automated assessment of rehabilitation exercises prescribed for patients with Parkinson's diseases. The focus of the paper is on the assessment algorithms instead of the description of the technical details of the system and efficacy of the system. The user interface is rendered on a mobile device, and the bulk of the computation is carried out on back-end servers. The assessment is divided into two steps. The first step focuses on the repetition recognition and the sub-action segmentation. The second step focuses on the determination of the quality of the movements. The assessment was developed around three exercises: (1) squat, (2) forward lunge, and (3) backward lunge. The features were manually extracted for the first step from the Kinect skeleton joint data and they include thigh angle, trunk angle, trunk-leg angle, knee angle normalized length of step, and shank angle. with these features, two Hidden Markov Models were constructed, one for a single repetition and the other for multiple repetition. For the second step, the correctness rules are defined by the physical therapist in terms of the features extracted in the first step. Then, the per-frame error for the actual movement was calculated with respect to the rules. The repetition error is defined as the average per-frame error. These type of errors are referred to by the authors as quantitative errors. In addition to quantitative errors, qualitative assessment was performed using a support vector machine (SVM)-based classification model. The input to the model includes the mean and maximum errors for each correctness criterion. Very interestingly, a task recommendation framework was proposed to automatically adjust the task difficulty based on the user's performance. The task recommendation is based on random forest using the mean and maximum errors as the input.

In [188], Bijalwan *et al.* presented their work on using hybrid deep learning to model the spatiotemporal features of 10 exercises for post-stroke rehabilitation (left/right shoulder flexion/extension/abduction/rotation) using Kinect data. The focus of the paper is on analyzing the data offline instead of evaluating the efficacy of the Kinect-based system. During the data acquisition process, it appears that no user interface is shown to the participant. The authors experimented with three deep learning models. The baseline model is a deep convolutional neural network (CNN), which is used to extract features from the pre-processed data for the classification of performed exercise. In the second model, the features extracted from CNN are then fed to a long short-term memory (LSTM) model, which is an architecture of the recurrent neural network (RNN), to capture temporal features. In the third model, the features extracted from CNN are fed to a gated recurrent unit (CNN–GRU). The modeling is evaluated in terms of the accuracy of recognizing the type of exercises, which has limited practical implications because it would be far more useful if the quality of an exercise performed can be assessed with specific feedback to the subject.

In [198], Cao *et al.* reported their effort on characterizing the severity levels of patients with spinal cord injury using data collected by a Kinect-based system. Again, the focus is on offline data analytics rather than the efficacy of the system to help patient rehabilitate. The participants were asked to perform nine upper limb exercises mostly involved with shoulder movements. The data were used to classify the functional capability of the participants into three levels according to the Brunnstrom Scale [218] using a deep learning model CNN-LSTM. The physical therapists manually observed the data to determine the levels as the ground truth. The authors showed that a deep neural model (convolutional neural network and long short-term memory network) performs significantly better than traditional machine learning algorithms.

In [203], Dorado *et al.* described a Kinect-based system (referred to as ArhriKin) designed for patients with rheumatoid arthritis to do rehabilitation at home. Four exercises were supported by the system, including shoulder abduction, shoulder extension, elbow flexion, and hip separation. Performance assessment and repetition count recognition are based on the measurement of the angle between the two limbs, which is common to all these exercises. More specifically, for each exercise, there are a maximum angle and a minimum angle, which is used to define the performance score and the delineate a repetition. The system provides both a graphical user interface and voice commands so that the user can use gestures and voice to navigate the system. The system is able to prompt the user if the arm or leg is in a wrong pose, and inform the user the number of repetition done. The authors performed a comprehensive evaluation of the system, including the evaluation of the motivation of physical therapists to use the proposed system (using the Intrinsic Motivation Inventory questionnaire [219]), the evaluation of the accuracy of the system used a reference system called Krypton K-400 (in terms of Pearson's correlation coefficient and Lin's Concordance correlation coefficient), the evaluation of the usability of the system used the ISO/IEC 9126-4 standard as the guideline, which consists of effectiveness, efficiency, and satisfaction. The effectiveness was computed as the completion rate (i.e., the number of tasks completed successfully vs. the total number of tasks undertaken). The efficiency was computed in terms of the time to complete the tasks. The user satisfaction was measured using the single ease questions [220] and the system usability scale [217].

In [187], Mortazavi and Nadian-Ghomsheh reported their work on a Kinect-based system for providing visual guidance and some form of feedback to patients doing stroke rehabilitation. The study focused on the recognition of eight exercises using k-nearest neighbor and the measurement of range of motion for relevant joints. The eight exercises include shoulder abduction, shoulder adduction, shoulder flexion, shoulder extension, shoulder horizontal abduction, shoulder horizontal adduction, elbow flexion, and elbow extension. The user interface consists of a 2D stick-figure denoting the skeleton of the subject with information on the recognized exercise being performed and the current range of motion. The study has an additional component on detecting compensatory movements and quantify them.

In [186], Lee *et al.* presented their work on the assessment of rehabilitation exercise performance based on Kinect data. The focus of the work is doing offline data analysis instead of the impact of the system to patients. Hence, the user interface is irrelevant. Three exercises for stroke rehabilitation were considered, including bring a cup to the mouth, switch a light on, move a cane forward. The features were extracted from the Kinect skeleton data with the input from the physical therapist. The authors hypothesized that the feature selection is key to the classification performance. Hence, the effort was on the selecting the most relevant features for each subject using reinforcement learning. The classification was regarding the quality of motion performed by the subject with binary outcomes: (1) correct and (2) incorrect. Two different models, one was referred to as prediction model, and the other was referred to as the knowledge model. In the predication model, the classification was done on all the features using traditional machine learning algorithms including decision trees, linear regression, support vector machine, and neural networks. In the knowledge model, manually crafted rules for each exercise are used for classification. Then, the results from the knowledge model and the predication model were then combined using a hybrid model by doing weighted average of the two basic models to achieve better result.

In [192], Osgouei *et al.* reported their work on analyzing the Kinect data to quantify exercise performance. Four exercises were considered, including shoulder abduction, hip abduction, lunge, and sit-to-stand. A total of 16 participants were recruited for the study. The Kinect-based system offers a rather polished user interface powered by the Unity framework. However, the quantification result does not seem to be displayed in real time on the user interface. Two methods, namely the dynamic time warping (DTW) and Hidden Markov Model (HMM), were used to determine the matching degree of the actual movements by the participants with respect to the recorded sequence of movements done by a physical therapist. The authors found the scores generated by the two methods were consistent with each other. Both methods were more sensitive to range of motion than duration.

In [196], Zhang *et al.* presented their study on objective assessment of postural abnormalities in patients with Parkinson's disease (PD) using Kinect skeleton data. Again, the focus is not on a system that improves the rehabilitation efficacy with good interface and real-time feedback to patients. Eight abnormal postures were identified in the study and Kinect data were collected with 70 participants with PD. For data collection, each subject was asked to stand in front the Kinect sensor for 5 s, and then attempt to correct the posture for another 5 s, and turn 90 degree and repeat. The objective of the study is to rate the participants using machine learning algorithms with respect to the eight abnormal postures and compare with the ratings from the physical therapist. Three machine-learning algorithms, namely, decision tree, support vector machine, and k-nearest neighbor were experimented as the classifiers. The accuracy, specificity and sensitivity, and statistical comparison with expert ratings (in terms of intraclass correlation coefficient) were provided.

We have also conducted this line of work. In [221], we proposed a set of extensive/adaptable rules (with error bounds to capture the fuzziness of requirements) for each exercise, such that specific feedback on rule violations can be provided directly to the patients. Unlike playing games, which a user normally only expects a total score, a patient who is carrying out a rehabilitation exercise expects to be informed exactly what is not done right to ensure proper recovery. The proposed rules have been incorporated into an at-home exercise monitoring system [193,194]. The system has a 3D user interface implemented using Unity3D. The left side of the user interface shows a 3D avatar demonstrating the correct movement, while the right side of the interface mirrors the patient's movement. A visual target ball for the leg in a hip abduction exercise is provided. The target ball changes color depending on whether or not any correctness rule is violated. If the patient does one iteration correctly, the ball shows the green color. Otherwise, it shows yellow and the specific rule that is violated is displayed in text on the interface. The target ball also shows the repetition count for correct iterations. In a separate paper, we also incorporated fuzzy logic to provide patients with categorical feedbacks [195].

5.3.3 Integration of Kinect and other sensing modalities

Even though the capability of tracking the skeleton joints fairly accurately, Kinect is incapable of tracking some movements, such as finger movements. Hence, other sensing modalities have been used in conjunction with Kinect such as smart gloves and IMUs. Kinect has also been used as a sensing modality to over occlusions in rehabilitation robotics [222].

Huang *et al.* reported a preliminary study on integrating Kinect and a smart glove to facilitate upper extremity rehabilitation with game interface [223]. The integrated system could track both arm and finger movements.

Bo *et al.* proposed a system that combines Kinect with IMU sensors for measuring human motion for rehabilitation purposes [224]. In this study, Kalman filtering was used to correct IMU sensor measurement using Kinect's skeleton data for long-term operations.

In [225], Tamei *et al.* reported a system that combines a Kinect sensor with a Wii balance board for in-home rehabilitation of patients with balance disorders. The Kinect is used to capture posture information and the Wii balance board is used to estimate the center of pressure. The system provides a rather simple but informative graphic user interface. The interface displays the user image with the skeleton superimposed on top, and cartoon-like avatars to highlight the anterior folding angle and the lateral bending angle. The center of pressure is displayed at the lower left area of the interface. The posture measurement accuracy of the system was validated by using marker-based optical motion capture system called MAC3D from Motion Analysis Corp.

In [205], Rahman and Hossain presented a multi-sensor framework for in-home therapy. They proposed to use Kinect sensor (for full body skeleton tracking), leap motion sensor (for finger tracking), and Myo armband sensor (for EMG tracking) together for motion tracking. A 3D graphical user interface is provided to the user. However, the validation is quite limited (a pendulum exercise), and it is not clear how the three sensors work together for the best motion tracking. The tracking parameters appear to be basic range of motion for different joints.

In [206], Avola *et al.* reported a system designed to help patients who suffer from neuromotor problems to rehabilitate. Instead of using the traditional display for visual help to patients, the participant would wear a virtual reality head goggle (Oculus). The system tracks the participant movements using both Kinect sensor and the leap motion sensor. The Kinect sensor is used to track the overall body movements, and the leap motion sensor is used to track primarily finger and hand movements. The data from the two sensors are joined using the elbow joint, which is tracked by both the Kinect sensor and the leap motion sensor. Three exercises were considered. The first one is to raise the knee while pinching the fingers. The second one is to march in-place while grasping with hands. The third one is to get up on heels/toes while cutting spheres displayed by Oculus by hands. A deep learning model (RNN-LSTM) was used to determine the capability of a participant with neuromotor issues compared with that of a healthy participant based on how similar the actions performed by a patient with that of a healthy participant.

We also used Kinect in conjunction with IMU sensors in an avatar-based system developed for rehabilitation at home for patients with total knee replacement [226]. Four different rehabilitation exercises, quad set, side-lying abduction, straight leg raise, and ankle pump, were examined to see if using Kinect alone could be feasible. While the first three exercises could be reliably tracked by Kinect alone, the ankle pump required the input from IMU for accurate measurement.

5.4   Beyond Kinect

In recent years, human pose estimation using images taken by a single regular color camera has advanced rapidly. Similar to the Kinect technology, numerous methods have been proposed and implemented to produce a human skeleton for each person in the image. This makes it possible to monitor and assess the quality of the exercise performed in rehabilitation. The advantage of using a regular camera for human motion tracking is obvious over the Kinect sensor because it is much more convenient and less costly to use a webcam in a desktop or laptop, and mobile devices, which typically come with a front camera and back camera. Indeed, we have seen this line of research getting started for rehabilitation [227–229] and other purposes [230].

Many studies are based on a python library called PoseNet. It is interesting to note that almost all papers that we have reviewed cited the wrong source, i.e., a 2015 paper with the term "PoseNet" as part of the title [231]. That paper was about camera pose estimation and has nothing to do with skeletonization of humans. The correct source of the PoseNet used in many studies on human pose estimation is actually a model as part of the TensorFlow deep learning library (https://github.com/tensorflow/tfjs-models/tree/master/pose-detection).

Chua, Ong, and Leow described a system they developed for supporting in-home rehabilitation via telehealth [227]. The system consists of a laptop computer with a built-in webcam as the sensing device. Data processing was done using a PoseNet-based algorithm. Two human subjects were recruit to perform a subset of actions defined in a whole-body human motion dataset. The focus of the work is to assess the measurement accuracy for angular movements of the left/right elbow and knee. Rick *et al.* reported their work on a similar PoseNet-based system in a short paper for geriatric rehabilitation [228].

Biebl *et al.* presented a study using a smartphone running an app to provide guidance and feedback to patients with knee or hip osteoarthritis [229]. Only the front camera is used to track the movements of the subject. Six exercises were experimented, including hip extension bent leg, knee flexion, strengthening hip extensors, strengthening hip adductors, strain front of thigh, and elongation of the hip flexors. The smartphone was placed on the ground about two meters away from the subject and was tilted slightly so that the subject can see the screen. The key on providing feedback is to examine the actual movements performed by the subject with respect to a set of predefined spatiotemporal constraints, which is rather similar to the rule-based approach that we proposed [194].

Chapter 6
Instrumented gloves

Hand is perhaps the most complex and amazing segment in the human body. Hand dexterity is essential for activities of daily living, and it may be impaired by neurological disorders such as stroke [9], spinal cord injury [9], and Parkinson's disease [232], and it could also be impacted by rheumatoid arthritis [1]. To help rehabilitation and monitor the progress of hand rehabilitation, it is important to measure hand dexterity. Hand dexterity consists of several key components, as summarized in [233]:

- Control of force. This refers to the degree of control of each individual finger in applying force when doing tasks such as power grip, precision grip, and grasp-and-lift.
- Finger independence. This refers to the degree of control of individual fingers independently of each other.
- Synchronization of finger movements. This refers to the degree of control of individual fingers on temporal movements.

The hand has many joints, as shown in Figure 6.1. At the bottom, there are carpals, which consists of several bones. Then, there are five metacarpal bones, one for each finger. Going upward, the thumb has proximal phalanx and distal phalanx. The other hand four fingers each has three bones called distal phalanges, middle/medial/intermediate phalanges, and proximal phalanges. Between the bones are the joints. Between the carpals and the metacarpals are five joints referred to as the carpo-metacarpal (CMC) joints. Between the metacarpals and the proximal phalanges are the five metacarpo-phalangeal (MCP) joints. Between the proximal phalanges and the intermediate phalanges are the five proximal-interphalangeal (PIP) joints. The four fingers other than thumb have the distal-interphalangeal (DIP) joints between the intermediate phalanges and the distal phalanges. Some papers refer the PIP joint of the thumb as the interphalangeal (IP) joint [234,235], and the MCP joint of the thumb as the MP joint [235]. Some work also considered the trapeziometacarpal (TM) joint for the thumb [234] (not shown in the figure).

Fingers have two primary types of movements, as shown in Figure 6.2. One is the movements within the hand sagittal plane, including abduction and adduction. Finger abduction happens when a finger moves away from the middle reference of the hand. Finger adduction happens for the opposite move when a finger moves towards the middle reference. The other type is the movements within the hand frontal plane, including flexion and extension. In flexion, the angle between the finger and the palm

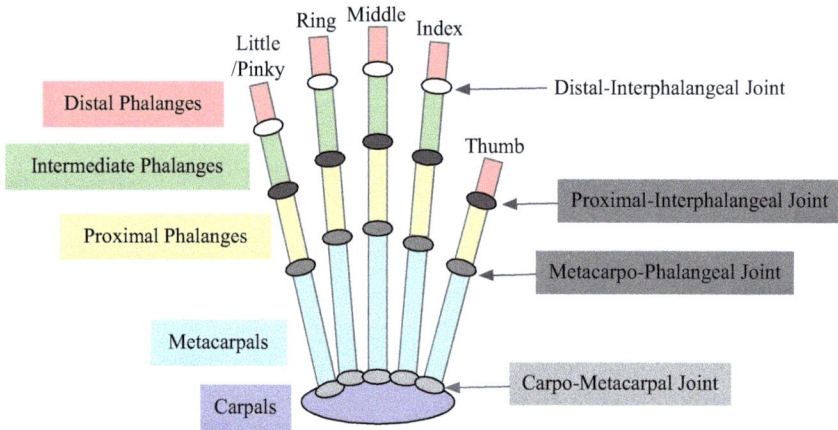

Figure 6.1   Hand anatomy with bones and joints of interest

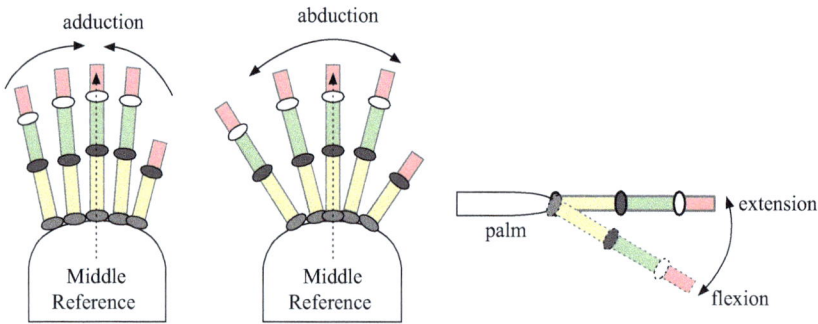

Figure 6.2   Major types of finger movements

becomes smaller. In extension, the angle becomes larger. Note that the definition of the thumb movements is different. Thumb flexion and extension would move within the hand frontal plane, where thumb flexion means that the thumb moves towards little finger and thumb extension means the thumb moves away from the little finger. Thumb abduction and adduction would move within the hand sagittal plane, where the abduction is an anterior motion (i.e., moving towards the front) and the adduction is the posterior motion (i.e., moving towards the back).

Evaluating the hand dexterity described above would require the measurement of kinematic characteristics of finger movements (such as a finger joint's flexion and extension), as well as the force applied by each finger [236]. Single-camera-based approach does not work well in tracking the hand movements due to self-occlusion caused by the intrinsic movement characteristics of the hand. A multi-camera motion analysis system can track hand movements highly accurately with markers placed on the hand. However, such systems are very expensive and not portable. The universal goniometer has also been used to manually measure the hand joint angle, but it is labor intensive and is also subject to inter-rater and intra-rater reliability problems [1]. Hence, sensor-based approach has been the dominating method. Research on instrumented gloves started fairly early and it is still very active to these days [1,2,4,235,237]. There are also several commercial products on instrumented gloves, although not all of them are intended for rehabilitation purposes. Instrumented gloves are also referred to as data gloves [1].

Sensors used in instrumented gloves include inertial measurement units (IMUs), flex sensors, optical sensors, and Hall effect sensors. Some gloves also combine with force sensors to measure the finger forces. These approaches all have their advantages and shortcomings. Most instrumented gloves can only measure finger (including thumb) flexion and extension.

Quite a few commercial instrumented gloves have been released, especially in the early days, usually not for the purpose of hand rehabilitation (e.g., for virtual reality applications or sign language interpretation). The technical details for these commercial gloves are typically not released to the public (except CyberGlove II from CyberGlove Systems, which does provide fairly detailed information at the company Website http://www.cyberglovesystems.com/cyberglove-ii/). In a review paper [238], the inner working of some commercial gloves are provided. However, it is unclear if the current version still follow the same design considering the rapid advancement of embedded and sensing technologies.

Some of the commercial gloves have been evaluated for hand rehabilitation. Data-Glove (based on optical sensors) from Visual Programming Languages was evaluated by Wise *et al.* [235]. HumanGlove (based on Hall effect sensors) from Humanware was evaluated in [239]. The 5DT (short for 5th Dimension Technologies) Ultra 14 glove was evaluated in [1] that showed that it has better usability (i.e., more comfortable to wear) and better repeatability results than the authors' research prototype glove. The 5DT Ultra 14 glove was said to be based on flex sensors in [238]. Cyber-Glove is based on the flex sensors, and it has been used in a couple of intervention studies on hand rehabilitation in a virtual reality setting [240,241].

Typically, an instrumented glove needs to be calibrated, often on an individual-basis due to the differences in hand and finger sizes. Some gloves, such as those instrumented with IMUs, may require periodical calibration. This calibration step might be cumbersome for some patients with hand disabilities. Gracia *et al.* [242] proposed a calibration procedure that aims to simply the process in the context of using a flex-sensor-based glove. The procedure requires the recording of a single hand position that could reflect the kinematics of the 16 finger/hand joints. Furthermore, they proposed a metric referred to as the "across-subject gains," which is obtained by averaging the gains obtained from calibrations made by multiple subjects over several days.

A comprehensive evaluation of an instrumented gloves would include accuracy, reliability (or repeatability), and feasibility (or usability) tests. The accuracy test typically compares the measured values from the glove and those obtained from another method that are accepted as highly accurate, such as using a goniometer or a marker-based multi-camera motion analysis system. Occasionally, other methods are used, such as a calibrated IMU. For the accuracy test, either a human subject wearing the glove, or a hand replica wearing the glove, would hold prefabricated mold of different shapes to ensure the finger joints are in specific set of angles.

The repeatability test is to check if the sensor output is the close to each other for the same joint angle in repeated measurements. This is to evaluate the reliability of the instrumented glove. Since a seminal paper was published in 1990 [235], the set of repeatability tests proposed by Wise *et al.* have been accepted as the standard for repeatability test of instrumented gloves. The set consists of four tests called test A, test B, test C, and test D. A plaster mold is used in test A and test B. In test A, the subject who wears the instrumented glove is required to clench his or her hand onto the plaster mold for 6 s and then release the hand for 6 s. The subject would repeat this clench-release cycle 10 times without taking the glove off. In test B, the subject is also required to clench the plaster mold for 6 s for a total of 10 times. However, in between two attempts, the subject would take off and put on the glove and there does not seem to have a predefined interval between two attempts except that the subject should complete all 10 times in the same day. In test C, the subject would wear the glove and place the hand with the glove flat on the table surface while the arm also remains flat on the table surface. The subject would stay flat for 6 s, and then clench his or her hand slightly for 6 s, which completes one cycle similar to test A. Without taking off the glove, the subject would repeat this cycle 10 times. Test D is rather similar to test B. The subject would take off the glove and put it back on in between two cycles. Some work performed reliability analysis by deriving the intraclass correlation coefficient (ICC) from the repeatability test results.

The feasibility (or usability) test typically would be done in the form of user surveys with a list of questions regard the use of the instrumented glove, such as whether or the glove impeded their finger movements.

6.1 Gloves based on IMUs

In IMU-based gloves, a special flexible printed circuit board (PCB) is designed to allow the mounting of IMUs and connectors on a flexible flat cable for data collection [1,2,4,237]. An IMU-instrumented glove would include a minimum of 16 IMUs according to [1,2]. As shown in Figure 6.3, on each side of the DIP and PIP joints has two IMUs. For the thumb, an additional IMU is placed below the MCP joint on the metacarpal bone. One more IMU is placed on the center of the metacarpal bone to detect the palm orientation and movement. Some glove used a different placement of 16 IMUs [3], where no IMU is placed on the metacarpal bone, but placed on the forearm instead. Some gloves use more than 16 IMUs. For example, in [4], 18 IMUs are used, where the two additional IMUs are used, one placed on the dorsal side of the metacarpal bone of the little finger and the other placed on the forearm.

Recent IMU-based gloves all use what is referred to as 9-axis IMU, which consists of a 3-axial accelerometer, a 3-axial gyroscope, and a 3-axial magnetometer [2]. Older IMU-based gloves have used IMUs that have only a 3-axial accelerometer and a 3-axial gyroscope [243]. While transiting to the 9-axis IMUs, Kortier *et al.* [237] experimented adding the magnetometer. In addition to the 6-axis 15 IMUs placed on

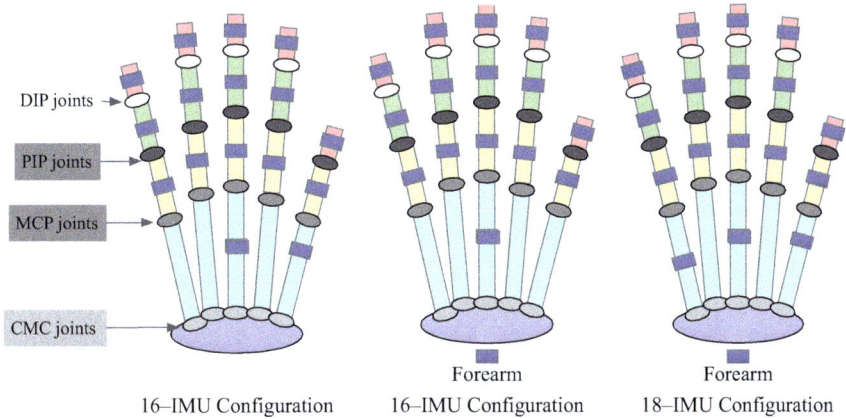

DIP joints →

PIP joints →

MCP joints →

CMC joints →

Forearm   Forearm

16–IMU Configuration   16–IMU Configuration   18–IMU Configuration

Figure 6.3   *Different IMU placements on a glove. The 16-IMU configuration on the left is used in [1,2], the 16-IMU configuration at the center is used in [3], and the 18-IMU configuration on the right is used in [4].*

each finger, they incorporated a magnetometer on each finger's tip and on the back of the hand [237]. They find that the use of magnetometer helps determine the heading of the movement [237]. What somewhat surprising is that despite the fact that IMU can measure 3D motions, all but one only reported the measurement of joint flexion and extension. Only the work by Kortier *et al.* [237] evaluated a form of finger abduction and adduction (index finger making circular movement).

In addition to IMUs, some gloves incorporated force sensors on the finger tips, such as [4], where the authors adopted a type of commercial force sensors called FlexiForce A101 from Tekscan. Force sensing for fingers is a necessary part of hand function assessment, in addition to hand kinematics.

6.1.1 Calibration

For IMU-based gloves, calibration is particularly important for several reasons. In [237], the IMU calibration was done to map the sensor coordinate frames to the coordinate frames of the segment to which the IMU is attached to. In [2], the calibration was done to remove the IMU sensitivity error and offset error. More specifically, in [2], the maximum and the minimum values of the accelerometer and the magnetometer are recorded. Then, the offset error is calculated based on the magnitude of the gravity and the geomagnetic field. The gyroscope offset error is identified by recording the gyroscope output at the stationary position. Lin *et al.* only reported the calibration for the magnetometer sensors in the glove [4]. In [237], an extended Kalman filter was used to calculate the relative finger kinematics and absolute hand kinematics.

In [1], the authors stated that the IMU sensors do not require complex calibration and did not elaborate any calibration step. Lin *et al.* [3] did not report any calibration procedure.

Table 6.1 Studies using consumer-grade devices containing IMUs

Reference	IMUs	Calibration	Human subject	Evaluation
Kortier [237], 2014	15+6	Mapping sensor coordinate frames to segment frames	1,1,5	Fingertip position validation, index finger and thumb pinching, dynamic range, repeatability
Connolly [1], 2017	16	None	9	Accuracy & Repeatability & Informal Usability
Chang [2], 2019	16	Offset & Sensitivity	None	Only roll and pitch angle
Lin [3], 2019	16	None	15+15	Classification between healthy and patients, and severity levels in patients
Lin [4], 2020	18	For magnetometer	1	Joint angle and fingertip force accuracy

6.1.2 Signal processing

For IMU-based gloves, a filtering algorithm is essential to remove the noise. Typically, a low-pass filter is used to remove high-frequency signals. The cutoff frequency is carefully determined. For hand motion, [2] uses 20 Hz as the cutoff frequency. Neither [3] nor [4] reported the use of any filtering algorithm. In [1], in addition to a low-pass filter, the authors used a complementary filter to fuse the output from the accelerometer and the gyroscope.

6.1.3 Reference systems for evaluation

In [237], the VZ-4000 system from PTI VisualEyez is used to validate their glove. In [4], Vicon Nexus 1.7.1 was used to establish the ground truth on joint angle measurement (in addition, the authors also verified the force sensor measurement via a force gauge, one finger at a time). In [1], the Vicon MX motion capture system was used to validate the joint angle accuracy measurement of the proposed IMU-based glove. All these systems use multi-cameras to perform motion analysis and they typically require the subject to wear multiple markers on the dorsal side of the hand.

6.1.4 Accuracy evaluation

The accuracy test could take several forms. Typically, the human subjects will be asked to perform a set of carefully designed tasks.

In [4], the subject was asked to perform two types of grip tasks, one to grip a horizontal cylinder and the other is to grip a vertical cylinder. The former is referred to as the horizontal flexion–extension movement, and the latter is referred to as the virtual flexion–extension movement. During this test, only the index finger IMU flexible board is activated. The subject was asked to perform each task continuously for 5 times. During each task, the subject is to grip for 3 s and fully extend the hand for 3 s. In addition to the two grip tasks, the subject is asked to perform random fast-paced and slow-paced flexion–extension movements continuously for 10 s. The accuracy of the glove is evaluated using the mean absolute error. The authors also tested the performance of the glove when both the joint angle and force sensing are integrated with one patient and a healthy subject [4]. The two subjects were asked to perform the same task to evaluate the difference between the patient and the healthy subject. The task is to hold the bottom filed with water, life it up, and move to a different position that is 50 cm apart from the original position.

In [1], the accuracy test itself did not use any human subject. Instead, the glove was placed on blocks of wood with predefined angles. Then, the measured joint angles from the proposed instrumented glove, the commercial 5DT Ultra 14 glove, a universal goniometer (manually measure the angles), and the Vicon MX optical system are all compared with the predefined angles in the blocks. They find that the universal goniometer actually gives the most accurate result.

Kortier *et al.* [237] performed several accuracy tests: fingertip position comparison, index finger and thumb pinching motion, finger dynamic range. Each test was performed by a single human subject. The fingertip position comparison consists of two tests and each test was repeated 10 times continuously. In the first test, the subject was asked to repeatedly flex the index finger to the maximum flexion angle of all three finger joints (i.e., MCP, PIP, and DIP). In the second test, the subject was asked to perform circular-like movements with the person's index finger in a way that the abduction and adduction angle of the MCP joint was maximized in each repetition. For the pinching motion, the subject was asked to move the index finger and the thumb closer until they touch each other and then move back to the original position as a cycle, and perform 10 cycles. For the dynamic range motion, the subject was asked to first play the hand flat on the table surface, raise forearm and open the hand fully and then close the hand at a tempo of 116 beat per minute for 30 times.

6.1.5 Repeatability and reliability evaluation

Both the work by *et al.* [237] and Connally *et al.* [1] performed two repeatability tests, one is referred to as the flat hand test and the other is called the paster mound test, which is a subset from the tests designed by Wise *et al.* [235]. Because in both "gloves" the sensors are not attached to any textile materials, the two original tests that would take off the glove were omitted. The flat hand test is to evaluate the glove's ability to maintain repeatability after the glove has been fully stretched. The plaster mound test is to evaluate the glove's ability to reproduce the same measured values for the same set of angles.

In [237], five healthy male subjects were recruited to perform the repeatability study. A subject would do each task continuously for 10 times as a group, and perform the group 10 times with a 1-min rest in between. The repeatability test was evaluated by the range and standard deviation.

In [1], the repeatability tests were done with nine patients that were suffering from rheumatoid arthritis. As part of the repeatability tests, the authors also performed a comparison study of their custom-built IMU-based glove and the commercial 5DT Ultra 14 data glove. User feedback shows that the 5DT data glove is more comfortable than the custom-made IMU-based glove, which is more bulky and the sensors do not fit well with the subject's hand. Therefore, the 5DT ultra 14 data glove have better repeatability in their tests.

Chang *et al.* [2] did not perform traditional accuracy and repeatability tests of the glove. They used the standard deviation under different pitch and yaw reference as the metric to estimate the measurement error.

6.1.6　*Classification of activities*

Instead of doing accuracy and repeatability tests, Ref. [3] used the data collected from three tasks performed by 15 patients with stroke, and 15 healthy subjects to class between the patients and the healthy subjects, and between different levels of severity. The severity of the patients with stroke was determined by physicians to have three levels, one of which is considered by the authors as indistinguishable from healthy subjects. Each subject was asked to perform three tasks:

- Thumb task: In this task, the subject would hold a vertical cylinder, press the button on top of the cylinder for 4 s and then release the button for another 4 s as one cycle. The subject would continuously perform 10 cycles. The purpose of this task is to assess the subject's thumb dexterity.
- Grip task: In this task, the subject would grip an elevated horizontal cycler for 4 seconds and release it for another 4 s as a cycle. The subject would continuously perform 10 cycles. The purpose of this task is to evaluate the entire hand's dexterity.
- Card-tuning task: In this task, the subject would turn a paper card as quickly as possible continuously for 20 times. The purpose of this task is to evaluate the hand mobility and stability.

The dimension of the raw measurement data was reduced using the principal component analysis for feature construction. The classification was performed using logistic regression.

6.2　Gloves based on flex sensors

This type of instrumented gloves typically uses commercial flex sensors. These sensors are also referred to as bend sensors. Saggio *et al.* provided a comprehensive review on resistive flex sensors [244], which are the most common sensors used in instrumented gloves. This type of sensors is thin film sensors. Each sensor consists of a flexible substrate (typically based on polymer) common one where the sensor film can be deposited onto, and electric contacts. The working principle of this type of sensors is that when the flex sensor bends, the resistance of the film sensor will increase due the presence of cracks caused by the bending, which is in a way rather similar to that of the piezoresistive pressure sensor we discussed in Chapter 2. Although there are several commercial flex sensor manufactures, it appears that the flex sensor from Flexpoint Sensor Systems has been embraced predominantly in custom-made instrumented gloves. These gloves use different number of flex sensor to measure the movements of different set of fingers, as shown in Figure 6.4.

Borghetti *et al.* [5] proposed an instrumented glove based on the flex sensors from Flexpoint Sensor Systems. Two flex sensors are used on each finger (i.e., a total of 10 sensors) to measure the flexion of the MCP and PIP joints. The validation of the glove is done using a 3-camera motion tracking system from OptiTrack (V100:R2) during a sequence of bending tests. Each test would bend every single joint of each finger between 0° and 90°.

Chuang *et al.* [7] reported their work on using an instrumented glove to detect four specific finger gestures. In each gesture, one or more fingers would move according

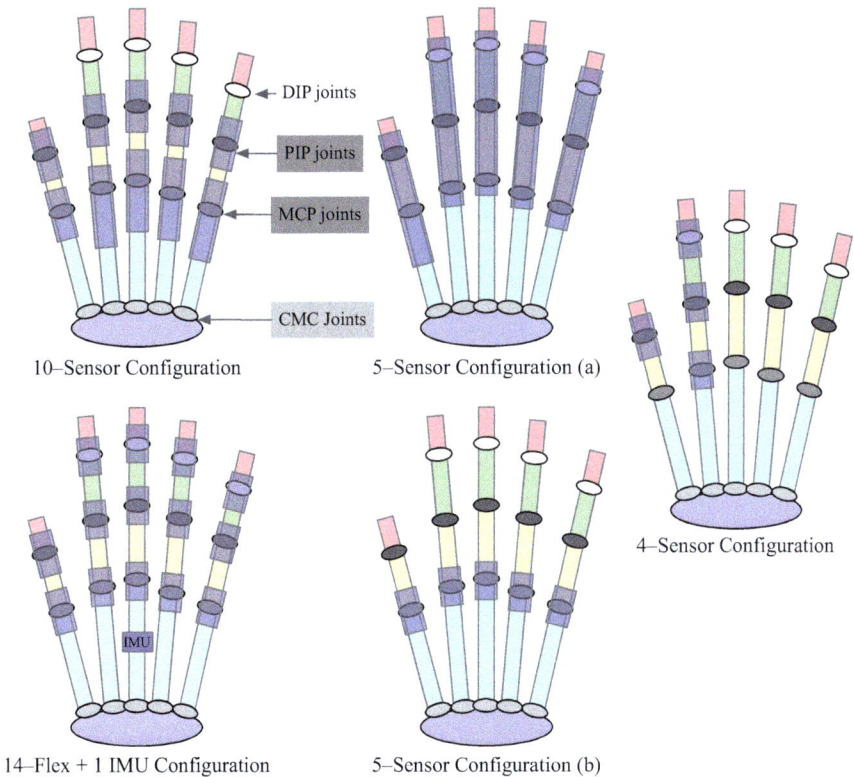

Figure 6.4   *Different flex sensor placements on a glove. The 14-sensor configuration on the top left is used in [5]. The 15-sensor (14 flex and 1 IMU) configuration on the bottom left is used in [6]. At the center are two 5-sensor configurations, where the top one is used in [7] and the bottom one is used in [8]. The 4-sensor configuration on the right is used in [9].*

to a predefined sequence. They used only a single 3.75-in long flex sensor for each finger in the glove (for a total of 5-sensors) (the manufacture of the flex sensor is not disclosed in the paper). Due to the objective of the research, they did not attempt to derive the joint angles. Instead, they used the measured resistance changes as input to a machine-learning algorithm called gated recurrent unit.

In [9], Oess *et al.* proposed a low-cost instrumented glove using 4 flex sensors from Flexpoint to measure the flexion of four joints, 3 at the index finger (MCP, PIP, and DIP joints) and 1 at the thumb (DIP joint) for the right hand. The authors' argument is that these four joints play the biggest role in performing main grasp tasks. The authors performed three types of evaluations on the glove, one on its reliability, the second on its validity, and the third one its feasibility. The reliability evaluation is done in terms of repeatability of the finger bending trajectories while the participants are doing three tasks that are common in activities of daily living (ADL): (1) pour water into a glass from a bottle; (2) unscrew the lid of a jar; (3) move a peg from one pegboard and to another pegboard. The validity evaluation is done by comparing the measurements of a set of predefined angles for each finger (at $10°$, $30°$, $50°$, $70°$, and $90°$) using the glove and a goniometer from Proteck AG. The feasibility evaluation included both user feedback on the glove (such as whether or not the glove reduces the touch sensitivity), and a quantitative correlation between the finger movement trajectories of the patients and the healthy subjects in the three ADL tasks. It is unclear why the latter is done because the way the patients performed the tasks is very different from that of the healthy subjects.

In [6], Sbernini *et al.* reported their work on using an instrumented glove to objectively assess the open surgery skills of the medical trainees instead of for purposes of rehabilitation. The glove uses 14 flex sensors from Flexpoint to measure the flexion of all the MCP, PIP, and DIP joints of the five fingers, and one IMU placed on the back of the palm. The sensor outputs are used to classify the expertise levels of the subjects while performing two surgery tasks: (1) a single interrupted suture (SIS) and (2) a simple running suture (SRS). Each task is classified as whether or not it is performed by a trained or an untrained user, and each subject is classified as an expert or a novice by considering the task performed by the subject. They experimented with three machine-learning algorithms, linear discriminant analysis (LDA), support vector machine (SVM), and artificial neural networks (ANN). They showed that overall, ANN performs the best.

In [8], Simone *et al.* proposed an instrumented glove for hand functional assessment. They designed the glove to be attached to the dorsal side of each finger using double-sided medical grade adhesive tape. The test version of the glove includes one flex sensor (from Flexpoint) for each finger covering the MCP joint (i.e., a total of five sensors). Prior to their testing, they first calibrated the sensors with respect to a set of joint angles (at maximum extension, $0°,10°$, $30°$, $60°$, $90°$, and maximum flexion). Then, they performed the repeatability testing of the glove by using a simplified set of tasks proposed by Wise *et al.* [235]. Based on the repeatability testing results, the authors performed reliability analysis. They also carried out a usability survey to collect user feedback, as well as the testing on the glove's wireless transmission and battery life.

The CyberGlove II is a commercial instrumented glove designed for motion capture in the graphic animation industry. It has been used in a couple studies for hand rehabilitation for post-stroke patients using a set of exercises delivered via a virtual reality interface, but without a systematic evaluation [240,241]. Unlike other commercial instrumented gloves, which barely publish any technical details, CyberGlove provides relatively detailed technical specification on its Website (http://www.cyberglovesystems.com/cyberglove-ii/). The CyberGlove family of instrumented gloves are based on their own patented flex sensors. CyberGlove II has two versions, one with 18 sensors, the other with 22 sensors. The 18-sensor glove consists of two sensors on each finger to detection flexion/extension, four sensors to detect abduction/adduction, and additional sensors to detect thumb crossover, palm arch, wrist flexion/extension, and wrist abduction. The 22-sensor glove consists of three sensors on each finger for measuring flexion/extension, and the remaining sensors are identical to the 18-sensor version.

6.3 Gloves based on optical sensors

Several different types of optical sensors have been proposed for use with instrumented gloves. One type is based on Fiber Brag Grating (FBG), which has also been used for pressure sensing and follows exactly the same working principle, although the use context for finger movement is significantly more challenging than if used in smart insoles for pressure sensing. In FBG optical sensor, the wavelength would be changed when the fiber is stretched or bent. Another type is based on the working principle that when the finger joint bends, so does the fiber, which will lead to the attenuation of the light at the light detector. Yet another approach is called optical linear encoder (OLE). It is drastically different from the previous two types in that OLE is device that measures linear movement using an optical scheme rather similar to that used in optical mouse for controlling the mouse pointer as a form of computer input.

6.3.1 FBG-based approach

This approach is becoming quite popular in recent years because of several advantages. By using optical fibers, a glove that could incorporate many FBG sensors (the largest number reported is 39 [245]) and yet very lightweight, and its operation is not affected by electromagnetic interference. However, using FBG for finger tracking has several issues. First, finger flexion/extension would cause both strains along the axial direction of the FBG sensor, but bending as well. Second, unless each FBG sensor is securely fixated at individual finger joints, some of the FBG sensors would move while some finger joint bends [245]. This means that even though the FBG sensors are placed exactly on the dorsal side of each finger joint, their positions could be moved away from their designated joints. Both factors would significantly complicate the algorithm needed to derive the joint flexion/extension based on the observed wavelength shift. Third, due to the need to analyze the wavelength shift for every FBG sensor on the same optical fiber, a multi-channel interrogator is required, which is very expensive and also makes the glove tethered to the equipment.

The most sophisticated FBG-based glove perhaps is the one proposed by Kim *et al.* [245]. The authors fully recognize the first two issues we mentioned above and designed highly complicated algorithms to accommodate the movements of the FBG sensors due to the flexion/extension of different finger joints and different hand sizes. The latter is because the movements of the FBG sensors would depend on the finger sizes. The glove is instrumented with 39 FBG sensors (in the paper, each FBG is referred to as a node) from FBGS Technologies GmbH. Each FBG sensor is 2-mm wide and 3-mm long. The spacing between adjacent FBG sensors (on the same finger) is 19 mm. For the index, middle, and ring fingers, nine FBG sensors are used each finger. For the litter finger, 7 FBG sensors are used. For the thumb, five sensors are used. Due the dense placements of the FBG sensors, multiple of them will be involved for each joint flexion/extension. In the study, an 8-channel interrogator from Smart Fibres Limited is used. The equipment operates in two bands (C and L) and each band has four channels. The glove is designed to measure only finger flexion/extension of the MCP, PIP, and DIP joints. The glove is evaluated using a 3D-printed hand replica at different angles for each finger joint (at $0°$, $20°$, $40°$, $60°$, $80°$, and $90°$). The ground truth is established by manually measuring the angle using a goniometer.

In [246], Jha *et al.* proposed an FBG-based glove that tracks the flexion/extension of 10 finger joints (the MCP and DIP joints for each finger). Unlike the design of Kim *et al.* [245], where a single sensor unit is used for each finger, Jha *et al.* used a separate sensor unit for each of the 10 finger joints. This design avoids the displacement of issue of the FBG sensors and only a single FBG sensor is needed to measure the flexion/extension of the target joint. The FBG sensor is first inserted and fixed in a PVC tube, and this tube is then inserted inside a spring to allow joint movement. Finally, the unit is packaged within a 3D-printed tube. Then sensor units are then secured to the glove using five strips of PVC latex foam. These sensor units are placed on the dorsal side of each MCP and PIP joint. Due to the use of only 10 FBG sensors, a 4-channel interrogator is used to analyze the wavelength shift of the sensors. The evaluation of the glove is done via comparison with a pre-calibrated IMU (MPU-6050 from InvenSense) in accuracy and repeatability testings. In the accuracy test, a subject would first wear the glove and place the hand flat on the table surface, then slowly bend each finger at about 1 degree per second. The repeatability test follows the four tests proposed by Wise *et al.* [235].

In an earlier work [247], da Silva *et al.* reported a relatively simple FBG-based glove. The glove is instrumented with a single optical fiber that has 14 FBG sensors. They aligned the fiber in a curvilinear layout on the top side of the glove so that the FBG sensors would be on top of the 14 intended finger joints (5 MCP, 5 PIP, and 4 DIP joints) and at the same time there are rooms for finger flexion. The glove uses a broadband light source from Luxpert in the C-band range, an optical circulator from Oplink Communications, and an interrogator (I-MON 80D) from Ibsen Photonics. The glove is evaluated on accuracy with a simple hand open and close motion. Initially, a subject would use two hand postures: (1) hand close and (2) hand open to calibrate. For subsequent movements, the output from the glove is compared with a reference

Figure 6.5   Optical fiber bend transducer

device called sliding T bevel device, which can be used for setting and transferring angles. The paper did not present repeatability and feasibility testing results.

6.3.2   Light-attenuation-based approach

While the several instrumented gloves covered in this section all follow the same basic principle of using light attenuation to detect finger flexion/extension, the actual techniques used different.

One of the earliest works on using an instrumented glove to objectively measure finger movements is by Wise *et al.* [235]. Wise *et al.* used a commercial instrumented glove called DataGlove from Visual Programming Languages. The DataGlove uses the light attenuation-based approach to detect finger flexion/extension. The basic idea is that when a finger bends, the loops of the specially treated optical fiber would attenuate the light transmission according to the degree of bending (the greater degree of bending the more attenuation). The technical details were not available. Being a commercial instrumented glove, the accuracy test has already been done by the manufacturer (1 degree of angular resolution and 5 degrees of static accuracy). Hence, the authors focused on doing repeatability testing and they introduced a set of tests that are later used as the standard for instrumented gloves. The DataGlove used in the study does not allow finger abduction and adduction.

Other gloves use various techniques to attenuate the light, including optical fiber bending transducer, hetero-core fiber–optic nerves, and an approach that utilize the Malu's Law on polarized light.

6.3.2.1   Bending transducer

In [248], Fujiwara *et al.* designed an optical fiber bending transducer to detect the finger flexion/extension. The term "transducer" refers to the conversion of finger movement to optical light attenuation. They used only four transducers on the MCP and PIP joints of the index finger and the thumb (one transducer on the dorsal side of each joint). The transducer design is illustrated in Figure 6.5. The design goal is to magnify the impact of finger bending on the attenuation of the light going through the fiber. The authors did this by using an array of silicon rods and thread the fiber over the rods. At the input end, the fiber loops are used. In fact, what is shown in Figure 6.5 is only one of the configurations they experimented. They chose the one illustrated here because this configuration has the best sensitivity. The rod width is 4 mm and height is 3 mm.

For calibration, the subject was asked to perform four hand postures in sequence:

- Remain in open hand position with all fingers tightly close to each other for about 7 s.

- Gradually abduct each finger to the maximum angular displacement with open hand.
- Grasp a prepared object with a preset angle so that the MCP and PIP joints are in an intermediary flexion position.
- Grasp another prepared object so that the MCP and PIP joints are in a more acute flexion position.

For accuracy evaluation of the glove, the subject is asked to grasp a set of objects with predefined angles. The repeatability test is done exactly as what have been defined by Wise *et al.* [235].

6.3.2.2   Hetero-core fiber–optic nerves

In [234], Nishiyama and Watanabe proposed an instrumented glove using hetero-core fiber–optic nerves. The principle is still to induce light attenuation when the finger flexes. In previous work, the authors established that the optical loss of the hetero-core structure linearly increases with its bending angles, which can be exploited to use as a sensor to detect finger flexes. The design of the glove is illustrated in Figure 6.6. The idea is to have two fixed points (using adhesive to secure the enclosure of the fiber to the glove) per finger. Figure 6.6 shows the placement of the fiber to track the index finger flexion. The fixed point A is located at the bottom center of the glove, and the fixed point B is located between the DIP and the PIP joint. From the PIP joint, over the MCP joint, to the back of the palm at point C are a sequence of short tubes to guide the fiber and allow the fiber to slide. The hetero-core sensor is located in between the point A and point C and the fiber is aligned curved sideways to allow the flexes of the finger, that is, when the finger flexes, the sensor will bend. This design has its advantage in that the measurement will not be affected by the random wrinkles in the glove at the joint because the sensor is actually not located directly on top of the joint of interest. However, this design can only reliably measure the joint angle when one joint flexes at a time (for the PIP or MCP joint).

The authors made two versions of the gloves. In the first type, three fiber lines are used to detect flexion of the thumb, index, and ring fingers. In the second type, the authors experimented with using two fibers with paired hetero-core sensors that cross each other as a single unit to detect the flexion of the ring and the little fingers. The authors presented the result on the light loss on the flexion of different finger joints, but did not carry out accuracy and repeatability testing. It is worth noting that the correlation between the joint angle and the light loss depends on the finger size, as shown in Eq. (1) in [234]. This implies that personalized calibration is required to use the instrumented glove.

6.3.2.3   Glove based on Malus's law

Yet another method to induce light attenuation when a finger flexes is to exploit Malus's law, as proposed by Wang *et al.* [10,11]. The idea is to first apply a polarizer to the unpolarized light, and then use a light analyzer prior to the detection of the light intensity. The polarizer will create a polarized light where the light wave "vibrate" within the single plane. In Figure 6.7, this plane is illustrated in a 2D view as the polarizer axis. The analyzer would allow the light that "vibrate" within its designated

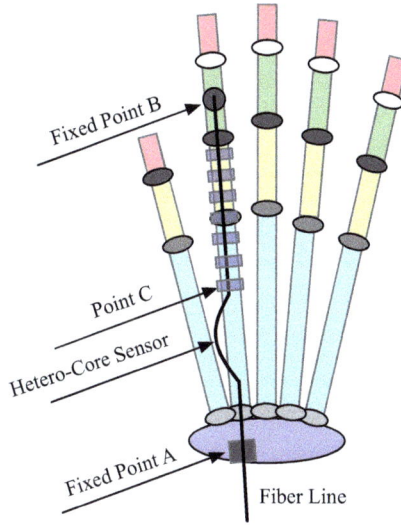

Figure 6.6   Hetero-core-based glove design

plane to go through. Initially, the axis of the analyzer aligns perfectly with the axis of the polarizer. When the finger flexes, it will cause the axis of the analyzer to misalign the axis of the polarizer. When there is an angle θ between the polarizer axis and the analyzer axis, only a portion of the polarized light could go through analyzer. Let the amplitude of the initially polarized light be E_0, the light can pass the analyzer would have an amplitude $E_0 \cos(\theta)$. The light intensity is proportional to the square of the amplitude. Hence, the light that has passed the analyzer would be $I_0 \cos^2(\theta)$, where I_0 is the light intensity prior to the light going through the analyzer.

If the analyzer and the polarizer are placed properly, the flexion angle would exactly be the angle between the polarizer axis and the analyzer axis. If so, the angle measurement will be independent of the size of the finger, which is highly desirable and could eliminate the need for the individualized calibration step. Wang *et al.* design a device that works exactly like this [11]. The device has two wings. Across the center of the device where the two wings are connected consists of a LED light source, a polarizer, an analyzer, and a photodiode as the light detector. In [10], the authors improved their design so that the device can detect both flexion/extension as well as abduction/adduction of the finger (i.e., the LED and the polarizer are located in a center piece of the device, and the analyzer and the photodiode are incorporated at the inner-end of each wing). Their device has the advantage of being untethered, unlike the FBG-based solution. However, how to adapt the device into an instrumented glove is unclear.

As shown in Figure 6.7, to track finger flexion/extension, it is attached on top of a finger joint and each wing will be placed on the dorsal side of the bones adjacent to the joint [11], and to track finger abduction/adduction, the device is placed on top of

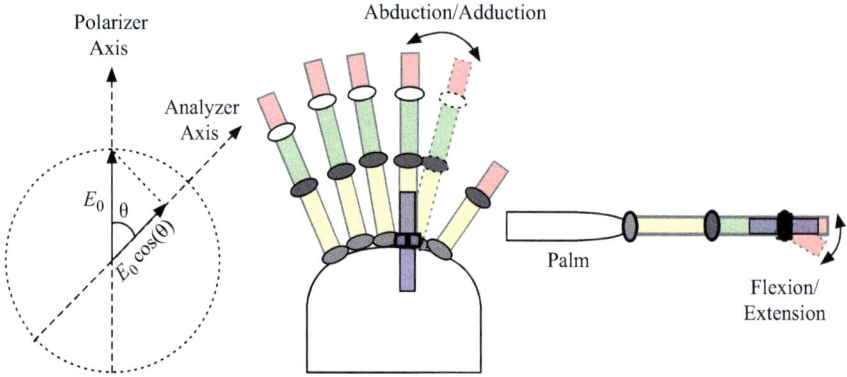

Figure 6.7   *The working principle of the Malu's law (left), and the device developed based on the Malu's law to track finger abduction/adduction in [10] and flexion/extension in [11]*

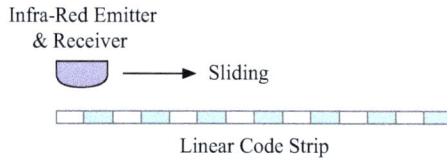

Figure 6.8   *The working principle the optical linear encoder*

the MCP joint of the index finger, with one wing on the dorsal side of the proximal phalanges, and the other wing on the dorsal side of the metacarpals [10].

The evaluation of the device is done with a motorized rotation stage (NR360S) from Thorlabs to systematically characterize the angle measurement in [11] and is done with human subject in [10]. In [11], the repeatability is established using statistical range and standard deviation.

6.3.3   Optical linear encoder

Drastically different from the previous two approaches, the optical linear encoder (OLE) is a scheme to measure linear distance traveled using an optical device instead of light attenuation or wavelength changes [249]. The working principle of OLE is illustrated in Figure 6.8. The OLE contains two components [249]: (1) a linear code strip that consists of parallel lines with 0.1 mm spacing, and (2) an infra-red light emitter and a receiver that together measure the distance traveled while moving along the linear code strip by emitting light to the strip and analyzing the reflected light collected by the receiver.

The OLE was first proposed to measure body movement [249] and was later adapted to measure finger flexion [250]. More specially, in [250], an inverted form of OLE is used where the emitter and receiver combination is fixed at the finger on one side of the joint, and the linear code string is fixed at the opposite side of the joint, and when the joint flexes, the strip will move, which can be detected. Also, in [250], the emitter and receiver combination is referred to as OLE. We will use the same notation from now on.

To deduce the flexion angle, the joint is modeled as a circular disk where the linear displacement is assumed to be due to the rotation of the disk. Apparently, the model would need the information on the thickness of the finger at the joint. Let R be the radius of the circular disk, when the joint flexes an angle ϕ in degree, the disk rotate on its edge for a distance of $L = 2\pi R\phi/360°$. Therefore, given the measured distance of L and assuming that R is known beforehand, we can calculate the flexion angle $\phi = \frac{L \times 360°}{2\pi R}$ (in degree). As can be seen, it is straightforward to track a single finger joint flexion tracking.

The authors developed an OLE-based glove for full hand joint tracking. The glove uses five linear code strips, one for each finger. In addition, two fixed OLEs are deployed for each finger (a total of 10 OLEs for the glove). In this design, any finger joint flexion would induce the detection of displacement by all the OLEs on that finger. This design makes necessary to develop a complex kinematic model for each finger. The OLE used in the glove actually is an optical mouse sensor product (ADNS-3530) from Avago Technologies. The OLEs are mounted to the glove using Velcro. For data collection and processing, each OLE has a ribbon wire connected to a microprocessor.

Like other instrumented gloves, one must go through a calibration step. The authors adopted four specific hand postures as proposed in [251]. The first posture is when the hand is flat when all joint angles are at 0 degree. The second posture is all other fingers are at 0 angle and the thumb's MCP joint at 45 degrees and PIP joint at 90 degrees. In the third posture, the MCP joints for all 4 fingers are at 90 degrees. In the fourth posture, the PIP joints of all four fingers are at 90 degrees. They used an accelerometer (LIS3LV02DQ) from STMicroelectronics as the reference to test the accuracy of the OLE. The accelerometer is attached to the second knuckle of the index finger. Hence, only the PIP joint flexion can be compared.

The glove's performance was evaluated with five health male subjects using the grip test and the flat test. In the grip test, a subject would grip a cylindrical metal bar for 6 s and release for 6 s. While the subject released the grip, he will place his hand flat on the table. The subject would do this grip/release cycle 10 times continuously as a set. The set is repeated five times. Hence, a subject would perform a total of 50 cycles, all without taking off the glove.

In the flat test, a subject would first place his hand flat on the table for 6 s, then raise his hand and slightly flexes his fingers before returning his hand flat on the table again. Again, the subject would continuously do 10 cycles of the movements as a set. The subject would do five sets total. Hence, the flat hand test is repeated for 50 times for each subject. Based on collected data, the repeatability is determined by the range and standard deviation. The reliability is calculated using intraclass correlation coefficient (ICC).

6.4   Gloves based on Hall effect

We are aware of only one instrumented glove that uses the Hall effect sensor to detect finger flexion/extension as well as abduction/adduction. It is called Human-glove and it was designed and released commercially in 1997 by Humanware (https://www.hmw.it/en/). It is unclear if the current version of the glove still uses the Hall effect sensor because no technical details are disclosed in the company website.

The Hall effect refers to a phenomenon that in the presence a magnetic field that is perpendicular to an electrical current, the electronics in the current will experience a Lorentz force sideways, as shown in Figure 6.9. This force will cause the built up of electrons on one of the conductors, which can be measured by the voltage level. The voltage V is linearly proportional to the magnetic field B if it is perpendicular to the current and the conductor surface. Hence, the stronger the magnetic field (along the right direction), the larger the voltage.

The presence of relatively strong magnetic field is needed for a Hall effect sensor to work, which is typically achieved by using a permanent magnet. Furthermore, the once the magnet is fixed at a particular location in the environment where the Hall effect sensor will operate, careful calibration is required because sensor output is proportional to the magnetic field only. For the sensor to be useful in measuring finger flexion/extension or abduction/adduction, we want to know the rotation angle of the finger instead of the magnetic field. The book on Hall effect sensor by Edward Ramsden provides an excellent description on the working principle of Hall effect sensors and their technical details [252]. In the following, we will follow the terminologies introduced in the book.

To understand how one can use the Hall effect sensor to measure rotation, we consider a simple scenario where the sensor is in a uniform magnetic field. The uniform magnetic field can be produced using two magnetic, as shown in Figure 6.10. A magnet has two poles, south pole and north pole. One can image that the magnetic

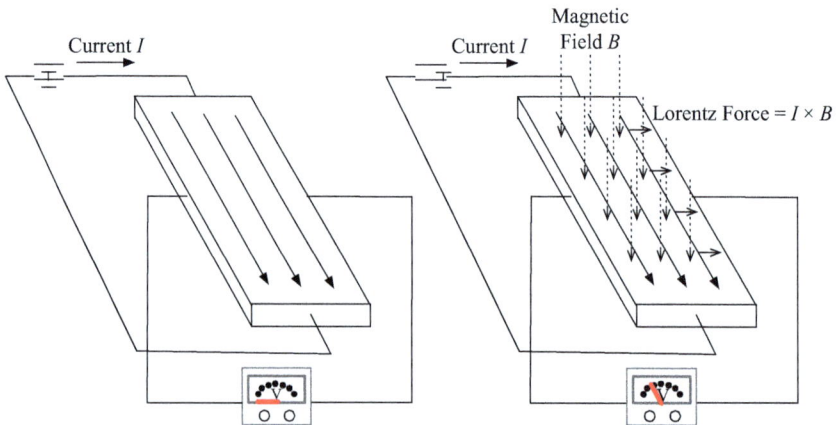

Figure 6.9   The working principle of the Hall effect sensor

field comes out of the north pole of the magnet and comes back in at the south pole. When two magnets are placed reasonably close to each other with one magnet's south pole directly opposite the other magnet's north pole. The region in between the two magnets can be considered uniform.

To calibration, the sensor is first oriented such that its conductor surface is parallel to the magnetic field, which should not lead to any voltage build-up across its surface. We define this orientation as angle 0°. The sensor output at this orientation is called offset V_{offset}. Then, the sensor is rotated so that its surface is perpendicular to the magnetic field. This orientation defines the 90° angle. The angle definition is shown in Figure 6.10. The sensor output is referred to as the gain V_{gain}. V_{gain} would be the maximum output and V_{offset} would be the minimum output among all orientations. The range of the angle can be measured by a single Hall effect sensor is listed to $-90°$ to 90°. Based on the offset and gain, we can normalize an arbitrary output V to between -1 and 1: $V_n = (V - V_{offset})/V_{amplitude}$, where $V_{amplitude} = V_{gain} - V_{offset}$. When either the sensor or the magnets rotate by θ around the sensor's current flow direction as the axis, the portion of the magnetic field that would be perpendicular to the sensor surface will become $B \sin(\theta)$. Furthermore, because the Hall sensor output is linearly proportional to the amount of magnetic field that is perpendicular to the sensor surface, we can derive the rotation angle $\theta = \arcsin(V_n)$.

A single Hall effect sensor can only measure rotation between $-90°$ and 90°. To measure all 360° rotation, two Hall effect sensors that are positioned perpendicular to each other, can be used together.

So far, we have assumed that the sensor is operating in a uniform magnetic field. In practice, often a single magnet is used, which means that the magnetic field is no

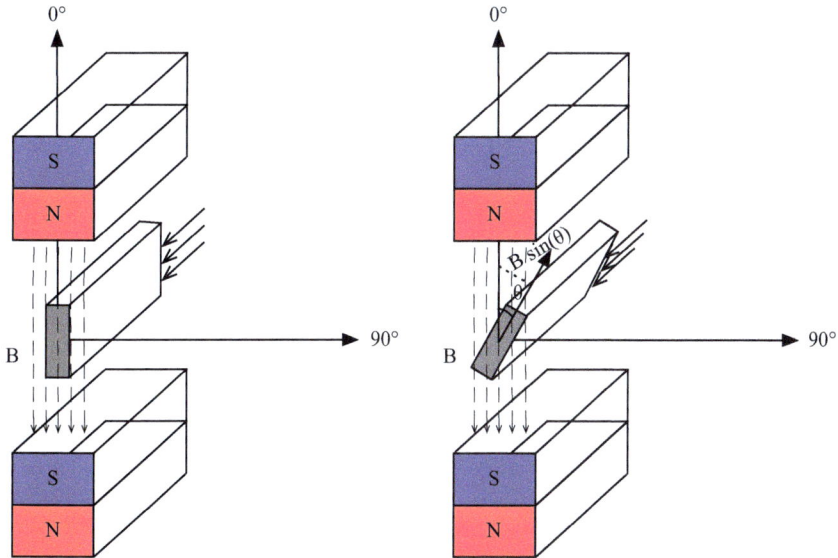

Figure 6.10   Hall effect sensor calibration and rotation angle determination

longer uniform. In fact, the magnetic field strength reduces sharply with the distance to the magnet. This could make the measurement of finger flexion/extension and abduction/adduction much more challenging because while the finger moves, the magnetic field would change even without rotation of the sensor. There is no publicly available information regarding how Humanglove measures finger flexion/extension and abduction/adduction (including the type of magnet used).

In [239], the authors reported their evaluation of the Humanglove on hand movement measurements. According to [239], the Humanglove has 20 sensors and it can measure the flexion and extension of all finger (including thumb) joints, and the abduction/adduction of all five MCP joints.

Prior to the tests, each subject would perform a calibration procedure. First, the subject would place his or her hand (with the glove) flat on the table surface. Then, the subject would make a hand posture that every finger joint has the maximum flexion. The repeatability test follows that proposed by Wise *et al.* [235]. From the data acquired using the glove, the authors calculated the range and standard deviation for each subject. Then, they did ICC analysis for each of the four repeatability tests.

Part II

Human motion recognition and exergames

Chapter 7
Measurement of basic parameters

In this chapter, we present some basic background information on human motion basic parameters, and on common statistical methods used to evaluate the feasibility of a new sensor-based measurement instrument (such as an IMU or Kinect-based system) to measure human motion, which would offer much lower cost and provide much better accessibility to patients, and ultimately facilitate practicing rehabilitation exercises at home.

7.1 Mechanics of body movements

To understand what basic parameters to measure, it is necessary to know some basic terminology related to the mechanics of body movements, which is often referred to as kinesiology [253].

7.1.1 Anatomical planes

Quite often, human movements are described in the context of anatomical planes. There are three different anatomical planes, one for each spatial dimension, as shown in Figure 7.1. The frontal plane is a vertical plane parallel to the human body from side to side assuming that the human is standing straight up. One can imagine that the frontal plane passes through the human body from one side to another side, and this plane would divide the body into the front (referred to as anterior) and back (referred to as posterior). The frontal plane is also referred to as the coronal plane.

The sagittal plane is a vertical plane perpendicular to the human body when the human is standing straight up. The sagittal plane would pass through the human body along the center line from front to back to divide the body into left and right halves.

The transverse plane is a horizontal plane parallel to the floor when the human is standing straight up on the floor. One can imagine that the plane passes through at the center of the body to divide the body into upper (referred to as superior) and lower (referred to as the inferior) halves.

Movements of human joints are often defined in terms of the three anatomical planes along certain axis. Movements in the frontal plane are about the anterior-posterior axis (which passes horizontally from front to back of the body). Movements in the sagittal plane are about the medial–lateral (also referred to as the bilateral) axis (which passes horizontally from side to side). Movements in the transverse plane are

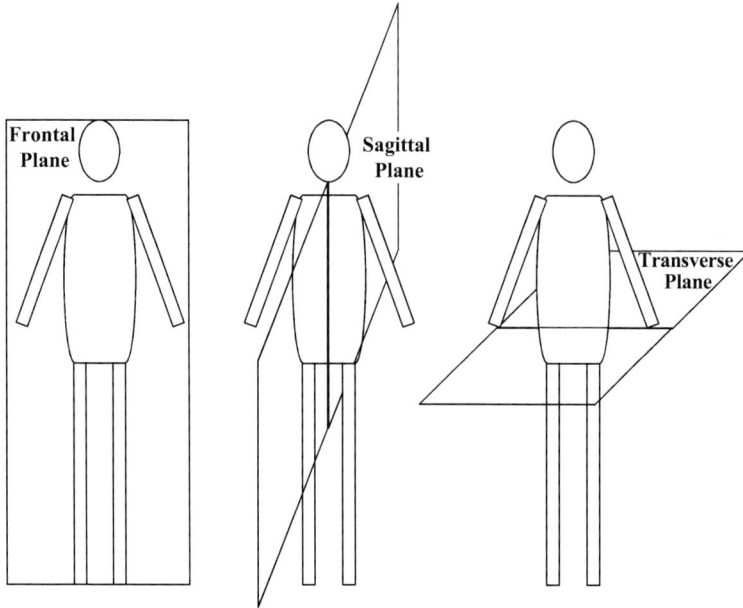

Figure 7.1   Three anatomical planes

about the longitudinal (also referred to as vertical) axis (which is perpendicular to the ground).

7.1.2   Joints and their movements

Joints are key to enable humans to move around. According to [253], joints can be categorized as nonaxial, uniaxial, biaxial, and triaxial. Nonaxial joints typically have very limited motion. In the context of sensor-based human motion tracking, this type of joints are rarely considered. A uniaxial joint allows the rotation around the joint in a single dimension. A biaxial joint allows the rotation around the joint in two dimensions. Similarly, a triaxial joint allows the rotation around the joint in three dimensions.

The most common joint movements include flexion, extension, abduction, and adduction. Lesser common movements include pronation, supination, medial and lateral rotation. Figure 7.2 illustrates examples of flexion, extension, abduction, adduction, pronation, and supination.

Flexion refers to the movement that decreases the angle between the two body segments connected by the joint. Extension refers the movement that increases the angle between the two body segments adjacent to the joint.

Abduction and adduction are movements relative to the midline of the body or the midline of the hand/foot. Adduction is the movement towards the midline, and the abduction is the movement away from the midline.

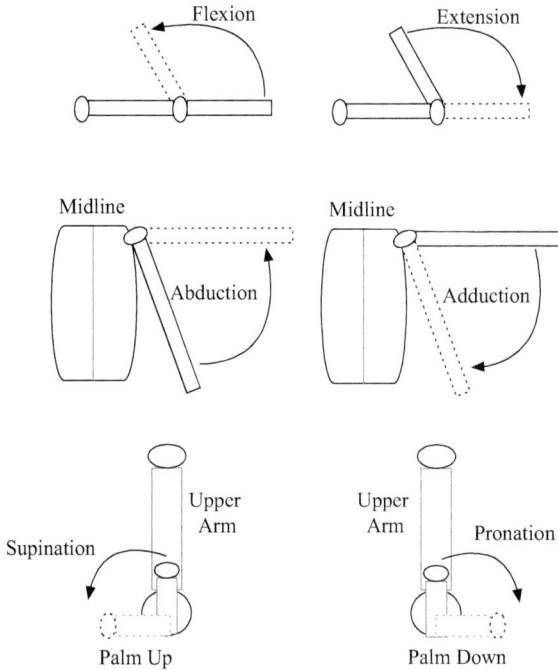

Figure 7.2   Some of the common joint movements

Pronation and supination are rotation movements of the hand or the forearm along the long axis. Pronation is the movement that turns the hand to the prone position (i.e., palm down). Supination is the movement that turns the hand to the supine position (i.e., palm up).

Medial and lateral rotations are movements of the limbs around their long axis. Medial rotation (i.e., also referred to as internal rotation) is the rotation of a limb towards the midline. Lateral rotation (i.e., also referred to as external rotation) is the rotation away from the midline.

Figure 7.3 shows the common types of movements for uniaxial, biaxial, and triaxial joints. All three types the joints except the forearm allow flexion and extension, which are movements within the sagittal plane. The forearm allows pronation and supination, which are movements within the frontal plane. Biaxial joints additionally allow abduction and adduction, which are movements in the sagittal plane. Triaxial joints allow horizontal adduction and abduction, which are movements in the transverse plane.

7.1.3   Range of motion

The range of motion (ROM) measures the amount of movement around a particular joint. There are three types of ROM: passive, active–assistive, and active. Passive

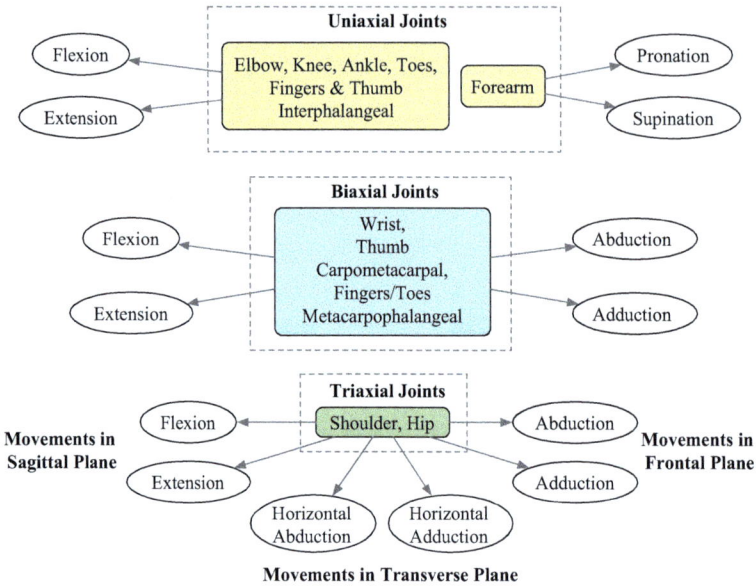

Figure 7.3　Uniaxial, biaxial, and triaxial joints and their movements

ROM is measured when another person (typically a physical therapist) or a specialized designed clinical device moves a person's body parts around a joint, for example, after a surgery of the joint when the person is not able to move his or her own muscles. Active ROM is measured when the person uses his or her own muscles to move the body part around a joint. During the recovery phases after a surgery, the active ROM would gradually increase to the normal level. Active–assistive ROM is measured when the person has gained some capability of moving his or her own muscle, but still requires another person or a special device to help.

To define ROM, a reference positive of the joint is necessary. Each joint has a neutral position. In this position, the ligaments and other non-muscular tissues around the joint have the least amount of support for the joint. Therefore, the muscles around the join are in the best position to control the movements around the joint. It is natural to use the neutral position as the reference point for defining ROM where the ROM is 0 degree at this neutral position, then the body part will move away from this position, either doing flexion or abduction, until the maximum range, as shown in Figure 7.4. Traditionally, the tool used to measure ROM is called goniometer, which is also included in Figure 7.4.

7.2　Joint angle measurement with various sensing modalities

As we can be seen in the first part of this book (i.e., the first six chapters), many different sensing modalities have been used to measure joint angle. The most pervasively

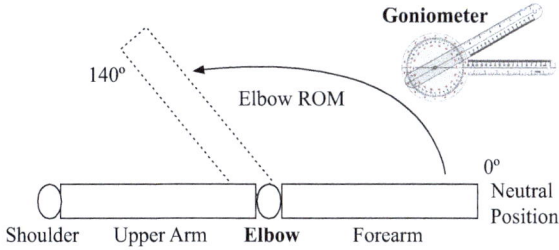

Figure 7.4   Elbow range of motion

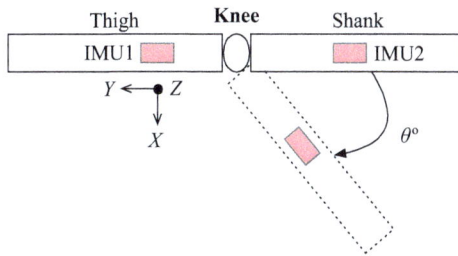

Figure 7.5   Using two IMUs to measure the joint angle using the knee joint as an example

used sensing modalities for rehabilitation include IMU and Kinect (i.e., computer vision with 3D skeletonization). Other sensing modalities either are less convenient to use or the measurement is highly dependent on the individuals (such as FSR and EMG). Hence, in this chapter, we provide technical details for the joint angle measurement using IMU and Kinect.

7.2.1   Joint angle measurement with IMU

In Chapter 1, we provided an overview on the studies that used IMU as the sensing modality to track human motion for rehabilitation. A significant portion of the studies focused on the measurement of ROM of one or more joints. Here we provide technical details on how to use IMUs to measure joint angle. A common approach is to use two IMUs, one IMU is placed on each of the two body segments around the joint, as shown in Figure 7.5 [26]. At the joint's neutral position, the two IMUs are aligned in parallel and the joint angle would be 0 degree. The IMUs' coordinate is also shown in the figure. When the joint flexes, the joint angle can be measured using the accelerometer readings of the two IMUs according to (7.1), where the acceleration measured by IMU1 is (x_1, y_1, z_1), and that by IMU2 is (x_2, y_2, z_2):

$$\theta = \arctan(-y_2/x_2) - \arctan(-y_1/x_1) \tag{7.1}$$

The above calculation requires strict alignment of the two IMUs: (1) both IMUs must be in the same sagittal plane, and (2) the y-axis for both IMUs must be in parallel. To be more robust, the joint angle can be calculated using (7.2) if at least both IMUs are in the same sagittal plane, where R_1 and R_2 are the magnitude of the acceleration from IMU1 and IMU2, respectively ($R_1 = \sqrt{x_1^2 + y_1^2 + z_1^2}$ and $R_2 = \sqrt{x_2^2 + y_2^2 + z_2^2}$):

$$\theta = \arccos\left(\frac{x_1x_2 + y_1y_2 + z_1z_2}{R_1R_2}\right) \tag{7.2}$$

It is also possible to use a single IMU to measure joint angle. For example, in [25], a single IMU is placed on the instep of the foot to track the ankle movements. The angle can be calculated in a way rather similar to (7.1). If the IMU is rotated along its x-axis, the y-axis is the pointing forward, and z-axis is the pointing upward, the angle of the IMU along the x-axis (i.e., the roll) would be defined by (7.3), where the acceleration is (x, y, z):

$$\theta = \arctan(z/y) \tag{7.3}$$

Accelerometer measurements are known to be noisy. This issue is often addressed by fusing the output from gyroscope, or even with the output from magnetometer as well. A simple way is to use a complementary filter [254] to fuse the output from the accelerometer and the output from the gyroscope, as shown in Figure 7.6. The filter takes two inputs. One input is the angle calculated using the accelerometer output (i.e., acceleration). The other input is from the gyroscope, which is angular velocity. The input from the gyroscope will have to be integrated to get the angle. Considering the nature of rehabilitation exercises, the accelerometer input will go through a low-pass filter to remove high-frequency noises. The integration of the gyroscope input will lead to drift, which is why the integrated value will be filtered using a high-pass filter to remove the drift. Then, the filtered accelerometer and gyroscope data will be combined to derive the current angle. Obviously, the parameters used in the low-pass and high-pass filters should reflect the nature of the movements. A simple formula shown in (7.4) has been used with good result in the context of rehabilitation motion tracking [226], where w is the angular velocity from the gyroscope, and δt is the time interval of the measurement, θ_a is the angle measured from the acceleration:

$$\theta = 0.98(\theta + w \times \delta t) + 0.02\theta_a \tag{7.4}$$

Kalman filter [255] is another filter used in several studies to fuse the accelerometer and gyroscope data [24,25]. Unlike the complementary filter, Kalman filter is much more sophisticated and more difficult to understand. It can use noisy input to estimate the state of the system variables. The algorithm requires the calculation of several matrices, including the state covariance matrix, state transition matrix, state-to-measurement matrix, measurement covariance matrix, and process noise covariance matrix, as well as the Kalman gain.

7.2.2   Joint angle measurement with Kinect

Unlike IMU, which measures acceleration and angular velocity, Kinect with its SDK could provide the 3D joint positions. This makes it straightforward to derive joint

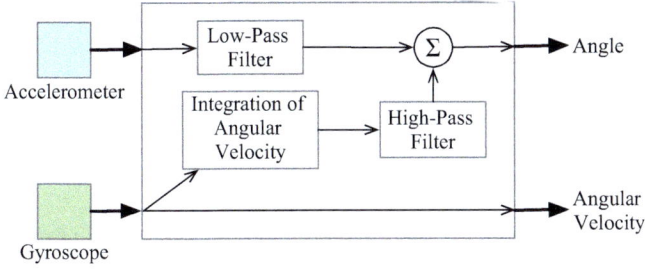

Figure 7.6 Complementary filter

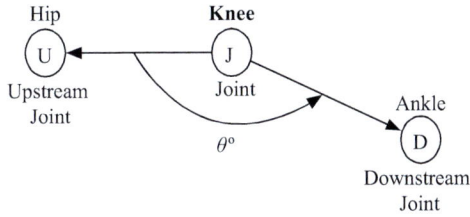

Figure 7.7 Joint angle calculation using Kinect skeleton joint data. If the joint of interest is the knee, then the upstream joint would be the hip joint and the downstream joint would be the ankle joint.

angle from the 3D joint positions. As shown in Figure 7.7, to calculate the angle of a joint, we use the positions of the upstream and downstream joints together with the current joint to form two vectors. Given a joint J with 3D position (x_J, y_J, z_J), its upstream joint U with 3D position (x_U, y_U, z_U) and its downstream joint D with 3D position (x_D, y_D, z_D), we can form two vectors $V_{JU} = (x_U - x_J, y_U - y_J, z_U - z_J) = (x_{UJ}, y_{UJ}, z_{UJ})$ and $V_{JD} = (x_D - x_J, y_D - y_J, z_D - z_J) = (x_{DJ}, y_{DJ}, z_{DJ})$. The angle between these two vectors can be calculated using the dot product of the two vectors: $V_{JU} \cdot V_{JD} = |V_{JU}||V_{JD}| \cos\theta$, where $V_{JU} \cdot V_{JD} = x_{UJ}x_{DJ} + y_{UJ}y_{DJ} + z_{UJ}z_{DJ}$, $|V_{JU}| = \sqrt{x_{JU}^2 + y_{JU}^2 + z_{JU}^2}$, and $|V_{JD}| = \sqrt{x_{JD}^2 + y_{JU}^2 + z_{JU}^2}$. Therefore, the joint angle can be defined by (7.5) (very similar to (7.2)):

$$\theta = \arccos\left(\frac{V_{JU} \cdot V_{JD}}{|V_{JU}||V_{JD}|}\right) = \arccos\left(\frac{x_{UJ}x_{DJ} + y_{UJ}y_{DJ} + z_{UJ}z_{DJ}}{\sqrt{x_{JU}^2 + y_{JU}^2 + z_{JU}^2}\sqrt{x_{JD}^2 + y_{JU}^2 + z_{JU}^2}}\right) \quad (7.5)$$

7.3 Measurement theories

Measurement is a process to determine the value of a variable using some instrument (such as an IMU or Kinect) or a rater (i.e., a trained human professional to observe something not yet possible or not convenient to measure using an instrument). Prior to the availability of modern instrument that enables objective assessment, various tools, such as goniometer, have been used to measure some human motion manually. During the measurement process, both the instrument (manually or automatically) and the rater could introduce errors. We hypothesize that the result of a measurement, r, would consist of the true value t of the variable, as well as some error e:

$$r = t + e \tag{7.6}$$

There are two types errors, systematic error and random error [256]. Systematic error could be due to the instrument itself, either the sensor issues or algorithm issues, and it could also due to the bias of the rater. Systematic errors in general can be compensated via calibration step. IMU often requires a calibration step. Random errors are introduced by chance. For a numerical variable, random errors could make the measurement result for the variable bigger or smaller with equal probability. Therefore, an effective way to mitigate random errors is to take multiple measurements of the same variable under the same condition within a short period of time. Then, the average of the measurement results would be closer to the true value of the variable because the random errors would be largely canceled out.

The quality of an instrument or a method of measurement is often evaluated by its reliability. Conceptually, reliability is the extent to which the measured result reflects the true value of the variable. Reliability can be quantified by taking measurements for a variable in a population (i.e., a group of subjects) under the same condition. These measurements form a sample. Then, the variance of the sample is calculated. The variance is a measure on the degree of dispersion among the measurements. The bigger the variance, the larger the disagreement of the measurements. The smaller the variance, the closer of the measurements with each other. In light of (7.6), we hypothesize that the observed variance are constituted by two parts: (1) the true variance due to the differences of the individual subjects within the population in the sample, and (2) the variance introduced by measurement errors. The reliability R can then be defined by (7.7):

$$R = \frac{true\ variance}{total\ variance} = \frac{true\ variance}{true\ variance + error\ variance} \tag{7.7}$$

Because there is no way we can know the true value of a variance and hence its true variance in a population, we can only estimate the reliability. The most popular method is to estimate reliability via the intraclass correlation coefficient (ICC). Modern ICC is positioned within the framework of analysis of variance (ANOVA) and in the framework of random effects models [257]. There are three models of ICC [258]. In the first model of ICC, often with a notation of ICC(1,1), each subject is rated by a different set of k raters, where the k raters are randomly selected from a population of raters. In the second model of ICC, k raters are randomly selected from the population of the raters, and each of these k raters would rate all subjects. In the third model of ICC, the k raters are not randomly selected from the rater population. The first model of ICC is also referred to as the one-way random effects. The second model of ICC is referred to as the two-way random effects. The third model of ICC is also referred to as the two-way mixed effects. Based on whether or not the result is taken from a single rater or the mean of k raters, each model can have two types. Thus, we will have six different ICC denoted as ICC(1,1), ICC(1,k), ICC(2,1), ICC(2,k), ICC(3,1), and ICC(3,k) (where the first element representing the model number) [258]. As can be seen, the ICC definitions are designed for subjective ratings by human beings instead of sensor-based instruments. It is not obvious how to translate the model into sensor-based instrumentation. When a sensor-based instrument is used to measure human motion for one repetition with dozens or hundreds of data points, should the measurement be regarded as using a group of randomly selected raters? If so, what is the fundamental reason? There is no universal agreement on how to evaluate the reliability of sensor-based measurement instruments [259].

7.4 Evaluating a new measurement instrument

To assess the feasibility of using a new sensor to measure human motion in the context of rehabilitation, it is a common practice to simultaneously measure the human motion using both the new sensor and an established method, such as a multi-camera human motion capture system. While many publications established the feasibility of using the sensor for human motion measurement by comparing the means of the two ways of measurements for key parameters, it is much more rigorous to use statistical methods for the comparison. For a new measurement instrument for human motion to be accepted, the method must prove to offer a good agreement with that of an established method (often referred to as a gold standard) [260]. In addition, the method should offer repeatability, reproducibility, and reliability on bar with that of the gold standard [260].

Agreement is to assess how close two methods of measurements for the same subject or the same set of subjects under the same condition. Several statistical methods could be used for this purpose, including root mean square error (RMSE), *t*-test, Pearson correlation coefficients.

Repeatability is to assess the variation in repeated measurement of the same subject under the same condition over a short period of time [260]. Reproducibility is to assess the variation in repeated measurement of the same subject under different conditions possibly over a long period of time [260]. In the former case, it is assumed that underlying variable remains unchanged if the measurement has no error. The latter assumes that the underlying variable could have gone through non-negligible changes. In many studies, no distinction was made between repeatability and reproducibility, and measurements from all subjects in the same cohort of the study were used in the repeatability calculation. Both repeatability and reproducibility are assessed with methods for determining agreement.

7.4.1 Root mean square error

RMSE has been used in a number of studies to compare a sensor instrument with a reference system [12,261–264]. The RMSE is calculated with respect to a regression model that would predict a sequence of n data points $(p_1, p_2, ..., p_i, ...p_n)$. Let the actual measured values (referred to as observed values) be $(o_1, o_2, ..., o_i, ..., o_n)$. The RMSE quantifies the error of the model as the square root of the mean square error of each data point, as defined in (7.8):

$$RMSE = \sqrt{\sum_{i=1}^{n} \frac{(o_i - p_i)^2}{n}} \tag{7.8}$$

While using RMSE to compare a new measurement instrument (such as IMU or Kinect) with a more established instrument, the results obtained using the established instrument can be considered as the predicted values. The rationale is that if the more established instrument has been regarded as the gold standard in measurement, then it must have proven to have high reliability and low measurement errors. Then, the RMSE of the new instrument compared with the gold standard would tell how well the new instrument performs (i.e., the lower RMSE the better). This RMSE is also referred to as between-system RMSE. Recall that in feasibility studies of a new instrument, it is common practice to measure the same subject using both the new instrument and the gold standard. The RMSE is calculated first for each subject, then the average of RMSE for all subjects is reported [12].

Some studies also reported the within-system RMSE [12]. However, no details were reported on what regression model was used, if at all. If the mean measurement results of all subjects for each variable are used as the predicted value, then RMSE becomes the standard deviation.

Example 7.1. To show how to calculate the between-system RMSE, we use the case presented in [12] where the sample consists of key gait parameters. Because the raw experimental data are not available, we use the mean values reported in the paper for the calculation, as shown in Table 7.1. Hence, it is not a surprise that the result differs from that presented in the paper. The parameters include the flexion angle F1 in the swing phase, extension angle E1 in the stance phase, flexion angle F2 in the stance phase, and extension angle E2 in the swing phase. The average measured values using IMU and the reference system are shown in Table 7.1. The difference between the two systems for each feature is listed as the error in column 4, and its square value is shown in column 5. By using (7.8), the RMSE is $\sqrt{1.962^\circ} = 1.4^\circ$.

Table 7.1   *Key gait features measured while walking by the IMU system and the reference system in [12]*

Joint angles	IMU system	Reference system	Error	Error^2
F1 (°)	56.1	53.7	2.4	5.76
E1 (°)	4.55	3.35	1.2	1.44
F2 (°)	16.5	15.7	0.8	0.64
E2 (°)	3.73	3.64	0.09	0.0081

7.4.2   *Student's t-test*

Student's *t*-test is a statistics method specifically designed for testing hypotheses about the mean of a small sample [265]. It is assumed that the sample was drawn from a normally distributed population and the standard deviation of the population is unknown. Considering that most studies in rehabilitation use a fairly small number of subjects in trials, *t*-test is a good fit for doing statistical comparison, for example, whether the measurements from a new instrument is indistinguishable from those made by a more established gold standard.

The *t*-test was invented by William Sealy Gosset using a pseudonym Student in his publication on *t*-test. A key part of the *t*-test is *t*-distribution. When the sample size becomes large (typically over 30), the *t*-distribution converges to the normal distribution. The *t*-distribution varies depending on the sample sizes. In fact, there are two versions of the *t*-distribution, one referred to as one-tailed and the other referred to as two-tailed. Which version to use depends on the alternative hypothesis to be tested. If the alternative hypothesis is that the sample mean is not equal to the comparison mean, then the two-tailed distribution should be used. If the alternative hypothesis is that the sample mean is greater or less than the comparison mean, then the one-tailed distribution should be used. *The null hypothesis for the t-test is that there is no difference between the sample mean and the comparison mean.*

There are three types of *t*-test, one-sample, two-sample, and paired *t*-test. All *t*-tests assume that the samples are normally distributed. The one-sample *t*-test should be used when comparing a sample mean against a standard value, for example, to compare a new measurement instrument against an established instrument regarded as the gold standard. The paired *t*-test should be used when comparing the means of two samples from the same population, for example, the performance of the same cohort of subjects before and after a rehabilitation program. The two-sample *t*-test should be used when comparing the means of two samples when the samples belong to different populations.

The procedure in doing a *t*-test is described as follows. First, calculate a *t*-value. Then, locate a cell in the *t*-distribution table based on the degree of freedom in the sample, which is the sample size minus one, and the confidence interval (also referred to as confidence level) *CI* or significance level $\alpha = 1 - CI$. Typically, a CI of 95% or $\alpha = 0.05$ is used in the *t*-test. The value in that particular cell is the critical value. If the *t*-value is positive and it is smaller than the critical value, then the null hypothesis holds. If the *t*-value is negative and it is larger than the critical value, then the null hypothesis holds.

The *t*-value is the ratio of the difference of sample means to the standard error of the difference. The method of calculating the *t*-value differs slightly for the three types of *t*-test. For the one-sample *t*-test, the *t*-value is calculated according to (7.9), where μ denotes the standard value, n is the number of data points in the sample $(x_1, x_2, ..., x_n)$, $\bar{x} = \sum_{i=1}^{n} x_i/n$ is the sample mean, and $\sigma = \sum_{i=1}^{n} (x_i - \bar{x})/(n - 1)$ is the standard deviation of the sample. The degree of freedom for one-sample *t*-test is $n - 1$:

$$t = \frac{\bar{x} - \mu}{\sigma/\sqrt{n}} \tag{7.9}$$

The method to compute the t-value for two-sampled t-test is slightly more complex. We assume that the first sample consists of n_1 data points $(x_{(1,1)}, x_{(1,2)}, ..., x_{(1,n_1)})$, and the second sample consists of n_2 data points $(x_{(2,1)}, x_{(2,2)}, ..., x_{(2,n_2)})$. The mean for the first sample is $\bar{x}_1 = \sum_{i=1}^{n_1} x_{(1,i)}/n_1$. The mean for the second sample is $\bar{x}_2 = \sum_{i=1}^{n_2} x_{(2,i)}/n_2$. Let the standard deviation of the first sample be σ_1, and that for the second sample be σ_2. There are actually two cases in two-sample t-test, one with the assumption that the populations of the two samples drawn from have identical variance, the other with the assumption that the two populations have different variances. The t-value for the two-sample equal-variance t-test is given in (7.10) where the denominator is the pooled standard deviation of the first and the second samples combined:

$$t = \frac{\bar{x}_1 - \bar{x}_2}{\sqrt{\left(\frac{(n_1-1)\sigma_1^2+(n_2-1)\sigma_2^2}{n_1+n_2-2}\right)\left(\frac{1}{n_1} + \frac{1}{n_2}\right)}} \tag{7.10}$$

The t-value for the two-sample unequal-variance t-test is given in (7.11):

$$t = \frac{\bar{x}_1 - \bar{x}_2}{\sqrt{\frac{\sigma_1^2}{n_1} + \frac{\sigma_2^2}{n_2}}} \tag{7.11}$$

For the equal-variance t-test, the degree of freedom is straightforward $df = n_1 + n_2 - 2$. For unequal-variance t-test, the degree of freedom is more complicated [266], as given by (7.12):

$$df = \frac{(\sigma_1^2/n_1 - \sigma_2^2/n_2)^2}{\frac{(\sigma_1^2/n_1)^2}{n_1-1} + \frac{(\sigma_2^2/n_2)^2}{n_2-1}} \tag{7.12}$$

The mathematical formula for calculating the t-value for the paired t-test are very similar to that for one-sample t-test, as shown (7.13). However, they are semantically different. The mean difference in the numerator is calculated by taking the average of the pair-wise differences. The standard error in the denominator is the standard deviation of the paired differences σ_d. Let the first sample be $(x_1, x_2, ..., x_n)$ and the second sample be $(y_1, y_2, ..., y_n)$. The differences are $(d_1 = x_1 - y_1, d_2 = x_2 - y_2, ..., d_n = x_n - y_n)$. Note that the standard deviation is calculated based on the set of pair-wise differences instead of either sample:

$$t = \frac{\bar{d}}{\sigma_d/\sqrt{n}} \tag{7.13}$$

An alternative method to performing t-table lookup is to use the p-value computed based on the t-value. The p-value is the probability that the null hypothesis is true. If the p-value is larger than the significance level, then the null hypothesis holds.

For the left-tailed (i.e., lower-tailed) t-test, the p-value is cumulative distribution function of the t-value $cdf(t)$, as shown in (7.14):

$$p\text{-}value = cdf(t) \tag{7.14}$$

For the right-tailed (i.e., upper-tailed) *t*-test, the *p*-value is $1 - cdf(t)$, as shown in (7.15):

$$p\text{-}value = 1 - cdf(t) \tag{7.15}$$

For the two tailed *t*-test, the *p*-value is $2 * \min\{cdf(t), 1 - cdf(t)\}$, as shown in (7.16):

$$p\text{-}value = 2 * \min\{cdf(t), 1 - cdf(t)\} \tag{7.16}$$

Example 7.2. To show how to do one-sample *t*-test, we again use the case presented in [12]. In this case, we want to see if the F1 angle measured by the IMU is indistinguishable with 95% confidence level (i.e., $\alpha = 0.05$) from that measured by the reference Vicon system. The alternative hypothesis is that the two instruments are not equivalent for F1 measurement. Hence, we will need to use the two-tailed t-distribution. According to [12], the sample size for IMU measurement is 10. Therefore, the degree of freedom is 9. The probability density function of the t-distribution for both the one-tailed and two-tailed versions for degree of freedom of 9 are shown in Figure 7.8.

We use the sample mean ($56.1°$) and the sample standard deviation ($8.4°$) provided in [12] for F1. The standard value for comparison is the sample mean measured by the Vicon system, which is $53.7°$. The *t*-value calculated using (7.9) is 0.0952. We should look up the two-tailed *t*-distribution table for degree of freedom of 9 for 95% confidence level, which is 2.262. Since the calculated *t*-value is smaller than 2.262, the null hypothesis holds. The *p*-value for the difference can be calculated using (7.16) (or using Excel or other software libraries such as R or python scikit-learn). $cdf(0.0952) = 0.537$ and $1 - cdf(0.0952) = 0.453$. Hence, *p*-value$= 2 * 0.453 = 0.926$. The illustration in Figure 7.8 shows how to interpret this result. Based on the alternative hypothesis, we should be looking at the two-tailed *t*-distribution probability density curve at the top. The red regions at the two tails of the curve have a total of 5%, which means if the *t*-value falls within those regions, the null hypothesis would be rejected. Otherwise, like this example, the null hypothesis holds. Alternatively, one could simply compare the *p* value for the *t*-value. As long as it is bigger than 5%, the significance level, the null hypothesis holds. Figure 7.8 includes the one-tailed *t*-distribution probability density function just in case the alternative hypothesis is to compare with the standard value in a particular direction (smaller or larger).

7.4.3 Pearson's coefficient of correlation

Pearson's coefficient of correlation is statistic method to determine the degree of correlation between two groups of paired data. It is also referred to as Pearson product–moment correlation coefficient, bivariate correlation, or simply Pearson's *r*. For example, in several studies on assessing the feasibility of using a new sensor instrument, the Pearson's coefficient of correlation was used to compare the data obtained using the new sensing modality against the established measurement method considered as the gold standard [12]. If the correlation is strong, it means that the new

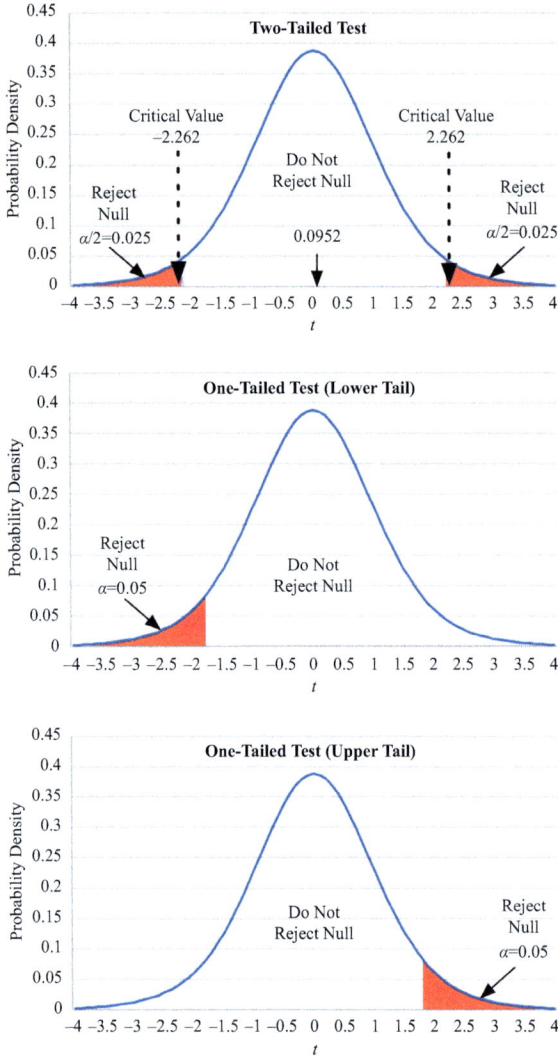

Figure 7.8   The probability density functions for t-distribution with freedom degree of 9 with annotations

measurement sensing modality would produce consistent measurement with respect to the gold standard.

For a population, the Pearson's coefficient of correlation ρ is defined as the ratio between the covariance of the two variables (x, y), $cov(x, y)$, and the product of the standard deviations of x and y, $\sigma_x \sigma_y$, as shown in (7.17):

$$\rho_{x,y} = \frac{cov(x, y)}{\sigma_x \sigma_y} \tag{7.17}$$

In actual studies, all we have are samples from a population. Pearson's coefficient of correction based on samples is represented using a symbol r. Given paired samples $(x_1, x_2, ..., x_n)$ and $(y_1, y_2, ..., y_n)$, the coefficient $r_{x,y}$ can be calculated using (7.18):

$$r_{x,y} = \frac{\sum_{i=1}^{n} (x_i - \bar{x})(y_i - \bar{y})}{\sqrt{\sum_{i=1}^{n} (x_i - \bar{x})^2} \sqrt{\sum_{i=1}^{n} (y_i - \bar{y})^2}} \tag{7.18}$$

The result is a normalized coefficient that ranges between -1 and 1. If the two variables measured by the two samples have no correlation, then $r_{x,y}$ would be close to 0. If the two variables have strong positive correlation, then then $r_{x,y}$ would be close to 1. If the two variables have strong negative correlation, then $r_{x,y}$ would be close to -1.

Example 7.3. We are going to again use the data in Example 7.1 shown in Table 7.1, where x sample (IMU data) consists of (56.1,4.55,16.5,3.73), and y sample (Vicon data) consists of (53.7, 3.35, 15.7, 3.64). The $\bar{x} = 20.22$, and $\bar{y} = 19.0975$. We plug these numbers in (7.18), we get $r_{x,y} = \frac{1755.83}{42.65 * 41.18} = 0.9998$. As can be seen, the correlation is extremely close to 1, which demonstrates that the IMU sensing instrument is as good as the gold standard.

7.4.4 Intraclass correlation coefficient

ICC has often been used as a statistical method to evaluate a new sensor-based measurement instrument compared with a gold standard measurement instrument. However, this practice is debatable [267] and there is no universal agreement on which ICC model should be used for the evaluation. Publications seldom report the type of ICC model used for the evaluation [12,176], but there are exceptions, such as [73] where ICC(3,1) is used.

In [176], test–retest reliability [268] was used to compare the agreement between a Kinect-based system with respect to a marker-based system considered as the gold standard. In [258], Koo and Li argued that test–retest reliability should be regarded as two-way mixed effect ICC, which is ICC(3,1) or ICC(3,k). Another article on test–retest reliability [269] seems to have confirmed this definition, although all forms of ICC are discussed. In [37], ICC(3,1) is used as the test–retest reliability evaluation. In [58], ICC(3,1) is used to compare the regression result using machine learning with the Fugl–Meyer Assessment scores provided by clinicians on upper-limb impairment level in patients who had stroke.

In ICC calculation, the reliability definition given in (7.7) is estimated using ANOVA. The total variance of the population is estimated to be the variance of measurements done to different subjects in the population, which is referred to as between-subject variance. In ICC(1), the variance of repeated measurements done to the same subject to estimate the error variance, which is referred to as the within-subject variance. This is intuitive because there would be no variation for repeated measurements done to the same subject under the same condition if there were no measurement errors. Because there are multiple subjects, the mean within-subject variance of all subjects will be used. In ICC(2) and ICC(3), it is possible to separate systematic errors from the random errors using ANOVA, and hence, the variance of the random error is used as the error variance.

We use the notations given by [257]. *BMS* denotes the between-subject mean square, *WMS* denotes the within-subject mean square, *EMS* denotes the random error mean square, n is the number of subjects, k denotes the number of raters in [257] and it was later reinterpreted as the number of trials taken [269] (here is also an implicit assumption that $k \geq 2$). ICC(1) and ICC(3) calculation are rather similar as shown in (7.19), (7.20), (7.21), and (7.22). The calculation for ICC(2) is omitted because it is irrelevant to comparing a new measurement instrument with respect to a gold standard:

$$ICC(1,1) = \frac{BMS - WMS}{BMS + (k-1)WMS} \tag{7.19}$$

$$ICC(1,k) = \frac{BMS - WMS}{BMS} \tag{7.20}$$

$$ICC(3,1) = \frac{BMS - EMS}{BMS + (k-1)EMS} \tag{7.21}$$

$$ICC(3,k) = \frac{BMS - EMS}{BMS} \tag{7.22}$$

In estimation of the reliability, the heterogeneity of the population in the sample plays a big role [260]. The higher the heterogeneity, the larger the between-subject variance, which will lead to a larger reliability. That is why Bartlett and Frost recommended to report not only reliability score, but the between-subject and within-subject variance as well [260].

ICC may also have an associated null hypothesis and the corresponding confidence interval. Originally, the null hypothesis is defined as the population reliability $\rho = 0$. The alternative hypothesis would be $\rho > 0$. The corresponding variable is BMS/WMS for ICC(1), and BMS/EMS for ICC(2) and ICC(3), and the associated distribution is the F distribution. The test of hypothesis is referred to as the row effects F statistics in [270]. McGraw and Wong pointed that the original null hypothesis is not meaningful for common use cases because one would be more interested in finding out if the reliability is greater than a particular positive value (between 0 and 1) [270].

The F-distribution has two parameters, i.e., the degree of freedom of the two groups of data in comparison, respectively. Given a significance level, and the two degree of freedom parameters, one can look up the F-table during the calculation of the lower-bound and the upper-bound of the population reliability ρ.

For ICC(1,1), let $F_0 = BWM/WMS$ and $F_{1-a}(i,j)$ be the $(1-a) \cdot 100$th percentile of the F-distribution with i and j degrees of freedom. Furthermore, we introduce two parameters F_U and F_L as defined in (7.23) and (7.24). The confidence interval for ICC(1,1) is defined in (7.25):

$$F_U = F_0 \cdot F_{1-1/2\alpha}[n(k-1), (n-1)] \tag{7.23}$$

$$F_L = \frac{F_0}{F_{1-1/2\alpha}[n(k-1), (n-1)]} \tag{7.24}$$

$$\frac{F_L - 1}{F_L + (k-1)} < \rho < \frac{F_U - 1}{F_U + (k-1)} \tag{7.25}$$

For ICC(1,k), the confidence interval is defined in (7.26), where F_L and F_U are defined in (7.23) and (7.24):

$$1 - \frac{1}{F_L} < \rho < 1 - \frac{1}{F_U} \tag{7.26}$$

For ICC(3,1) confidence interval calculation, $F_0 = BMS/EMS$, F_L and F_U are redefined in (7.27) and (7.28). The interval has the same form as that defined in (7.25) in terms of F_L and F_U:

$$F_U = F_0 \cdot F_{1-1/2\alpha}[(n -)(k - 1), (n - 1)] \tag{7.27}$$

$$F_L = \frac{F_0}{F_{1-1/2\alpha}[(n - 1)(k - 1), (n - 1)]} \tag{7.28}$$

For ICC(3,k), again the confidence interval formulation has the same form as that in (7.26) in terms of F_L and F_U (note that $F_0 = BMS/EMS$).

7.4.5 Bland–Altman limits of agreement

Bland and Altman proposed an amazingly simple statistical method to assess the agreement between two measurement methods. The idea was first published in 1983 [271], and it was a later paper published in a medicine journal *Lancet* in 1986 [13] that got the medical field to widely accept the method for agreement assessment. Bland and Altman argued that correlation-based statistical methods are not suitable for medicine because obviously correlation does not mean the two methods would necessarily have good agreement.

Here we assume that n subjects were measured by two different measurement instruments (e.g., for new instrument and another established instrument considered as the gold standard). Let $(x_1, x_2, ..., x_i, ...x_n)$ be the sample obtained using the new instrument, and $(y_1, y_2, ..., y_i, ...y_n)$ be the sample obtained using the gold standard. The mean and standard deviation of the mean for pair-wise difference of the two samples are then calculated

$$\bar{d} = \frac{\sum_{i=1}^{n} (x_i - y_i)}{n} \tag{7.29}$$

$$s = \sqrt{\frac{\sum_{i=1}^{n} (x_i - y_i - \bar{d})^2}{n}} \tag{7.30}$$

Then, a plot will be made based on the two samples. The horizontal axis is plotted against the average of each data pair (one from the new instrument, and the other from the gold standard), i.e., $((x_1 + y_1)/2, (x_2 + y_2)/2, ..., (x_i + y_i)/2, ...(x_n + y_n)/2)$. The virtual axis is plotted against the difference of the paired data from the two samples. The plots also include three horizontal lines. One indicates the mean of the difference $\bar{d}$. The other two are the upper limit and lower limit of agreement as defined by $\bar{d} - 2s$ and $\bar{d} + 2s$. Therefore, the larger the standard deviation for the difference, the large range of the limits of the agreement. For the new instrument to be useful, the standard deviation is expected to be sufficiently small.

*Table 7.2   Measurement data for two different instruments with 17 subjects
provided in [13]*

Subject	Instrument1	Instrument2	DataDiff	DataMean
1	494	512	−18	503
2	395	430	−35	412.5
3	516	520	−4	518
4	434	428	6	431
5	476	500	−24	488
6	557	600	−43	578.5
7	413	364	49	388.5
8	442	380	62	411
9	650	658	−8	654
10	433	445	−12	439
11	417	432	−15	424.5
12	656	626	30	641
13	267	260	7	263.5
14	478	477	1	477.5
15	178	259	−81	218.5
16	423	350	73	386.5
17	427	451	−24	439

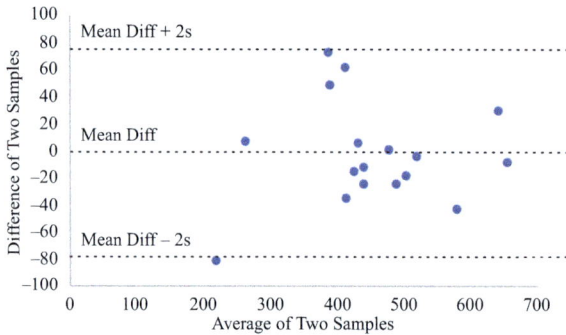

Figure 7.9   The Bland–Altman plot in comparison of the two instruments

Example 7.4. We use the example provided in [13], which is shown in Table 7.2.

The corresponding Bland–Altman plot is shown in Figure 7.9. As can be seen, the distance between the upper and lower limits of agreement is quite large, indicating the new instrument is not acceptable for medical use.

As can be seen in the above example, the Bland–Altman plot not only shows the agreement between two instruments clearly but also let one inspect the entire range of data. Other statistical methods, including root mean square error, Student's *t*-test, Pearson's coefficient of correlation, and ICC reduce the assessment into a single value. The simpleness and intuitiveness of the Bland–Altman agreement method have made it the most widely used statistical method in the health and medical field [12,73,176,259].

Chapter 8
Machine-learning-based activity recognition

Rehabilitation exercises are complex human activities. While automating the measurement of basic parameters such as joint angle is relatively straightforward using various sensing modalities, it is a challenge to perform automated recognition and assessment at the entire exercise level. In this chapter, we review the studies that have used various machine-learning algorithms and models to perform various levels of recognition and assessment. Major steps in machine-learning-based activity recognition is shown in Figure 8.1. In the context of rehabilitation exercises, machine learning has been used in tasks of different levels of challenges as shown in Figure 8.2, from recognizing the exercise being performed (which could be useful to perform automated repetition count), to assessing the quality of the exercise being performed (correct/incorrect or multi-categorical such as excellent, very good), to performing automated scoring according to some well-established clinical assessment scales.

The efficacy of machine learning heavily depends on the quality of data. Because it is inevitable for the raw data to contain various forms of noises or systematic errors, the raw data are typically pre-processed prior to doing machine learning, which could include filtering, data synchronization, and fusion in the presence of multi-modality of sensors.

The data collected for rehabilitation activity recognition are naturally in a temporal sequence. It is often not practical to use the entire sequence as the input for machine learning simply due to the large dimension of the data. Hence, the sequence is often segmented. Considering that rehabilitation exercises are repetitive, ideally, each segment would contain exactly one repetition. For some exercises, it is possible to identify the occurrence of each repetition such as with zero-crossing in velocity [272]. If this is not feasible or convenient, a common practice is to experiment with a sliding window to produce segments of different sizes. The window size that coincides with the repetition would lead to the best machine learning outcome.

Traditional machine-learning algorithms require features being manually extracted from the segmented data. Extracting features and selecting the most distinctive features for the classification with machine learning is often referred to as feature engineering [272]. Machine learning can make categorical classification or support regression over a range of numbers.

Deep learning models could take segmented data directly and infer the most relevant features automatically, and have been proven to be more powerful. Typically much larger training data are required for deep learning to work well.

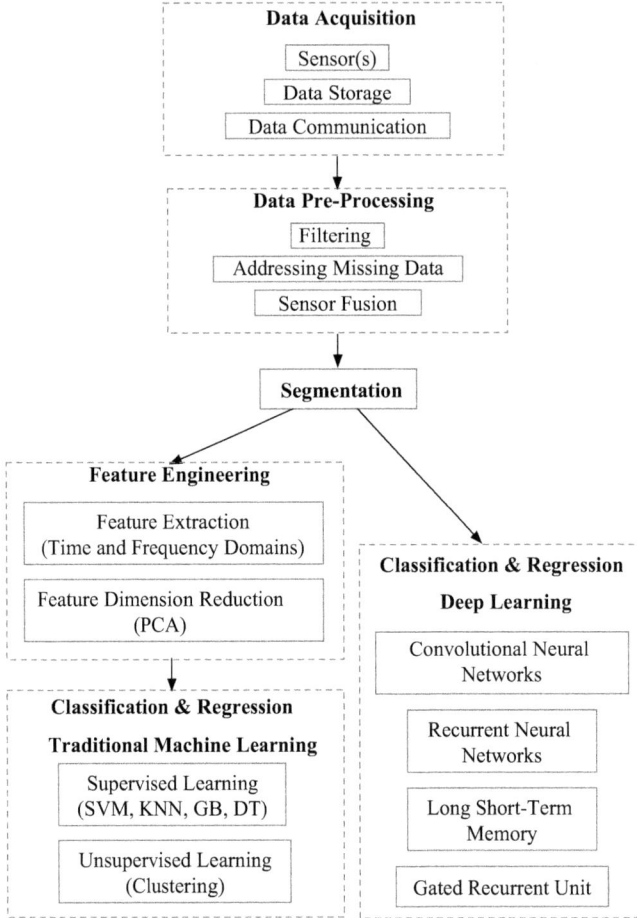

Figure 8.1   Major steps in machine-learning-based activity recognition

While most studies have used supervised machine learning, which requires the availability of accurately labeled training data, some studies chose to use unsupervised learning, often to differentiate exercises performed by healthy subjects from patients, and/or to identify different levels of recovery or disease severity.

There are various metrics on evaluation the machine learning performance. We will also briefly introduce these metrics in this chapter.

8.1   Data pre-processing

Data pre-processing is often necessary prior to feature extraction. As shown in Figure 8.3, there are three major tasks in pre-processing: (1) filtering to remove

Machine Learning-Based
Classification and Regression

Activity Recognition
Identify Rehab Exercise
Identify ADL
Identify Repetition
Segmentation

Performance Assessment
Correct/Incorrect
Categorical
Determination

Clinical Assessment
MFT, ARAT, ARAT
5-Sit-to-Stand
FMA, OGM, NIHSS

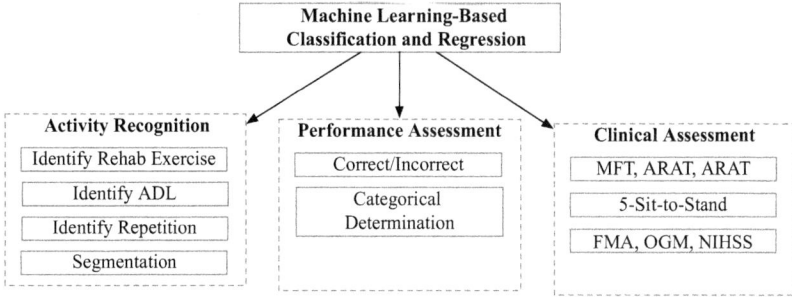

Figure 8.2   Various tasks that machine learning has been used for rehabilitation exercises

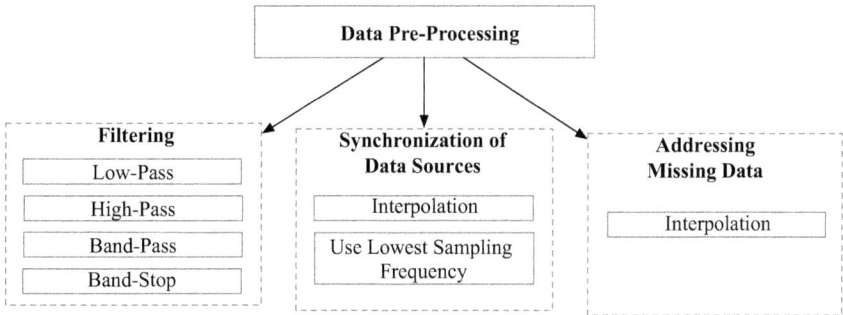

Data Pre-Processing

Filtering
Low-Pass
High-Pass
Band-Pass
Band-Stop

Synchronization of Data Sources
Interpolation
Use Lowest Sampling Frequency

Addressing Missing Data
Interpolation

Figure 8.3   Three main tasks in data pre-processing

noises and errors in data acquisition or processing (such as integration), (2) the synchronization of multiple data source, and (3) addressing the missing data if present. Some system would use multi-modality sensing, which obviously would require synchronization so that the same time basis can be established to facilitate sensor fusion. It also occurs often to compare a new sensor-based measurement method against an established method, where the two methods must be synchronized due to different sampling rate used.

Some low-frequency or high-frequency noises can be easily removed using simple low-pass or high-pass filters. For example, the movements in rehabilitation exercises are typically slow, rarely more than a new times per second. Hence, a low-pass filter can be used to remove noises that are higher than a few Hz. When using gyroscope data to derive the rotation angle, the drift introduced during integration could be easily removed by applying a high-pass filter. It is also conceivable to use a band-pass filter to remove both low-frequency and high-frequency noises, or use a band-stop filter to remove a particular frequency range in the samples.

All these filters originated from resister capacitor electric circuits, where the product of the resister (R) and the capacitor (C), RC, is referred to as the time constant $\tau = RC$, which is the time it takes to charge the capacitor through the resister from an initial charge voltage of 0 to about 63.2% (more precisely $1 - e^{-1}$, where e is

Euler's number) of the value of the applied voltage. This time constant determines the low-pass or high-pass cutoff frequency $f_c = \frac{1}{2\pi\tau} = \frac{1}{2\pi RC}$.

Considering that the sensor measurement results would form discrete sequences, given the sampling interval of δt, the ith raw data point x_i, the low-pass filtered value y_i is defined by (8.1):

$$y_i = \alpha x_i + (1 - \alpha)y_{i-1} \tag{8.1}$$

where $\alpha = \frac{\delta t}{\tau + \delta t}$. Because α is considered a parameter in the low-pass filter, it is often arbitrarily set in some studies. We should note that this parameter must be set according to the nature of the movements during data acquisition.

Similarly, a simple high-pass filter is defined by (8.2):

$$y_i = \alpha y_i + \alpha(x_i - x_{i-1}) \tag{8.2}$$

For better filtering performance, the Butterworth filter is commonly used. The Butterworth filter can be used for low-pass, high-pass, band-pass, and band-stop filtering. The mathematics behind the filter is complex, hence, it is out of the scope of this book. Many software packages are available to perform Butterworth filtering. The scikit-learn Python library provides Butterworth filtering and has application programming interfaces (API) for many traditional machine learning algorithms. The Butterworth filter operates in different orders, where the higher the order the more effective in filtering (and also the higher complex in mathematics computation). Figure 8.4 shows the results of order 3, 6, and 9 of the Butterworth band-pass filter (produced using the scipy library in Python). As can be seen, the higher order the shape resembles more a square wave with sharper drop off over the cutoff frequencies. The order of 6 appears to be a good comprise. For example, in [58], Oubre *et al.* used a Butterworth band-pass with 0.1Hz as the low cutoff frequency and 8Hz as the high cutoff frequency to filter the raw data.

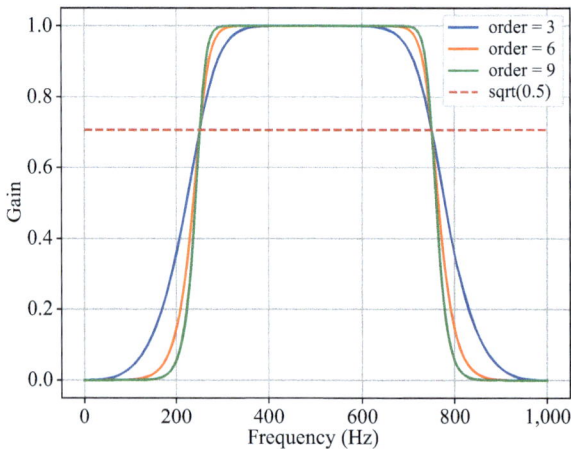

Figure 8.4   Butterworth band-pass filter frequency response for orders 3, 6, and 9

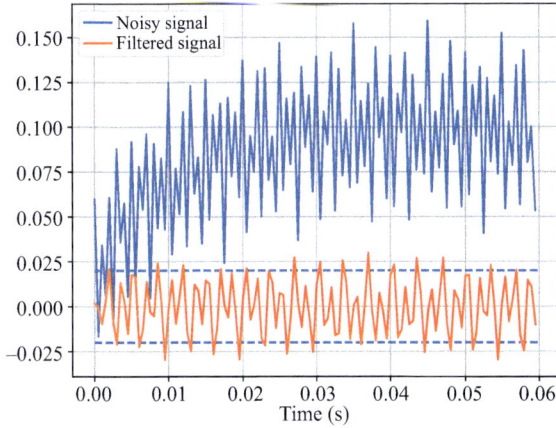

Figure 8.5 The original noisy signal and the filtered result with band-pass Butterworth filter with order of 6

To show how the Butterworth band-pass filter works, we will show the result of applying the filter to a generated discrete noisy sequence according to (8.3), where $t_i = i/2,000, i = 1, 2, \ldots n$ (i.e., the sampling rate is 2,000 Hz). The noisy signal has several frequency components at 2 Hz, 312 Hz, 600 Hz, and 1,000 Hz:

$$x_i = 0.1 \sin{(2\pi\,1.2\sqrt{(t_i)})} + 0.01 \cos{(2\pi\,312t_i + 0.1)} + \\ 0.02 \cos{(2\pi\,600t_i + 0.11)} + 0.03 \cos{(2\pi\,1,000t_i)} \tag{8.3}$$

We would like to remove the frequencies that are below 250 Hz and higher than 750 Hz using the Butterworth band-pass filter with order 6. The filtered signal should only have two frequency components left at 312 Hz and 600 Hz. As can be seen from Figure 8.5, indeed, this is the case.

Besides filtering, pre-processing of data may also include the synchronization of data sources if multiple sources are present [273]. This is to make sure that the same time basis can be used to sensor fusion, data analytics, and comparison of different measurement methods. Either the source with lowest sampling frequency is used as the time basis [181], or the linear interpolation is used to insert more data points to the lower frequency sources [274]. Missing data in the raw data could also be addressed via interpolation [274].

8.2 Data segmentation

The simplest way of doing data segmentation is to use a sliding window to separate the raw temporal data into a sequence of segments [275,276]. A very short window might not contain enough information to enable feature extraction, while a large window might contain several repetitions of the exercises, which might or might not be helpful in extracting distinctive features for machine learning. Some studies chose to use a heuristic window size for segmentation. For example, in [276], Lee *et al.* used 2.5 s as the window size without any overlap in between two consecutive windows to segment the IMU data in the context of stroke rehabilitation. In [277], Yang *et al.* used a 20 ms window size with 50 ms overlap for EMG data. For IMU-based data, typically a window of 2–10 s is used [273]. In [56], Laudanski *et al.* chose to use a 2-s window with 50% overlap between consecutive windows such that the window is big enough to capture at least one gait cycle while not limiting the number of windows. In [74], Kyritsis *et al.* used a 5-s window with 4-s overlap between consecutive windows for gait measurement where each step is about 1-s. In other studies, the sliding window size, and the amount of overlap in two consecutive windows have been experimented to find one combination that would lead to the best machine-learning performance [278].

It is also possible to exploit the nature of the raw measurement results to perform segmentation. A common practice is to identify zero-crossing of the velocity in the raw data (often the angular velocity) and use this as the way to segment the raw data [58]. More specifically, zero-crossing refers to the points where the velocity value changes sign (from positive to negative or from negative to position), which indicates that the movement is changing directions. Sometimes zero-crossing also include the points where the average velocity within a small window is below a predefined threshold [279]. A potential issue with this method is that it tends to over-segment given noisy data [278].

In [58], Oubre *et al.* first converted the IMU data into a sequence of velocity along with each anatomical axis. The nature of the exercise dictates that the starting and ending velocity would be zero. Hence, this observation enables them to segment the data reliably with each segment containing a single repetition.

Another line of approach to segmentation is to use a template for doing segmentation. The template is a sequence of pre-recorded data that has been manually inspected to contain one or more repetition of the same exercise. Common methods used in the template-based approach include dynamic time warping and the hidden Markov model [278] where zero-crossing of velocity has often been used as features for matching. However, this approach inevitably also includes the recognition of the activity itself, which goes beyond the general data segmentation domain, and sometimes it is referred to as the template-based approach for activity recognition [169,195,272].

8.3 Feature engineering

Traditional machine learning heavily depends on the features used as the input for training and testing. Furthermore, the dimension of the features must be reduced to avoid the overfitting problem in machine learning. Typically, the first step is to extract various features from the filtered and segmented data (feature extraction). To reduce the dimension of the features, the most distinctive features will then be identified, often through statistical analysis, and used for the training and testing with machine-learning algorithms.

8.3.1 Feature extraction

Normally the measurements for human motion are done over a period of time. Hence, the data acquired are naturally in the time domain. Features are most often extracted from the time domain. But some studies (using IMU, EMG, other sensing modalities) have extracted features from the frequency domain as well as the time–frequency domain using various transform methods. As shown in Figure 8.6, in the time domain, the mean of the segment is one of the most popular features used because it reflects the main characteristics of the human motion during this time window. The variances of the segment, root mean square of the segment, and correlation with other segment or other axis of segment could also be used as features [273].

Another common practice is to extract frequency domain features by transforming the time domain data into the frequency domain using fast Fourier transform [56, 74,280] or discrete cosine transform [273]. It is also possible to extract features that reflect both frequency characteristics and the time characteristics using discrete wavelet transform [281].

8.3.2 Feature selection

If there are a large number of extracted features, typically the set of features will need to be reduced, either manually based on experiences or more preferably automatically using some statistical methods. The popular statistical methods are shown in Figure 8.7. Regardless of the method used for feature selection, the goal is to identify those features that are the most distinctive for classification or regression. As shown in Figure 8.8, not all features are helpful in making classification or regression. Those that are not helpful are called irrelevant features or useless features. The features that are helpful in making classification or regression are referred to as revenant features. Among relevant features, some of them are actually redundant to each other. Statistically, features that are redundant are strongly correlated. Such correlation can be detected via statistical methods. We want to select the minimal set of features that are all relevant but containing no redundancy.

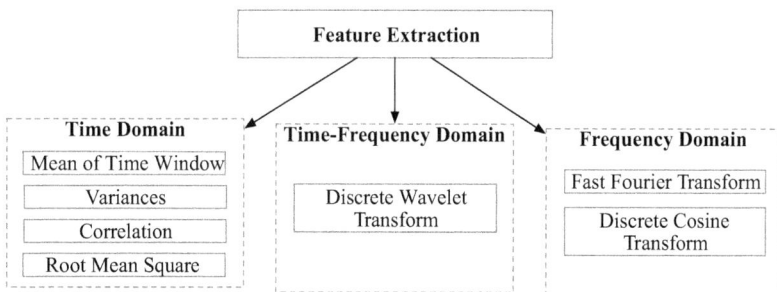

Figure 8.6 Feature extraction has been done in the time domain, the frequency domain, and the time–frequency domain

Manifold Learning
(Nonlinear Dimensionality Reduction)

| Isomap |
| Diffusion Map |
| Laplacian Eigenmaps |
| Hessian Eigenmaps |
| Spectral Embedding |
| Semidefinite Embedding |
| Locally Linear Embedding |

| Principal Components Analysis (Nonlinear Dimensionality Reduction) |

| Correlation-Based Feature Selection |

| Minimal Redundancy Maximal Relevance |

| Least Absolute Shrinkage and Selection Operator |

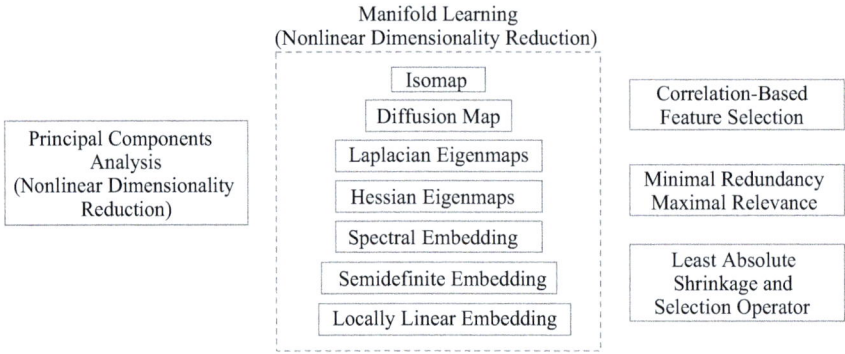

Figure 8.7   Popular statistical methods for feature selection

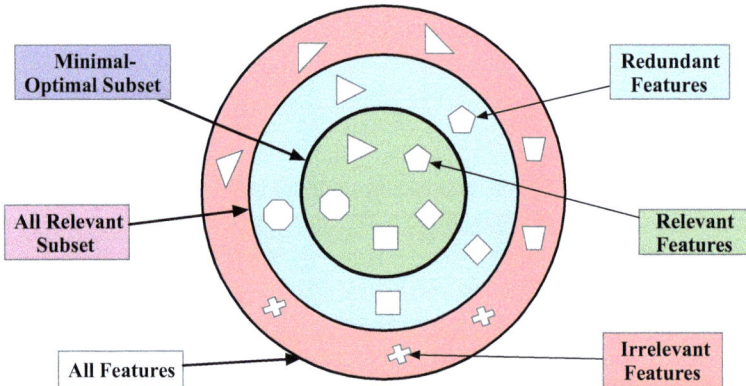

Figure 8.8   Different types of features

To reduce the dimension of the features extracted, Principal Component Analysis (PCA) is perhaps the most widely used method to identity a subset of the most distinctive features [30,277]. The fundamental idea of PCA is that the set of features are regarded as being interrelated variables, and they are transformed to a new set of uncorrelated and ordered set of variables that retain most of the variances of the original feature set [282]. However, PCA works well only if the structure of the features are linear [283]. Manifold learning (also referred to as nonlinear dimensionality reduction) [283] extends PCA to look for the underlying uncorrelated variables that are located beyond a two-dimensional plane (e.g., in some arbitrary lower-dimension surface, i.e., manifold). Many algorithms have been proposed for manifold learning, such as isomap [284], diffusion map [285], locally linear embedding [283], Laplacian eigenmaps [286], Hessian eigenmaps [287], spectral embedding [288], and semidefinite embedding [289]. The scikit-learn Python library provides the implementation of several manifold learning algorithms.

Correlation-based feature selection [290] has also been used to select the most distinctive subset of the features for machine learning in the context of rehabilitation exercise assessment [37,58]. The basic idea is that the subset of features should be highly correlated to the class they predict, but the features in the subset should have low correlation with each other. The best first search is used to find this subset of features. A blog post provides excellent explanation of this algorithm and a python implementation based on the scikit-learn library at https://johfischer.com/2021/08/06/correlation-based-feature-selection-in-python-from-scratch/.

Another very influential feature selection algorithm is the minimal-redundancy-maximal relevance (mRMR) method [291]. The intuition behind this method is rather similar to the correlation-based feature selection [290] because redundant features would have strong correlation with each, and the most relevant features to a class would appear to be highly relevant to the class. It appears that the authors of mRMR are not aware of Hall's work because the earlier study was not cited. The theoretical foundations for the two methods are drastically different. The most up-to-date python implementation for the mRMR method seems to be https://github.com/smazzanti/mrmr. The mRMR method was used in [276] to select the most relevant features.

Least absolute shrinkage and selection operator (lasso) [292] is yet another method for feature selection. The goal of lasso is to preform variable selection and regularization to improve prediction accuracy. It learns the relationship between the features and the target class. More specifically, lasso finds the best subset of variables (i.e., features) such that the sum of the absolute value of the regression coefficients of these variables is less than a predefined value. This would force some coefficient to be set to 0, which in effect will exclude correlated variables in the subset. In [293], lasso was used to select features for measuring patient progress after knee or hop replacement surgery.

Besides using statistical methods to automatically select the most distinctive features, it is still a common practice to manually handcraft a small feature set for machine learning based on domain knowledge and experiences.

8.4   Supervised machine learning

Supervised machine learning refers to the task of inferring the function that reflects the relationship between a set of input variables and a set of output variables [294]. Supervised machine learning can be used to predict both categorical variables and continuous variables. The former is referred to as classification and the latter is referred to as regression. A training dataset, where each input data sample is labeled with the correct output value, is used to infer the function, or rather, the parameters for the function. A challenge is that the function learned using the training dataset would be able to perform well with unforeseen input data. How well the machine-learning algorithm can generalize from the training dataset to other previously unknown data is used to evaluate the performance of the learning algorithm.

As we elaborated in the previous section, a feature vector is extracted from the training dataset, and the feature vector is used as the input to the machine learning algorithm with the corresponding labels in the training step. To evaluate how well the machine-learning algorithm perform, or to select the best learning algorithm, the training dataset is divided into two parts, one part is used for training and the other part is used to test the performance.

8.4.1 *Mathematical model for supervised machine learning*

Let x_i be the feature vector of the ith sample in the training set, and y_i be the corresponding labeled output, and the set of training data with n samples be $\{(x_1, y_1), ..., (y_n, y_n)\}$. A supervised machine-learning algorithm attempts to find a function g that maps the input space X to the output space Y, i.e., $g : X \rightarrow Y$. The function g can also be represented as a scoring function f such that g is the one that has the highest score, as shown in (8.4):

$$g(x) = \text{argmax}_y f(x, y) \tag{8.4}$$

While seeking for g, the training dataset is assumed to consist of samples of independent and identically distributed, and the performance of a candidate function is evaluated using a loss function $L(y, g(x))$. Additionally, the risk of function g, $R(g)$, is defined as the expected loss of g given the training dataset, as given in (8.5):

$$R(g) = \frac{1}{n} \sum_i L(y_i, g(x_i)) \tag{8.5}$$

In general, supervised learning would aim to find the function g that minimizes the risk as defined in (8.5). This approach is referred to as the empirical risk minimization and many optimization algorithms could be used to find the most optimal g. However, when there is insufficient training data, or the space for searching for g is too large, this approach could lead to poor performance in generalization. This issue is referred to as overfitting.

This overfitting issue can be addressed by applying a regularization penalty $C(g)$ into the optimization process such that the goal is to find the function g that minimizes $J(g) = R(g) + \lambda C(g)$, where λ is a parameter that controls the weight of the regularization penalty. The objective for applying a regularization penalty is to favor simpler functions over complex ones. This approach is referred to as structural risk minimization.

By tuning the parameter λ, one can make a tradeoff between bias and variance. Both bias and variance are errors. Bias refers to the difference between the expected value and the true value of the output. Variance refers to the fact that the output may be drastically different for inputs with small fluctuations. The larger the λ, the larger the bias and the lower the variance. A small λ could lead too high variance when overfitting occurs even though it would lead to low bias.

A high bias is an indication that the model failed to capture the relationship between the input variables and the output variables from the training set. This phenomenon is referred to as underfit. A high variance is an indication of overfit because although the model can fit the training set well, but it generalizes poorly for unseen data. Although it is desirable to have both low bias and low variance, it is generally impossible to achieve both simultaneously.

8.4.2 Cross validation

Cross validation is commonly used to evaluate how well a supervised machine-learning model performs given a dataset. It resamples the dataset by using different portions of the data to train the model and the test the performance of the model in multiple rounds. The validation results for these rounds are then averaged as the performance of the model.

Cross-validation methods can be divided by whether or not the validation method is exhaustive. Leave-k-out cross validation is an exhaustive cross validation method. In this method, each time k samples are selected from the n total samples in the dataset as the validation set and the remaining samples are used the training set. This method would require C_k^n rounds to exhaustively test the model with all possible combinations. For large dataset, it is impractical to use this method due to sheer number of rounds.

The most common non-exhaustive cross-validation method is the k-fold cross validation. In this method, the dataset is randomly divided into k partitions. Then, in each round, one of the partitions is used as the validation set and the remaining $k - 1$ partitions are used as the training set. There will be a total k rounds so that each partition is used as the validation set once.

8.4.3 Common supervised machine-learning models

Support-vector machine (SVM) was originally proposed to perform linear classification [295]. Linear classification means the classification is done based on the linear combination of the feature vector. For binary classification, SVM aims to find a hyperplane that would lead to the maximum separation of the samples into two groups. Given a n-dimension feature vector, the hyperplane will be an $n - 1$ dimension hyperplane. The hyperplane is also referred to as the maximum-margin hyperplane. SVM can also perform non-linear classification by using a kernel to map the feature vector into high-dimensional feature spaces. Common kernels include homogeneous and inhomogeneous polynomial, Gaussian radial basis function, and sigmoid function. SVM can be extended into multiclass classification by transform a multiclass problem into multiple binary classification problems [296], or into a single optimization problem [297].

Naive Bayes refers to a family of classifiers that are based on the assumption that each feature is independent from other features, and the classification decision is based on Bayes' theorem. Gaussian naive Bayes is a commonly used classifier when the samples consist of continuous values. Given each feature, the assumption is that the observed samples for the feature follow the normal (i.e., Gaussian) distribution.

The classification is done based on the normal distribution with the mean of the feature, and the Bessel corrected variance.

The k-nearest neighbors (KNN) algorithm can be used for both classification and regression. Given a test sample, KNN determines the output value by taking the majority vote of the k-nearest neighbors on their classes (for classification) or by taking the average of the values of the k-nearest neighbors. KNN depends on some distance metric to find the k-nearest neighbors of the test sample. The parameter k should be tuned using the training dataset. In general, a larger k could minimize the impact of noise on the classification or regression, but the tradeoff is that the boundaries between classes become less distinctive.

The decision tree is a method to perform classification or regression by constructing a tree using the feature vector [298]. Each internal node of the tree is labeled with a different feature, and the leaf nodes are class labels. The tree is constructed recursively. In each step, some algorithm is used to evaluate the best feature to split the features.

Artificial neural network is a method of learning via a network of artificial neurons inspired by the biological neural networks in the human brain [299]. A neural network typically consists of an input layer, an output layer, and one or more hidden layers in between the input and output layers. The user would have to choose the network architecture, including the number of layers, the number of neurons in each layout (the number of neurons in the input layer should be the same as the number of features in the training data, and the number of neurons in the output layer should be the same as the number of classes), and how the neurons in adjacent layers are connected. Every input edge to a node has a weight, and each node has an activation function to derive an output based on the sum of the input weights. Using the training dataset, the weight of each edge is adjusted such that the final output is closest to the label (i.e., smallest error). A fixed learning rate could be specified or an adaptive learning algorithm could be used.

A multilayer perceptron (MLP) is a common neural network. It consists of a minimum of three layers of neurons, including an input layer, a hidden layer, and an output layer. Furthermore, an MLP is a fully connected network, i.e., a node is connected with every node in the next layer. MLP also uses a non-linear activation function. MLP uses backpropagation during the learning steps.

8.4.4 *Performance evaluation for classification*

We first go over the evaluation metrics used for classification. The most common way to assess the performance of a machine learning model is accuracy, which is defined as the ratio of the number of correct predictions and the total number of predictions made. Typically, the classification is binary, for example, the output is either a rehabilitation exercise is done correctly or incorrectly. In the context of binary classification, a correct prediction can be further divided into true positive (TP) and true negative (TN). TP means that if the test sample is labeled as correct, and the machine learning model predicted the sample as correct as well. TN means that if the test sample is labeled as incorrect, and the machine learning model predicted the sample as incorrect as well. If a sample's label is correct, but the predication result is incorrect, this is referred to as false positive (FP). When a test sample is labeled as correct, but the prediction result is incorrect, this is referred to as false negative. Therefore, the total number of correct predictions is the sum of true positive and true negative, and the total number of predictions is the sum of TP, TN, FP, FN. The accuracy is defined in (8.6) in terms of TP, TN, FP, and FN:

$$\text{Accuracy} = \frac{TP + TN}{TP + TN + FP + FN} \tag{8.6}$$

In medicine, precision, sensitivity, and specificity are often used instead of accuracy. In this context, a positive classification means that a patient has or is diagnosed as having some disease, and the negative means the patient is healthy. Precision refers to the proportion of correct positive predictions among all positive results (i.e., sum of true positives and false positives), as defined in (8.7). Precision is also called positive predictive values:

$$\text{Precision} = \frac{TP}{TP + FP} \tag{8.7}$$

Sensitivity is the true positive rate (also referred to as recall), i.e., the ratio of true positives versus the total positively labeled positives samples (sum of the true positives and false negatives), as defined by (8.8). Sensitivity is a measure how sensitive the machine learning model is to actual positives, i.e., if the patient has certain disease, sensitivity means how likely the model can make the correct diagnosis:

$$\text{Sensitivity(recall)} = \frac{TP}{TP + FN} \tag{8.8}$$

Specificity is the true negative rate, which is defined as the ratio of true negative versus the total negatively labeled negative samples, as defined in (8.9). Specificity means that if the patient is healthy, how likely the model can confirm that the patient is indeed healthy:

$$\text{Specificity} = \frac{TP}{TP + FP} \tag{8.9}$$

To achieve higher sensitivity, one would minimize the number of false negatives, which means that the model should avoid mistakenly predicting a patient that has the disease as healthy. To achieve higher specificity, one would minimize the number of false positives, which means that the model should avoid mistakenly predicting a patient that is healthy as having the disease. Minimizing false positives and minimizing false negatives are often conflicting. Hence, one typically needs to make a trade-off between the two.

To help make the tradeoff, F1-score is introduced because it combines precision and recall. More specifically, F1-score is the harmonic mean of precision and recall, as defined in (8.11). We note that the number of true negatives in the sample plays no role in the F1-score. This is an advantage compared with accuracy because in the presence of large TN, the accuracy could give a false sense of efficacy:

$$\text{F1-Score} = 2\frac{\text{Precision} \times \text{Recall}}{\text{Precision} + \text{Recall}} \tag{8.10}$$

Furthermore, one can use the precision–recall (PR) curve and the receiver operating characteristic (ROC) curve to visually compare different machine-learning models and decide on the best threshold. The PR curve displays the relationship between the precision (as the vertical axis) and the recall (as the horizontal axis). An example PR curve using the breast cancer dataset as part of sklearn (a Python machine-learning library) is shown in Figure 8.9.

The ROC curve displays the relationship between sensitivity (as the vertical axis) and (1-specificity) (or false positive rate) (as the horizontal axis). Sensitivity is also referred to as true positive rate ($= \frac{TP}{TP+FN}$). 1-specificity is the same as the false positive rate ($= \frac{FP}{FP+TN}$). Given a ROC curve, the larger the area under the curve, the better the performance. Hence, when comparing with different models, the model that gives the largest area should be selected. Using the same breast cancer dataset as an example, the ROC curve is shown in Figure 8.10. The diagonal dotted line indicates the worst-case scenario. The threshold should be the point that is furthest away from this diagonal line.

For multi-class classification, confusion matrix is often used to evaluate the model performance visually. Each row of the confusion matrix represents the instances in a class, i.e., that total number of instances in each row is the total number of labeled samples in the test set for this particular class. Each column represents the instances in a predicted class, i.e., the total number of instances in each column is the total number of instances that have been predicted to be in a particular class. If the model makes perfect classification, then only the cells in the diagonal from upper left to the bottom right should have non-zero instances. The confusion matrix provides an

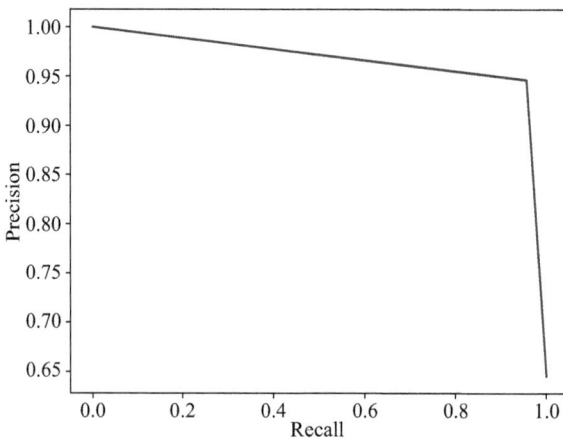

Figure 8.9   An example precision–recall curve

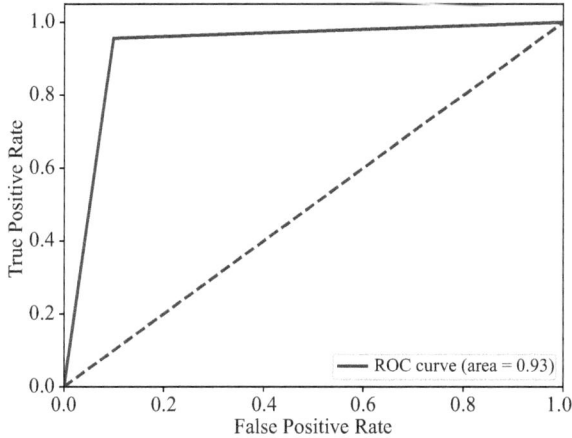

Figure 8.10   An example precision–recall curve

intuitive visual representation of how often an instance of test sample is confused as another class by the model. In Figure 8.11, we show an example confusion matrix using the iris dataset as part of sklearn and SVM as the machine learning model. In this example, there are three classes labeled as 0, 1, and 2. The first row in Figure 8.11 is 13, 0, and 0. Class 0 samples are clearly highly distinguishable from classes 1 and 2 because no confusion is made by the model. The second row in Figure 8.11 is 0, 10, and 6. This revealed that the model confused 6 of class 1 samples as class 2 samples. The third row in Figure 8.11 is 0, 0, and 9. It is somewhat surprising that the model makes no mistakes in classification for class 2.

8.4.5   Performance evaluation for regression

A machine-learning model for regression can be evaluated by the error rate of the predication compared with the labeled dataset. Statistical metrics for regression include mean absolute error (MAE), root mean square error (RMSE), coefficient of determination (R2), and adjusted R2.

The mean absolute error is defined in (8.11), where y_i is the labeled sample i output, and $\hat{y}_i$ is the predicated output for sample i, and n is the total number of samples. The less the error, the better performance of the model:

$$\text{MAE} = \frac{1}{n} \sum_{i=1}^{n} |y_i - \hat{y}_i| \tag{8.11}$$

In Section 7.4.1, we have introduced RMSE before. Here we redefine RMSE in a different context in (8.12). Similar to MAE, the less the RMSE, the better performance

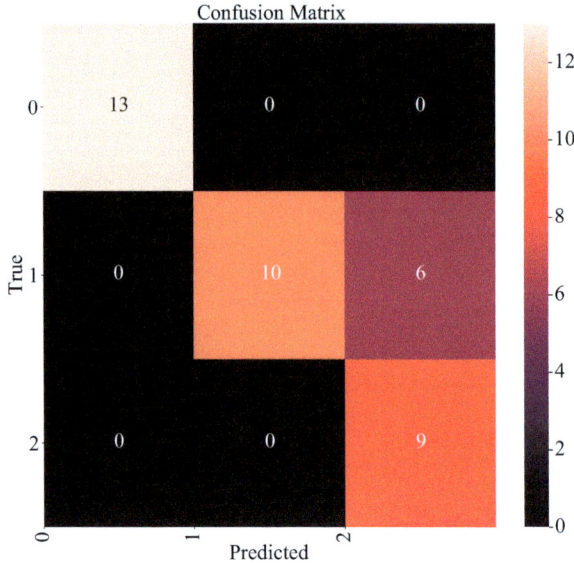

Figure 8.11   An example confusion matrix

of the model. Different from MAE, RMSE sums up the square of the error for each sample. Hence, RMSE imposes a heavier penalty on errors (especially large errors):

$$\text{RMSE} = \sqrt{\frac{\sum_{i=1}^{n} (\hat{y}_i - y_i)^2}{n}} \tag{8.12}$$

Coefficient of determination (R^2) is defined in (8.13) in terms of the residual sum of squares (RSS) and the total sum squares (TSS), where $\bar{y}$ is the average sample output values (i.e., $\bar{y} = \frac{1}{n} \sum_{i=1}^{n} y_i$). Statistically, R^2 reflects the proportion of variation in the output variable (i.e., dependent variable) that is predicted by the model from the input variable (i.e., independent variable). If the prediction is perfect, then $\hat{y}_i = y_i$, and RSS would be 0, which in turn would lead to $R^2 = 1$:

$$R^2 = 1 - \frac{\text{SSR}}{\text{SST}} = 1 - \frac{\sum_{i=1}^{n} (\hat{y}_i - y_i)^2}{\sum_{i=1}^{n} (y_i - \bar{y})^2} \tag{8.13}$$

R^2 could also be defined in terms of the regression sum of squares and the total sum of squares, where the regression sum of squares is defined as $\sum_{i=1}^{n} (\hat{y}_i - \bar{y})$, as shown in (8.14). In this case, we assume that the total sum of squares equal to the sum of the residual sum of squares and the regression sum of squares. The regression sum of squares is also referred to as the explained sum of squares, emphasizing how much the regression model could explain the variance:

$$R^2 = \frac{\sum_{i=1}^{n} (\hat{y}_i - \bar{y})^2}{\sum_{i=1}^{n} (y_i - \bar{y})^2} \tag{8.14}$$

A known issue with R^2 is that the more independent variables used for doing regression (each independent variable is often referred to as a regressor), the better R^2 value. The remove the impact of the additional regressors, adjusted R^2 is introduced, as defined in (8.15), where n is the number of samples, the k is the number of independent variables:

$$R^2_{Adj} = 1 - (1 - R^2)\frac{n-1}{n-k-1} \tag{8.15}$$

8.5   Unsupervised machine learning

Unsupervised machine learning is to discover patterns exist in the unlabeled data. There are many algorithms for unsupervised machine learning. Clustering is one of the most popular unsupervised machine-learning approaches.

The k-means clustering aims to divide the samples into k number of clusters. The goal is to minimize the squared distances between each paired point within the same cluster. The standard k-means algorithm would first randomly select k samples as the initial means. Then, each additional sample is assigned to one of the k clusters that have the smallest squared Euclidean distance, and then the cluster mean is recalculated accordingly. In [51], Seiter *et al.* compared the results using a rule-based approach and the k-means clustering to discover patterns of daily life activities routine in hemiparetic rehabilitation patients. They showed that the rule-based approach outperformed the clustering approach by about 10%.

Hierarchical clustering is another approach to clustering. There are two types of hierarchical clustering, agglomerative and divisive. Agglomerative clustering starts with forming individual clusters each with a single sample, and then merge a pair of existing clusters at each step until a single cluster is formed. In each step, the two clusters that have the smallest distance is chosen to merge. Divisive clustering starts with a single cluster that contains all samples, and then spilt to have one extra cluster at a time until each sample forms its own cluster. Some metric will be used for the merging or splitting decision. The metric is typically some distance between each pair of samples or clusters, such as Euclidean distance, and squared Euclidean distance. When comparing distance between two clusters, a linkage criteria is needed. Popular linkage criteria include complete-linkage clustering and single-linkage cluster. The former takes the maximum pairwise distance among the samples in the cluster, and the latter takes the minimum pairwise distance among the samples in the cluster. Hierarchical clustering results are typically illustrated in terms of dendrograms. One axis of a dendrogram shows the sample numbers, and the other axis shows the distance. In [23], Nguyen *et al.* performed pattern recognition of the movements in healthy and patients who are doing post-surgical rehabilitation using hierarchical agglomerative clustering.

Example 8.1. We use a simple example to show how k-means and hierarchical clustering work. Assume that we have the performance scores for 5 healthy subjects and 5 subjects who are rehabilitating. The scores for the 5 healthy subjects are 99, 95, 93,

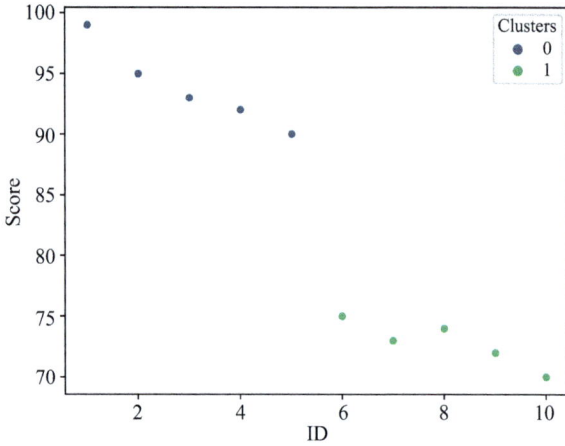

Figure 8.12   An example k-means clustering result

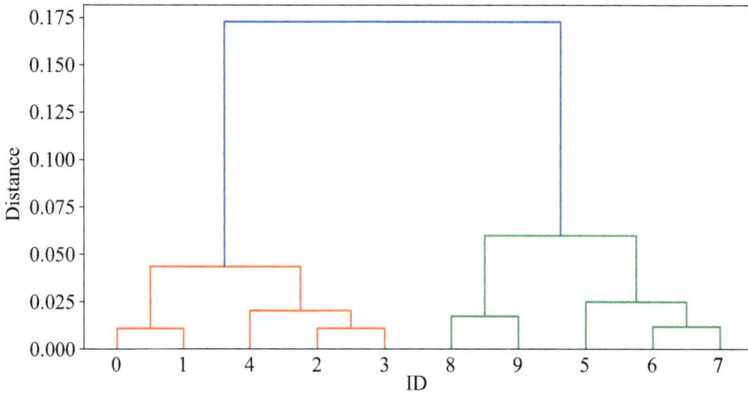

Figure 8.13   An example dendrogram for hierarchal agglomerative clustering

92, and 90. The scores for the 5 rehabilitating subjects are 75, 73, 74, 72, and 70. A sequence number of 1,2,...,10 is used as the ID for each subject. We use the sklearn python library for both k-means and hierarchical clustering. The k-means clustering result is shown in Figure 8.12. The hierarchical clustering result is shown in Figure 8.13. We note that hierarchical clustering can be used to determine the optimal number of clusters. This could be done by setting a threshold along the vertical axis (i.e., distance). One can imagine a horizontal line as this threshold. The number of lines that this threshold line cuts into would be the optimal number of clusters. For example, we set the threshold to be 0.075, then the number of clusters would be 2.

8.6 Deep learning

Deep learning is an artificial neural network that uses multiple layers. A particular benefit of deep learning is its capability of extracting features from the raw input data. This is done over multiple layers progressively. There are several types of deep learning architectures, including convolutional neural networks, recurrent neural networks, deep belief networks, and deep reinforcement learning. Here we briefly introduce convolutional neural networks and recurrent neural networks.

Convolutional neural network (CNN) is a regularized multilayer perceptron network. Identical to MLP, a neuron is connected to all neurons in the next layer. An issue with this fully connected architecture is overfitting. To cope with this issue, regularization is used in CNN. A CNN typically consists of one or more convolutional layers, pooling layers, and fully connected dense layers. A convolutional layer convolves the input and generates a feature map. A pooling layer is used to reduce the dimension of the data by combining the inputs from several neurons in the previous layer as a cluster and generating one output to a single neuron in the next layer. Each cluster is merged by taking the maximum value in the cluster or taking the average of the cluster.

Recurrent neural network (RNN) uses a particular network architecture that facilitates the processing of time series or other sequential data. In the RNN architecture, an output from a neuron could feed back to itself, as shown in Figure 8.14. Such an architecture could unfold according to the time sequence (or sequentially). The long short-term memory (LSTM) is one of the most popular RNN [300]. LSTM is capable

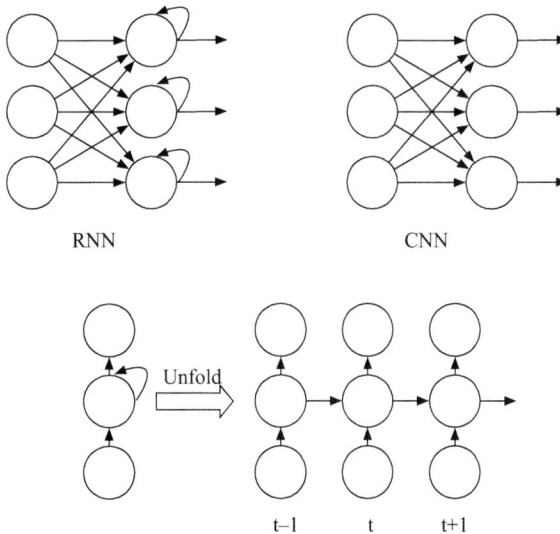

Figure 8.14 The characteristics of RNN in comparison to feed-forward networks such as CNN

of learning tasks that require memories of events in the past in the sequence. Gated recurrent unit (GRU) is another RNN [301].

Deep learning has been used in the context of general human activity recognition, which could be applicable to evaluating the quality of rehabilitation exercises. CNN has been used for human activity recognition in [302–306]. RNN has been used for human activity recognition in [307–310]. It is also possible to combine CNN and RNN for human activity recognition, such as [311].

8.7   Assessment of rehabilitation exercises

We follow the taxonomy proposed by Boukhennoufa *et al.* [273] with some minor modifications. Assessment of rehabilitation exercises can be roughly divided into three levels. At the lowest level is to recognize the activities performed, including the activity type, pose, and repetition count. At the intermediate level is to assess the quality of the exercise performed, i.e., how accurately the patient is adhering the requirements of the rehabilitation exercise. At the highest level is to assess the performance at the clinical level, often using established clinical routines and scales.

8.7.1   Activity recognition

Examples of activity recognition include the recognition of standing, sitting, sit-to-stand, and other activities of daily living. Activity recognition also includes the recognition of gestures and the type of rehabilitation exercises performed by the patient.

In [312], naive Bayes, SVM, and decision tree were used to make several levels of classification. At level 1, the classification is whether or not the subject is in the mobile or immobile states. At level 2, the classification is between the sit and stand poses. At level 3, the classification is between three poses: sit, stand, and lie. At level 4, the classification is between stair climbing movements and other large movements. At level 5, the classification is the recognition of five different movements, including ramp up, ramp down, stairs climbing up, stairs climbing down, and other large movements. At level 6, the classification is to recognize some small movements, such as slightly moving while seated or standing. At level 7, the classification is about the transitions between different movements.

In [313], 19 activities of daily living (such as brush teeth, mixing powders, eating with hands, eating with utensils) in stroke patients were classified using decision tree, random forest, SVM, and XGBoost. SVM and XGBoost exhibited the best classification results.

In [314], deep learning (CNN) is used to recognize four rehabilitation excesses designed for upper limbs in patients with chronic stroke, including bilateral shoulder flexion, wall push exercise, active scapular exercise, and towel slide exercise.

In [276], logistic regression is used to detect goal-directed and non-goal-directed activities of daily living in stroke patients.

In [274], KNN, SVM, and decision tree were used to recognize the type of three arm exercises (elbow flexion/extension, forearm supination/pronation, and wrist extension/flexion). Furthermore, the repetition count was determined by the MATLAB® findpeaks function.

In [315], KNN is used to classify a set of activities of daily living using IMU and barometer sensors. The set of activities includes walk, run, stand, sit, lie down, stair descent, star ascent, etc.

8.7.2 Performance quality assessment

In [276], random forest is used to assess the quality of the rehabilitation exercises performed by stroke patients. The quality is determined based on the accuracy, smoothness, speed of the movements, and whether or not any compensatory movement is detected.

8.7.3 Clinical assessment

In [316], Kaku *et al.* aimed to measure the impairment levels using deep learning algorithms. The subjects where divided into several cohorts based on their impairment levels using the upper extremity Fugl–Meyer Assessment Scale (FMA) (the higher score the less impairment) [317]. More specifically, the authors aimed to assess five functional primitives, including (1) motion to contract (referred to as reach), (2) motion to move an object (referred as transport), (3) motion to close to an object (referred to as reposition), (4) motion to keep an object still (referred to as stabilize), and (5) stand without any motion (referred to as idle). CNN, LSTM, and a modified version of CNN is used for deep learning. In addition, random forest is also used as a classifier. They showed that scores for more severely impaired subjects indeed have lower scores.

Some studies aimed to directly determine a score based on the FMA scale. In [318], linear regression and random forest regression are used to determine the impairment score and the quality of movement score based on the FMA scale [317], and the Functional Ability Scale [319]. In [320], ensemble regression is used to determine the FMA score based on motion data of the upper limb, wrist, and fingers. In [58], a regression model is used to estimate the FMA score of stroke survivors with different degrees of impairment.

National Institutes of Health Stroke Scale (NIHSS) [321] and the Medical Research Council (MRC) scale [322] are also common scales for motor weakness measurement. In [323], SVM and an ensemble classifier are used to classify five different scores based on NIHSS-0, NIHSS-1, MRC-7, MRC-8, and MRC-9.

In [280], SVM is used to make a binary determination on the limb impairment level for stroke patients. The impairment level is used on the Oxford Grading Motor Scale [324], which has six motor scores. Score 5 indicates the normal muscle strength, and score 0 indicates the most serious impairment where the muscle cannot move the arm at all. The authors used score 3 and the threshold to create a binary classification.

Several studies focused on more diagnosis of patients who are suffering from a particular diseases based on analysis of motion data. In [325], SVM (and majority voting) is used to identify subjects who are post stroke, who are with Huntington's disease, and who are healthy elderly based on the gait data. In [326], random forest, decision tree, Gaussian naive Bayes, MLP, and Adaboost are used to differentiate subjects with stroke from those who are with other neurological disorders. In [327], deep neural networks are used to identify stroke patients from healthy subjects from the gait data. The goal of the study is to develop a tool that can be used by clinicians to diagnose different abnormal gaits. In [328], random forest is used to differentiate upper extremity functional and nonfunctional movements.

In [37], the authors argued that a rehabilitating patient might gain movement capacity that is not reflected by the performance scales such as FMA. Support Vector Regression is used to estimate the hand capability.

Chapter 9
Rule-based activity recognition

Unlike the learning-based approach, where one does not need to know the details of a rehabilitation exercise, the rule-based approach requires a detailed description of each exercise to be recognized and evaluated in terms of a set of rules. Exactly because of this, it is difficult for the learning-based approach to provide specific feedback to patients regarding exactly what was performed correctly and what requires improvements. The rule-based approach could provide much more specific feedback to the patients. Studies on rule-based activity recognition can be roughly divided into two categories: (1) ad hoc rules that are applicable to a particular rehabilitation exercise; and (2) general purpose rules that are designed to define a broader range of exercises or activities. Another perspective related to rehabilitation exercises is the tolerance on movements within which are considered correct.

For both types studies, such uncertainty can be captured explicitly by a threshold in the measurement as part of the rule, or by using the fuzzy interference using some membership definition. Fuzzy inference has also been used to detect activities.

9.1 Ad hoc rule-based studies

For rehabilitation exercises, it is a common practice to define the most critical rules to each exercise. The rules might not be distinctive enough to identify a particular exercise among a group of exercises. For example, in [172,173], gait retraining rules are defined in terms of the trunk flexion angle, trunk lean angle, and the distance between some joints for postural control traverse. In [224], the rules for sit-to-stand and squat exercises are defined in terms of the knee angle and the ankle angle, and the rules for shoulder abduction/adduction are defined in terms of the shoulder angle. In [329], the rules for knee rehabilitation exercises are defined in terms of the knee angle. In [330], the rules for the sit-to-stand exercise are expressed in terms of the minimum hip angle and the movement smoothness of the head.

In [66], five different activities are detected using joint angles and some kinematic data. For the sit-down and the stand-up activities, the trunk and sacrum acceleration along the z-axis, thigh acceleration along the y-axis, and the hip angle are used for detection. For the reach up and the reach mid activities, the shoulder angle, the knee angle, and the hip angle are used for detection. For the reach ground activity, the shoulder angle, the knee angle, the hip angle, and the shoulder acceleration along the *x*-axis are used for detection. For the turn activity, the angular velocity of the sacrum, thigh, and trunk are used. For the walk activity, the magnitude of the linear acceleration of the shin, thigh, trunk, and sacrum are used for detection.

In [331], a fuzzy inference system is used to classify several activities of daily living, including lie, sit, stand, and walk. The fuzzy inference system uses six fuzzy variables, including the activity type before and after the current activity, the strength of the transaction and the type of transition at the beginning and the end of the current activity period. A total of 14 fuzzy rules are used.

In [332], a dual-stage hierarchical fuzzy inference system is developed to recognize several activities of daily living, including sitting, standing, walking, lying, climbing stairs, and taking an elevator. Stage-1 processor is referred to as the event fuzzy inference system. Stage-2 processor is referred to as the behavior fuzzy inference system. The input to the stage-1 processor includes the previous activity that has been recognized, the current activity (usually unknown), the detection probability for postural transition, the classification probability for postural transition, and the altitude change (if any) before and after the postural transition. The output for the stage-1 processor is the current activity type. The output of the stage-1 processor is fed into stage-2 processor for further processing. Stage-1 processor uses 6 membership for its input variables, and rules for lying-to-sitting and lying-to-standing. Stage-2 processor uses a set of fairly complex rules to improve the accuracy of the detection of the current activity.

9.2   General-purpose rule-based activity recognition

General-purpose rules have been proposed for classifying hand gestures [333] and whole body activities [334]. In [334], each activity (or gesture) is described by a set of rules according to a Gesture Description Language (GDL). GDL consists of basic rules and the final rule. The key to defining the rules is the identification of key frames as part of the activity. Once the key frames are identified, then the basic rules are defined based on set of joint positions in these key frames. The final rule consists of a sequence of basic rules. Completely defining the rules based on key frames alone could miss requirements that depend on the entire trajectory of a gesture or an activity. Another issue with GDL is that no guideline is offered on how to identify key frames.

Previously, we proposed a rule-based framework for describing rehabilitation exercises and for providing real-time feedback to patients [194]. The approach is inspired by [333], where a hand gesture is defined by a sequence of movement segments in terms of a monotonically increasing or decreasing key parameter, such as the angle between two fingers. Each segment is referred to as a monotonic segment. In our framework, we also monotonic segments to define dynamic rules. The frames that delineate two consecutive segments are referred to as reference configurations. Therefore, the reference configuration corresponds to the particular frame that the variable reaches a maximum if the variable increases in the segment, or a minimum if the variable decreases in the segment of movement.

Unlike [333], where only dynamic rules were defined, we also define invariance rules and static rules [221]. The invariance rules specify the rules by which the movement must abide by throughout the entire repetition. The static rules specify the poses of segments that must remain stationary. The invariance rules and static rules are critical for rehabilitation exercises, but they may not be important for gesture/activity recognition.

More specifically, the rules in our frameworks consist of three types: dynamic, static, and invariance. Considering that the rehabilitation exercise is repetitive, the rules are defined for each repetition:

- A dynamic rule defines a sequence of consecutive positions for a moving body segment.
- A static rule specifies the position and/or orientation of body segment that should remain stationary.
- An invariance rule describes the condition that a moving body segment must adhere to.

All rules are described in terms of one or more reference configurations. A reference configuration could be defined in terms of the joint angle in between two adjacent body segments, such as the angle between a moving arm and a stationary segment. A reference configuration could also be defined in terms of the orientation with respect to an anatomical plane, such as the frontal, sagittal, or transverse plane. In some cases, the reference configuration may be defined in terms of the distance between some joints or relative positions of different joints.

9.2.1   Rule encoding method

Rules are encoded in the eXtensible Markup Language (XML) for their readability and extensibility. Listing 9.1 shows a template for encoding the rules. The rules start with an ExerciseRules element where the exercise name should be given as an attribute for identification purpose. This is followed by a list of dynamic rules enclosed in a DynamicRules element, one or more static rules in a StaticRules element, and one or more invariance rules in an InvarianceRules element.

The DynamicRules element consists of one or more DynamicRule elements. Each DynamicRule element consists of two or more Configuration elements. If there is only a single DynamicRule, then the structure can be compacted to have the DynamicRule

Listing 9.1   The template structure on rule encoding for an exercise

```
1 <ExerciseRules  Name="Exercise Name">
2     <DynamicRules> ... </DynamicRules>
3         <DynamicRule> ... </DynamicRule>
4         ...
5         <DynamicRule> ... </DynamicRule>
6     <StaticRules> ... </StaticRules>
7     <InvarianceRule> ... </InvarianceRule>
8 </ExerciseRules>
```

directly under the ExerciseRules element. The StaticRules element and the Invariance element are rather similar to the DynamicRule element.

Each reference configuration is expressed in a Configuration element. It starts with a Type element so that the element can be parsed properly. As shown in Listing 9.2, the JointAngle configuration uses the joint angle between two adjacent body segments. The JointAngle configuration uses several additional elements:

- The CenterJoint element defines the common joint of the two segments.
- The DownstreamJoint and UpstreamJoint elements specify the other endpoints of the two segments.
- The Angle and MaxAngleDeviation elements define the target value and the tolerance value. The tolerance value is determined heuristically and empirically with the following constraints:
  - The tolerance value must not lead to the overlapping of the current configuration with any other configuration to avoid confusion.
  - The tolerance value must not result in an unsafe posture for the user.
- Some exercise may require the patient to complete a repetition not too fast and not too slow with a target duration. This is captured by the Duration element. The tolerance on the duration is specified is the MaxDurationDeviation element.

As shown in Listing 9.3, the JointDistance configuration is defined in terms of the distance between a moving joint and a stationary joint, encoded as MovingJoint and StationaryJoint elements. The tolerance value is provided in the MaxDistDeviation element. Duration and MaxDurationDeviation may also be specified if there is a requirement on the exacted duration for each repetition.

Listing 9.2   A configuration for joint angle

```
1 <Configuration>
2    <Type>"JointAngle"</Type>
3    <CenterJoint>"JointName"</CenterJoint>
4    <DownstreamJoint>"JointName"</DownstreamJoint>
5    <UpstreamJoint>"JointName"</UpstreamJoint>
6    <Angle>"ExpectedAngleValue"</Angle>
7    <MaxAngleDeviation> "AngleTolerance" </MaxAngleDeviation>
8    <Duration>"ExpectedDurationValue"</Duration>
9    <MaxDurationDeviation> "DurationTolerance" </MaxDurationDeviation>
10 </Configuration>
```

Listing 9.3   A configuration for distance between two joints

```
1 <Configuration>
2    <Type>"JointDistance"</Type>
3    <MovingJoint>"JointName"</MovingJoint>
4    <StationaryJoint>"JointName"</StationaryJoint>
5    <Distance>"ExpectedValue"</Distance>
6    <MaxDistDeviation> "DistanceTolerance"</MaxDistDeviation>
7    <Duration>"ExpectedDurationValue"</Duration>
8    <MaxDurationDeviation> "DurationTolerance" </MaxDurationDeviation>
9 </Configuration>
```

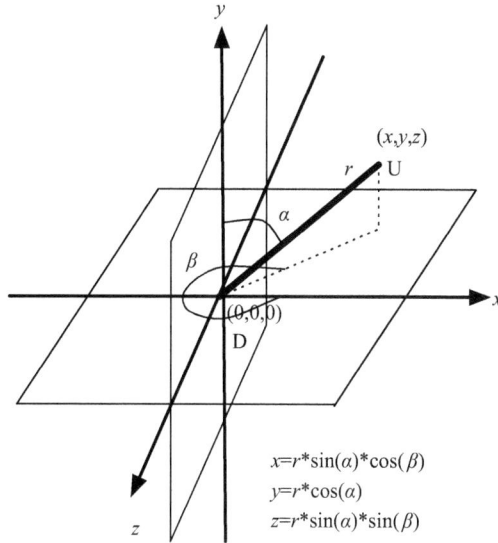

$x = r * \sin(\alpha) * \cos(\beta)$
$y = r * \cos(\alpha)$
$z = r * \sin(\alpha) * \sin(\beta)$

Figure 9.1   The spherical coordinate system used to define the BoneOrientation element

Static and invariance rules often require the use of a BoneOrientation element to specify a configuration regarding the orientation of a body segment. The body segment is defined using two joints represented by the DownstreamJoint and the UpstreamJoint elements. The segment orientation is represented in terms of a spherical coordinate system where the DownstreamJoint is used as the origin, as shown in Figure 9.1. Assuming that the subject is facing the Kinect sensor, the frontal plane is the plane determined by the x–y axes, the transverse plane is determined by the x–z axes, and the sagittal plane is determined by the y–z axes. The two end points of segment shown in Figure 9.1 are the DownstreamJoint (with a Cartesian coordinate of (0,0,0)) and the UpstreamJoint (with a Cartesian coordinate of (x,y,z)).

In Figure 9.1, we assume that the reference plane is the transverse plane defined by the x–z axis, and the reference axis is the x-axis. In this context, the spherical coordinates of the UpstreamJoint are (r, α, β), where α is defined as the angle between the body segment and the y-axis, and β is defined as the angle between the projected segment in the x–z plane and the y-axis. Hence, the Cartesian coordinates for the UpstreamJoint can be expressed in terms of r, α, and β: $x = r * \sin(\alpha) * \cos(\beta)$, $y = r * \cos(\alpha)$, $z = r * \sin(\alpha) * \sin(\beta)$.

The BoneOrientation reference configuration would need to define a reference plane (i.e., "Frontal," "Sagittal," or "Transverse"), a reference axis (i.e., "X," "Y," or "Z"), and the α and β angles, as shown in Listing 9.4. If the movement of a body segment is to be within an anatomical plane in an invariance rule, the α value will be 0 with no β value and no axis specified, and the reference plane should be that particular anatomical plane.

Listing 9.4   A configuration for bone orientation

```
 1  <Configuration>
 2      <Type>"BoneOrientation"</Type>
 3      <DownstreamJoint>"JointName"</DownstreamJoint>
 4      <UpstreamJoint>"JointName"</UpstreamJoint>
 5      <Plane>"RefPlane"</Plane>
 6      <Axis>"RefAxis"</Axis>
 7      <AlphaAngle>"ExpectedAlphaAngleValue"</AlphaAngle>
 8      <BetaAngle>"ExpectedBetaAngleValue"</BetaAngle>
 9      <MaxAngleDeviation> "AngleTolerance" </MaxAngleDeviation>
10      <Duration>"ExpectedDurationValue"</Duration>
11      <MaxDurationDeviation> "DurationTolerance" </MaxDurationDeviation>
12  </Configuration>
```

Example 9.1. We use the hip abduction/adduction exercise as an example to show how to define the rules using our framework. Hip abduction/adduction requires that the abducting leg moves away from the body to about 45 degrees within the frontal plane, and this is followed by hip-adduction so that the leg would move back to the initial configuration. This exercise has one dynamic rule with two reference configurations, one for the initial pose and the other for the final pose where the abducting leg reaches the out-most position. This exercise also has an invariance rule includes two requirements: (1) the abducting/adduction leg must move within the frontal plane; and (2) the knee of the abducting/adduction leg must be straight. The rules for the hip abduction/adduction exercise are provided in Listing 9.5.

9.2.2   Real-time motion tracking

Tracking motion with respect to the static rules and invariance rules is relatively straightforward because it can be done at the individual frame-level. As soon as a violation is detected, feedback regarding the violation of the particular rule can be provided to the subject immediately.

Tracking against the dynamic rules is performed by a combination of two-levels analysis. At the first level is at the individual frame level. Here we assume that the movement is captured in terms of a sequence of frames, such as those provided by the Kinect software development kit. The second level is at the system level where the state of the movement is tracked. The recognition of each repetition is also done via the interpretation of the dynamic rules.

More specifically, the tracking against dynamic rules is guided by a finite state machine, as shown in Figure 9.2. The states are determined by the reference configurations, which delineate a sequence of monotonic segments. The first reference configuration is the initial state of the finite state machine. An exercise is modeled to have k reference configurations as defined in the dynamic rules. Each reference configuration constitutes a state, i.e., C_i where $i = 1, 2, ...k$. Hence, there are $2k - 1$ states in the finite-state machine for the exercise. The machine enters a state C_i on detecting the corresponding reference configuration and will stay in that state until the detection of the next reference configuration as shown in Figure 9.2.

Listing 9.5   The rules for the hip-abduction exercise

```
1 <ExerciseRules Name="Hip Abduction">
2   <DynamicRule>
3       <Configuration>
4               <Type>"JointAngle"</Type>
5               <CenterJoint>"HipCenter"</CenterJoint>
6               <DownstreamJoint>"RightAnkle"</DownstreamJoint>
7               <UpstreamJoint>"LeftAnkle"</UpstreamJoint>
8               <Angle>"0"</Angle>
9               <MaxAngleDeviation> "15" </MaxAngleDeviation>
10      </Configuration>
11      <Configuration>
12              <Type>"JointAngle"</Type>
13              <CenterJoint>"HipCenter"</CenterJoint>
14              <DownstreamJoint>"RightAnkle"</DownstreamJoint>
15              <UpstreamJoint>"LeftAnkle"</UpstreamJoint>
16              <Angle>"45"</Angle>
17              <MaxAngleDeviation> "10" </MaxAngleDeviation>
18      </Configuration>
19  </DynamicRule>
20  <InvarianceRule>
21    <Configuration>
22        <Type>"BoneOrientation"</Type>
23        <DownstreamJoint>"HipCenter"</DownstreamJoint>
24        <UpstreamJoint>"RightAnkle"</UpstreamJoint>
25      <Plane>"Frontal"</Plane>
26      <AlphaAngle>"0"</AlphaAngle>
27      <MaxAngleDeviation> "15" </MaxAngleDeviation>
28      </Configuration>
29      <Configuration>
30        <Type>"JointAngle"</Type>
31        <CenterJoint>"RightKnee"</CenterJoint>
32        <DownstreamJoint>"HipCenter"</DownstreamJoint>
33        <UpstreamJoint>"RightAnkle"</UpstreamJoint>
34        <Angle>"180"</Angle>
35        <MaxAngleDeviation>"15"</MaxAngleDeviation>
36      </Configuration>
37    </InvarianceRule>
38 </ExerciseRules>
```

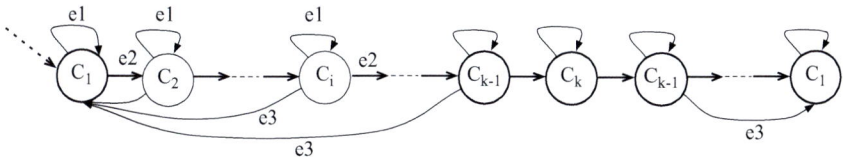

Figure 9.2   The finite state machines for the dynamic rule

Upon the arrival of a new data frame, the rule execution engine treats it as a new event. It will determine the event type as follows, as shown in Figure 9.2. Assuming the current state is C_i:

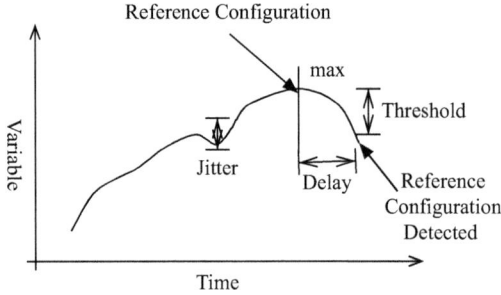

Figure 9.3　The mechanism to determine the next reference configuration that can tolerate measurement jitters

- Event $e1$: The frame does not match the next expected reference configuration C_{i+1}. Hence, the finite-state machine stays in the current state C_i.
- Event $e2$: The frame matches the next expected reference configuration C_{i+1}. The finite-state machine transition to the next state C_{i+1}.
- Event $e3$: The current frame satisfies the initial configuration C_1. The finite-state machine then switches to the initial state C_1.

The reason for event $e1$ to occur is because there are many frames in between the two reference configurations. The reason for event $e3$ is because the subject might decide to stop in the middle of a repetition, for example, the subject has received feedback that one or more rules have been violated.

In practice, a subject may not execute an exercise exactly as described, hence, we cannot assume the next reference configuration will take place all the time. Therefore, we must track each actual monotonic segment dynamically as the subject is performing the exercise. This is implemented by tracking the variables defined in the rule (such as the joint angle, the body segment orientation angle, or the distance between two joints). The current monotonic segment ends when the variable reaches a peak (if the variable is increasing) or a valley (if the variable is decreasing).

To cope with measurement jitters, the tracking for a monotonic segment with an increasing (or decreasing) value, it is not terminated until the current value is smaller (or greater) than the last seen maximum (or minimum) value by a threshold, as illustrated in Figure 9.3. The threshold depends on the magnitude of jitters. Inevitably, this mechanism would a cause some delay in the detection of the next reference configuration.

9.2.3 Fuzzy interference extension

In [195], we extended the rule-based framework for fuzzy interference so that an overall categorical feedback can be provided to the subject. It is relatively straightforward to define the set of fuzzy interface rules in terms of the dynamic, static, and invariance rules (which we will refer to as kinematic rules in this section). To do fuzzy interface, we will need to provide a mechanism to perform similarity calculation on the adherence of each of the kinematic rules with a numerical score. The score is then being categorized based on the membership function.

9.2.3.1 Similarity calculation

Given a reference configuration and the corresponding key frame, the similarity between the two is determined by the distance between them. The smaller the distance is, the more similar between the two, and therefore, the better performance. The overall distance is the weighted sum of the distance for each of the variables defined in the configuration. For each variable, the distance is calculated based on the difference between the value defined in the reference configuration and the observed value. The variable could be angle, distance, or duration.

Given a pair of values for a variable X, let X_r be the value defined in the reference configuration, X_o be the observed value, E_X be the maximum deviation from the tolerance value defined in the reference configuration (e.g., the MaxAngleDeviation, or MaxDistDeviation, or MaxDurationDeviation), the distance between the two, D_X is defined in (9.1):

$$D_X = \frac{|X_r - X_o|}{E_X} \tag{9.1}$$

The distance is normalized against the maximum tolerance value for each variable so that the distances can be aggregated in a meaningful way. If the subject moves within the maximum tolerance value E_X, then D_X would range between 0 and 1.

Given a set of n variables defined in a reference configuration C, $X_1, X_2, ..., X_n$, the total distance between the reference configuration and the observed configuration is defined as the weighted sum of individual distances for the n variables, as shown in in (9.2):

$$D_C = \sum_{i=1}^{n} w_i D_{X_i} \tag{9.2}$$

where w_i is the weight for the variable X_i, and the sum of all weight must be 1 (i.e., $\sum_{i=1}^{n} w_i = 1$). By default, the weight is the same for all variables, i.e., $w_i = 1/n$. For static and invariance kinematic rules, every frame observed will be compared against each of the reference configurations defined in each such rule. Therefore, we should aggregate the individual distances calculated for each frame at the end of the repetition for each reference configuration, and further aggregate them together for the entire rule. For a dynamic kinematic rule consisting of several reference configurations, the distances for each individual configuration would also be aggregated.

Let m be the number of frames to be aggregated, and D^i_{Cj} be the distance calculated for frame i (where $i = 1, 2, ...m$) for reference configuration j in a rule. The aggregated distance for the entire repetition for the reference configuration is defined in (9.3). Note that this step is needed for the static rule and the invariance rule only:

$$D_{Cj} = \frac{\sum_{i=1}^{m} D^i_{Cj}}{m} \tag{9.3}$$

Assuming that there are k reference configurations in the particular rule, the total distance for the rule is given in (9.4):

$$D = \frac{\sum_{j=1}^{k} D_{Cj}}{k} \tag{9.4}$$

9.2.3.2   Fuzzy inference

The similarity calculation is used for the fuzzy inference system to evaluate the execution quality of the current repetition for each fuzzy rule. The overall structure of the fuzzy system is illustrated in Figure 9.4. For simplicity, we assume that there are three kinematic model: (1) one dynamic rule, (2) one static rule, and (3) one invariance rule.

Each input variable corresponds to the similarity calculation result for one kinematic rule for the current repetition, i.e., the distance between the ideal execution of the repetition according to the rule and the actual execution. For each input variable, a membership function is defined so that the degree to which it belongs to the corresponding fuzzy set can be determined. In this study, we used the following classes in the fuzzy set: "excellent," "very good," "good," "fair," and "poor." The input variable value is limited between 0 and 1 inclusive. The distance calculated from the similarity calculation might be larger than 1 if the subject deviates from a kinematic rule by more than the specified tolerance, in which case, the distance value is rounded down to 1.

The fuzzy inference rules are based on the kinematic rules, and in linguistic terms they can be expressed as:

1.   If the evaluation result as determined by the similarity calculation for every kinematic rule is excellent, then the current repetition is excellent.

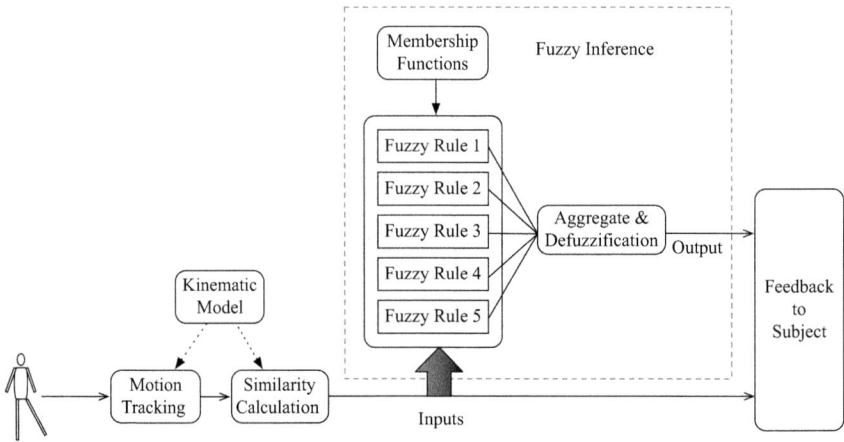

Figure 9.4   Overall structure of the fuzzy interference system

2. If the evaluation result as determined by the similarity calculation for every kinematic rule is very good or excellent, with at least one rule is evaluated to be very good, then the current repetition is very good.
3. If the evaluation result as determined by the similarity calculation for every kinematic rule is good, very good or excellent, with at least one rule is evaluated to be good, then the current repetition is good.
4. If the evaluation result as determined by the similarity calculation for every kinematic rule is fair, good, very good or excellent, with at least one rule is evaluated to be fair, then the current repetition is fair.
5. If the evaluation result as determined by the similarity calculation for any kinematic rule is poor, then the current repetition is poor.

The five linguistic rules are expressed in terms of $125 = 5 \times 5 \times 5$ fuzzy rules because we have three inputs and each input has five categories. In the fuzzy inference, we assume that the rules have equal weight. The output from each fuzzy rule evaluation also belongs to a fuzzy set, and we choose to use the same membership function as that of the input. To provide categorical feedback to the subject, the output value is fuzzified again based on the membership function for the output.

Example 9.2. We again use the hip abduction/adduction exercise as an example. The upper right corner in Figure 9.5 shows the measured hip angle, off-frontal-plane angle, and knee angle of a run of five hip abduction/adduction repetition using Microsoft

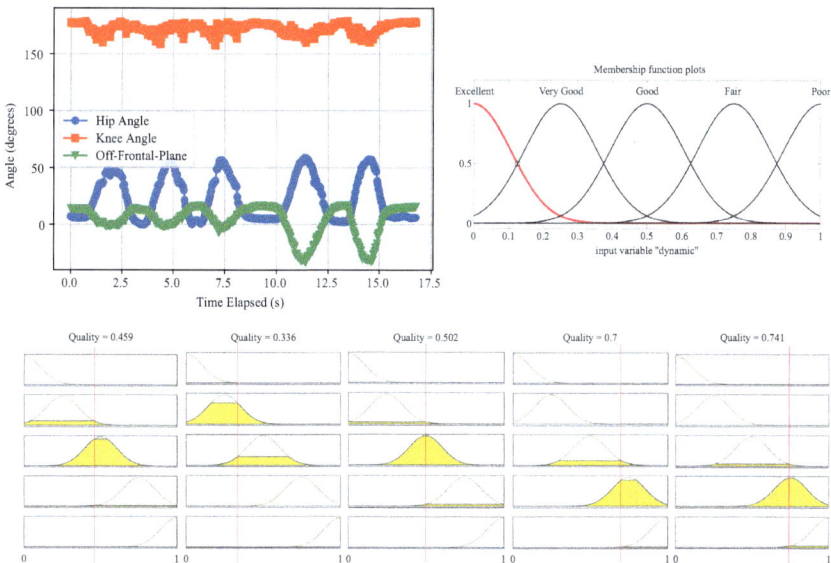

Figure 9.5 *Upper left corner: the raw motion data for a run of five repetitions of the hip abduction/adduction exercise. Upper right corner: the membership function for input/output. Lower row: the outputs for the five repetitions.*

Table 9.1 The inputs and the output for five repetitions of hip abduction/adduction

	R1	R2	R3	R4	R5
D-C1	0.357	0.003	0.019	0.237	0.123
D-C2	0.131	0.509	0.563	0.670	0.604
$D\text{-}C_{dynamic}$	0.244	0.256	0.291	0.454	0.354
I-C1	0.329	0.276	0.328	0.256	0.396
I-C2	0.578	0.395	0.676	1.159	1.202
$I\text{-}C_{invariance}$	0.453	0.336	0.502	0.707	0.799
Output	0.459	0.336	0.502	0.700	0.741

Kinect. These variables are needed to evaluate the dynamic rule and the invariance rule for hip abduction/adduction. In this run, the first three repetition were performed correctly, and the last two repetitions were performed intentionally wrong by moving the abducting leg out of the frontal plane. The hip angle is defined as the angle between the two legs. The knee angle refers to the angle formed between the upper leg and the lower leg. The off-frontal-plane angle refers to the angle formed between the abducting leg and the frontal plane.

The distance between the observed hip angle and that defined in each of the two reference configurations (one for the initial configuration and the other for the target final configuration) is calculated based on (9.1). Because it is the only parameter in the configuration, it is also the distance for the configuration itself. The total distance with respect to the dynamic rule can then be computed using (9.4). The invariance rule is defined by two reference configurations, each with only a single parameter. The similarity calculation can be done by following Equations (9.1), (9.2), (9.3), and (9.4).

The membership function is identical to all input and output variables and it is shown in the upper right corner of Figure 9.5. The results of the similarity calculation for the dynamic rule and the invariance rule are used as the two inputs to the fuzzy system defined in Section 9.2.3.2. The output of the fuzzy system is obtained via MATLAB®. The inputs and the outputs for the three correct repetitions and two wrong repetitions are summarized in Table 9.1. The outputs are then fuzzified again so that we can provide categorical feedback to the subject. As shown in the bottom row of Figure 9.5, the first three repetitions are categorized as "Good" or "Very Good" and the last repetitions are classified as "Fair."

Note that the inputs and consequently the output and its fuzzy classification depend on the maximum tolerance parameters specified in the kinematic rules. A large value might categorize a wrong iteration as "Good," while a small maximum deviation value might classify a correct exercise as "Fair." For example, if we set the maximum deviation to be 20 for the parameters in the invariance rule, the last two iterations would be classified as "Good." On the other hand, if the maximum tolerance is set to be 10, two of the three correct iterations would be classified as "Fair."

Chapter 10
Exergames

Traditional prescriptive rehabilitation exercises would work best in the clinical setting under direct supervision of a clinician. However, often it is necessary for a patient to perform additional repetitions of the prescribed exercises at home. A typical practice is for the patient to be given a written description of the exercises and perhaps video demonstrations as well. It is not surprising that the rate of adherence to the prescribed amount of repetitions is low. The low adherence rate may have many reasons, but lack of motivation to follow through with the prescribed exercises is obvious. The lack of motivation is caused by several factors:

- Written instruction on how to perform a prescribed exercise is hard to follow. A video instruction would be better, but it still leaves a lot to be desired.
- The patient receives no feedback on the exercise performed and the patient would have to perform repetition count.
- Doing the prescribed exercise, which is intrinsically repetitive, could be boring without any entertaining elements.
- The patient will not be held accountable for not adhering to the prescribed exercises because there is no way for a clinician to know the truth.

By incorporating some game mechanics, the prescribed exercises would be made much more attractive to the patients. This would require some form of motion tracking so that the patient's movement can be detected and evaluated, and some form of game-like user interface, such as a 3D avatar, displayed on a computer screen or the TV, with real-time instruction and feedback. This is a simple form of virtual reality because it provides a virtual environment during the game play so that the game environment is to some degree immersive to its user. This type of exercises is referred to as exergames or serious games. Some also used the term active video games. The model we use for exergame is shown in Figure 10.1.

The integration of the game mechanics with prescriptive exercises provides a compelling solution to the patient adherence problem faced by clinicians and physical therapists. The benefits of exergames can be roughly divided into three categories: (1) education, (2) psychosocial, and (3) clinical. The educational benefits and the psychosocial benefits are facilitated by the game mechanics that will help achieve the clinical benefits, which are what ultimately matter to the speedy rehabilitation of the patients.

Exergame

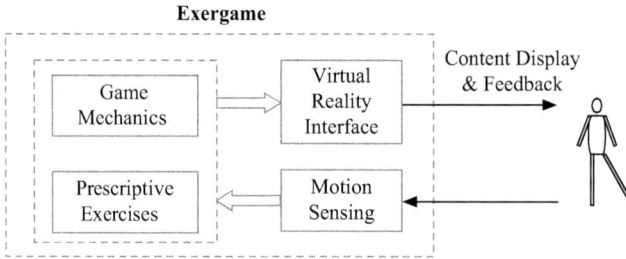

Figure 10.1   A model for exergame

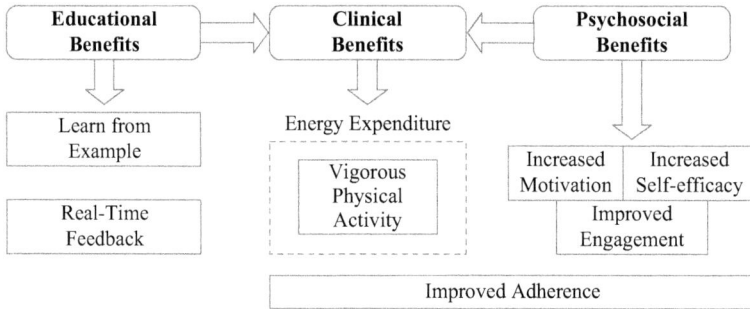

Figure 10.2   A taxonomy for the benefits of exergames

As shown in Figure 10.2, educational benefits include providing the patient with step-by-step video instruction on how to perform each task, and providing real-time feedback on the quality of each task performed by the patient. Furthermore, an important game mechanics is the design of a sequence of tasks that have progressively more difficulty levels so that the patient who is at different stages of recovery could find tasks that are sufficiently challenging and yet could still manage to complete in a satisfactory manner.

The game mechanics and the virtual reality interface directly provide a pleasant and fun environment for the patient to engage in the physical activities in the form of completing carefully planned tasks with detailed instruction and real-time feedback. Due to the range of difficulty levels provided, the patient could always complete some of the easier tasks, which boosts the patient's self-efficacy in recovery and the motivation for doing physical activities. This in turn would lead to improved engagement in performing the prescribed physical activities. Ultimately, this would lead to improved adherence both in terms of the amount of physical activities and the quality of performing rehabilitation exercises. There are also reports on reduced stress and depression levels as side benefits of exergames, which can also be regarded as psychosocial benefits.

The clinical benefits are determined based on the outcome of the intervention using exergames. One apparent clinical benefit is increased exercise expenditure (sometimes also referred to as energy expenditure). For cardiac rehabilitation, which is to improve the health of patients with cardiovascular disease and patients with low cardiorespiratory fitness, maintaining sufficient level of exercise expenditure (moderate to vigorous physical activity) is necessary [335]. This is also the outcome of good adherence to the prescribed rehabilitation program. That is why improved adherence is listed under both psychosocial and clinical benefits.

10.1 Commercial game-console-based exergames

Many studies have used commercially available games running on game-consoles as exergames, especially prior to the availability of the Microsoft Kinect sensor. The first motion tracking accessory to PlayStation 2 (a game console from Sony) is called EyeToy. EyeToy was released on October 2003 and is essentially a webcam. The supported games would use computer-vision algorithms to perform gesture recognition based on the color frames captured by EyeToy. However, its support is limited.

In 2006, Nintendo released its Wii game console, which came with a wireless remote controller integrated with inertial sensors (referred to as Wiimote or later Wiimote Plus). Wiimote enables the tracking of hand movements of the player during game play. In addition to Wiimote, Nintendo introduced Wii Balance Board for its Wii Fit game, which has been used as a low-cost device to measure center of pressure for balance study. In 2010, Sony released its PlayStation 3 game console, which came with a similar motion sensing remote controller called Move Motion Controller.

In 2010, Microsoft released an add-on motion sensing gadget called Kinect to its Xbox game console. Instead of using inertial sensor, initially Kinect uses a depth-sensing computer-vision technology for full-body motion tracking. A number of Kinect games were commercially released, such as Kinect Sports season one and season two.

10.1.1 Wii

In [336], the Wii Fit game was used in conjunction with Wii balance board in a 6-week intervention program for senior adults with balance deficits. Some participants demonstrated significant improvements as measured by the Berg Balance Scale [337]. In [338], the Wii Fit game and the Wii balance board were used in a 6-week intervention program for patients with Parkinson's disease (PD). The outcome was assessed using a number of standard clinical tests, including the sit-to-stand test, timed-up-and-go, Tinetti performance-oriented mobility assessment [339], 10-m walk test, activities-specific balance and mobility assessment, and unimodal stance duration, and with significant improvement. In [340], the Wii Fit game was used in a 12-week intervention program for a single patient with CREST syndrome and multiple other chronic conditions. The program demonstrated improved results in the 6-min walk test (6MWT), timed up and go (TUG) [341], 30-s sit to stand, hand grip strength, ankle plantarflexion and dorsiflexion, Tinetti performance oriented mobility assessment, resting systolic blood pressure, and oxygen saturation.

In [342], the Wii Fit game and the Wii balance board were used in an 8-week intervention program for patients with PD. The outcome was evaluated using the Berg Balance Scale [337] and the Dynamic Gait Index [343] with statistically significant improvements. However, the study reported that there is significant changes on patient ratings for the activities-specific balance confidence [344] and the geriatric depression scale [345]. In [346], the Wii Fit game and the Wii balance board were used in a 3-month intervention program to improve balance in order adults. The outcome of the program confirmed the effectiveness of the program with improved score using Berg Balance Scores. In addition, the walking speed has also been increased.

In [347], the Wii Fit game and the Wii balance board were used in a 6-week intervention program to reduce fall risk for a single 76-year-old patient with bilateral peripheral neuropathy. The patient was observed to have improved scores on Berg Balance Scale, activities-specific balance confidence, and TUG.

In [348], the Wii Sports game was used in an 8-week intervention program for patients with PD. The primarily outcome of the program was assessed using the Nottingham Extended Activities of Daily Living Test (NEADL) [349], and second outcome was evaluated using several other clinical scales, including Unified Parkinson's Disease Rating Scale (UPDRS) [350], the 9-hole peg test [351], the Purdue Pegboard Test [352], a timed tapping test, TUG, Hamilton Depression Scale (HAMD) [353], PD quality of life scale (PDQ-39) [354]. The observed improvements in these scales show that the intervention program is effective in improving both motor and non-motor functionalities and the quality of life of the participants with PD.

In [355], the Wii Sports game was used in a 2-week (6 sessions) intervention program for stroke patients. The outcome of the program was evaluated using the Fugl-Meyer Assessment of upper Limb Motor Function [356], Motricity Index, Modified Ashworth Scale (MAS) for assessment of muscle tone of the shoulder, elbow and wrist [357], and visual analogue scale for upper limb pain.

In [358], the Wii Sports game was used in a 6-session (1 h each) intervention program for a single 89-year-old patient with balance disorder. The outcome shows improvements in Berg Balance Score, the Dynamic Gait Index, TUG performance, and the Activities-specific Balance Confidence Scale.

In [359], the Wii Sports game was used in a 12-week intervention program with a goal to increase the exercise capacity and daily physical activity for patients with heart failure. The post-test shows increased exercise capability in slightly over half of the patients compared with the pre-test result. However, there is no observed increased daily physical activity.

Although the pre- and post-assessment for many studies have demonstrated improvements in several clinical scales, randomized trials demonstrated conflicting results regarding the efficacy of Wii-based exergames. This is probably heavily depends on the activities arranged for the control group or the lack thereof.

In [360], a 6-week trial was conducted with 41 order adults to test the efficacy of using the Wii Fit Plug game as the intervention. Each participant in the experimental group (with 19 participants) was provided with a Wii game console, a Wii balance board, and the Wii Fit Plug game. The group was encouraged to conduct three unsupervised intervention sessions per week (each session approximately lasts about

30 min). The control group (with 21 participants) was expected to live their life as usual without any specific intervention. The outcome was assessed with nine clinical tests (TUG, functional reach, lateral reach left/right, single leg stance left/right, 30-s chair stand, gait speed, and the Iconographical Falls Efficacy Scale [361]), and the experimental group exhibited better improvements than the control group in five of the nine tests.

In [362], an 8-week randomized trial was conducted to test the effectiveness of using Wii Fit as an intervention to improve the balance and gait function. A total of 30 older adults participated in the study and they were randomized assigned to the experimental group and the control group. The experimental group conducted 30-min exercise three times per week using the Wii Fit game (Soccer Heading, Snowboard Slalom, and Table Tilt). The control group engaged in a physical ball exercise for the same number of sessions and the same duration in each session. Both groups showed improvement in post-trial assessment with decreased sway length, average sway speed, and TUG time. However, the experimental group exhibited more significant reduction.

In [363], an 8-week trial was conducted with 32 healthy older adults to assess the effect of using the Wii Fit game as the intervention for balance training. The experimental group performed three 30-min sessions of Wii Fit training per week. The control group received no intervention. The outcome was assessed using Romberg Test [364], and the experimental group demonstrated significant better performance in balance than the control group.

In [365], a 10-week randomized trial was conducted to test the effectiveness of the Wii Fit-based training for community-dwelling older adults. The experimental group consists of 28 participants and participated two sessions of training per week. Each session has two components of intervention, one is for postural balance and another is for muscle conditioning. For postural balance training, a participant could choose one from five balance exercises, i.e., table tilt, slalom ski, perfect 10, tight rope tension, and penguin slide. For muscle conditioning training, the standing rowing squat was used. A training session lasted for about 70 min. The participants in the control group (with 30 participants) were asked to wear the ethylene vinyl acetate copolymer insoles during the period of trial. The post-trial evaluation showed that the experimental group had higher maximal voluntary contraction strength than the control group. The experimental group also obtained better performance in the rate of force development, TUG test, the short form of the Falls Efficacy Scale-International [366], and 30-s repeated chair stand test [367]. However, the two groups have similar center of pressure velocity moment.

In [368], 40 senior adults (between 60 and 95 years of age) were recruited to participate in a randomized 6-week trial. The experimental group received Wii Fit balance training, and the control group received no intervention. The outcome of the trial is evaluated using the 8-foot Up and Go Test [369] and the Activities-Specific Balance Confidence Scale [344]. The experimental group demonstrated significant improvement than the control group.

In [370], a 12-week trial was conducted to test the efficacy of using Wii games to improve the physical and cognitive function in older adults (one additional week was used to do pretest and another week was used to perform posttest). The experimental group completed two 1-h session per week. In each session, pairs of participants played Wii Tennis or Wii Boxing (part of Wii Sports), and this was followed by practicing with the Wii Soccer Headers, Wii Ski Jump, and Will Marbles games (part of Wii Fit). During the final period of the training session, participants played four games, including Wii Ski Slalom, Wii Hula Hoop, Wii Trampoline, and Wii Tennis Return of Serve (Wii Fit and Mario & Sonic on Olympic Games). The control received no intervention. A set of clinical tests were conducted after the trial. The experimental group demonstrated better performance than the control group in measures of physical function and cognitive measures of executive control and processing speed. However, the two groups did not differ on visuospatial measures.

In [371], 12 healthy senior adults were recruited to participate in a randomized 3-week trial. The experimental group received Wii Fit training, and the control group received no special training (i.e., the group continued their normal activities). The outcome was evaluated using Berg Balance Scale, Fullerton Advanced Balance Scale, Functional Reach, and TUG. The experimental group exhibited significant improvement on the Berg Balance Scale. However, no significant changes were observed for both groups on other measures.

In [372], the Wii Fit game was used to study the efficacy of using the exergame with respect to traditional physical therapy to improve balance control for older adults. The study recruited a total of 17 participants and they were divided into three groups: (1) a group that receives both physical therapy and Wii Fit training, (2) a group that receives only Wii Fit training, and (3) a group that receives only physical therapy. The intervention outcome was evaluated using the Berg Balance Scale and the Bubble Test. Wii Fit training failed to show any superiority than physical therapy. Several other randomized trials [373–376] using Wii games also showed no statistical difference in the improvements after the trial for the experimental group using exergames and those for the control group.

10.1.2   Xbox

In [377], a 12-week randomized trial was conducted to test the efficacy of using the Kinect Sports game for improving the memory function in older adults. The experimental group (with 23 participants) engaged in two 1-h training sessions per week using Kinect Sports. The control group (with 22 participants) also participated two 1-h training sessions with low-intensity conventional multimodal exercises under supervision. In each session, the participants played each of the six games in Kinect Sports. The study identified specific fitness components for each game. The 10-pin bowling could improve shoulder flexibility balance, the boxing game could improve bilateral arm flexibility, agility, endurance, and balance. The track and field game could improve all-round flexibility, endurance, agility, and the lower body strength and balance. The table tennis game could improve all-round flexibility and agility. The soccer game could improve the lower body strength and agility. The beach volleyball could improve upper body strength and flexibility. The efficacy was measured using several clinical tests, including 6-min walk, TUG, dynamic balance, functional reach, mini-mental state examination, n-back task, and the modified troop task. The experimental group exhibited significant improvement in the 6-min walk, TUG, dynamic balance, and functional reach. The experimental group also provided higher number of total number of correct responses on the Stroop task than the control group.

In [378], a 4-week single-blind randomized controlled trial was conducted with 20 post-stroke patients using two commercial Xbox Kinect games, one is the Kinect Sports game (only Bowling was used in the study) and another is the Dr Kawashima's Body and Brain Exercises game (only Mouse Mayhem was used in the study). Both games would require the use of the upper extremities. Both the experimental group and the control receive would engage in conventional rehabilitation for one hour a day and 5 days a week for 4 weeks. The experimental group would participate Kinect game practices for an additional one-hour a day and 5 days a week for 4 weeks. The efficacy of the intervention program was evaluated using several clinical scales, including Box and Blocks Test (BBT) [379] for gross manual dexterity, Wolf Motor Function Test (WMFT) [380] for the motor function of the upper extremity, Functional Independence Measure (FIM) [381] for upper extremity related functional independence level, and Brunnstrom motor recovery stage (BMRS) [218] to determine the motor recovery stage. Although both groups exhibited improvements after the intervention program, the experimental group demonstrated more significant improvements.

10.1.3   PlayStation

In [382], a 4-week randomized trial was conducted with 20 hemiparetic inpatients where the experimental group and the control group each has 10 subjects. The experimental group used the EyeToy Kung-Foo, Goal Attack, MrChef, Dig and Home-Run games to help practice flexion and extension of the paretic shoulder, elbow, and wrist, and the abduction of the poetic shoulder. The intervention outcome was evaluated using the Functional Independence Measure [383], and the experimental group demonstrated better results.

10.2   Custom-developed exergames

In this section, we highlight several studies on using custom-developed exergames in the intervention. Although there are a large number of publications reporting on systems that use various sensing modalities for motion tracking, and provide some degree of virtual reality interfaces, most of them lack sufficient game mechanics. Furthermore, the maturity of such systems is still far from that of commercial health-related games. This line of publications are mostly about the technical details of the design and implementation of the systems, assessment of the measurement accuracy of the systems, and feasibility and usability studies. Hence, we intentionally exclude these publications. We also note that randomized clinical trials using custom-developed exergames are rare.

10.2.1 IMU

In [384], four rehabilitation games were used to test the efficacy of using exergames to improve the motivation and exercise intensity. The games were developed based on an IMU-based commercially available arm rehabilitation system called Bimeo. One game is a single-player game where the player would play against a computer opponent. Two games are two-player cooperative games where two players would help each. One is a two-player game where two players would play against each other. The study recruited 29 participants with chronic arm impairment. The study consists of a single session where each participant played all four games. The Intrinsic Motivation Inventory questionnaire was used to capture the participants' subjective experience after each game. A final questionnaire about game preferences was provided to the participants at the end of the session. The exercise intensity was measured in terms of the hand velocity in each game. The study found that both motivation and intensity are positively correlated with rehabilitation outcome.

In [385], two rehabilitation games were used to facilitate self-directed arm therapy at home for out-patients rehabilitation after stroke. The study is an open-label single group 6-week study with 11 stroke patients. The rehabilitation games were developed by the authors based on the commercial ArmeoSensor system, which uses IMUs to track the arm movements [386]. One of the games is called Meteors and it was designed to improve arm workspace and reaching velocity. In the game, the participant would use his or her arm to catch meteors that were falling towards a planet. The other game is called Slingshot and it was designed to train arm coordination and to improve the precision of arm pointing and reaching movements. In the game, the participant would control a slingshot to shoot stones at static or moving targets by extending the elbow. The participants used the games at home without supervision during the 6-week. The outcome of the invention was measured using the Fugl–Meyer Assessment of the upper extremity, WMFT, kinematic metrics (i.e., arm function, training intensity, and trunk movement) derived from the motion data obtained from the IMUs. The arm function of the participants improved significantly, but there was no significant change in WMFT.

10.2.2 Kinect

Kinect and the corresponding software development kit from Microsoft are the most common tools for developing custom exergames with functionalities beyond those offered by commercial-off-the-shelf games, such as the collection of raw and processed motion data, more specific feedback to the user regarding the quality of the execution of each repetition, and the reporting to the clinicians. The user interface is often developed using the 3D Unity game engine. In general, the user interface is not as polished as those of the commercial games. The algorithms used to evaluate the user performing during game play vary, from single angle measurement, to rule-based, to machine-learning based. The feedback offered to the user also varies, from an overall single score, to more specific guidance such as areas of improvement and repetition count.

Although there have a large number of publications on the design, implementation, and occasionally human subject tests on Kinect-based custom-made systems. Almost all of such systems can be regarded as assistive technology for conducting physical therapy rather than exergames due to the focus on providing guidance to the patients and the lack of game mechanics. In general, their maturity is far from that of the commercial health-oriented games. We are not aware of any randomized trials for intervention that use custom-made Kinect systems.

In [387], a custom-made Kinect game called wipe out was used a new assessment tool for upper limb function instead of as an intervention for rehabilitation for patients with Friedreich ataxia (FRDA). The rationale for the study is that traditional clinical assessment tools are no longer effective for patients with advanced FRDA, especially for wheelchair-bound patients. The game would require the patient to clean the screen in a scene covered by fog using a tissue controlled the wrist movement of the patient. To move the wrist, the patient would make medio-lateral and inferior–superior displacements of the upper limb relative to the trunk. Such movements of the patient were recorded by the game. During the test, the patient was asked to play the games three times and the average of the three repetitions was used for statistical analysis. The study recruited 27 FRDA patients and a group of healthy subjects of similar age. Significant differences were found for the time it takes to complete the game and the accuracy between FRDA patients and the healthy subjects.

In [388], a Kinect-based exergame was developed for post-stroke rehabilitation. The exergame consists of six activities for the subjects to practice: (1) shoulder adduction/adduction, elbow and wrist extension; (2) hip flexion; (3) shoulder abduction, horizontal adduction, elbow and wrist extension; (4) hip abduction/adduction; (5) shoulder flexion; (6) hip and knee flexion. Two forms of user interfaces were experimented. One is a 3D smart TV and another is the Oculus Rift headset. A questionnaire for the two user interfaces was issued to the subjects to collect their feedback. No clinical assessment was performed.

In [389], a Kinect-based exergame was also developed for post-stroke rehabilitation. The exergame is a shooting game referred to as duck duck punch where the participant would move his or her arm (with shoulder and elbow rotation). The performance was evaluated based on the measured elbow angle and the arch percentage. Six post-stroke patients were recruited to test the efficacy of game as an intervention via five supervised therapy sessions. During each session, a participant was required to make 100 reaching movements using the duck duck punch game. After the intervention program, the participants demonstrated reduced abnormal trunk displacement, greater shoulder flexion, and increased elbow extension.

10.2.3 Wii balance board

The Wii balance board was initially released as an accessary for the Wii Fit game in late 2007 in Japan and in 2008 elsewhere. Subsequently, more Wii games were released that utilize the Wii balance board. The Wii balance board is a low-cost, wireless, and portable device to measure the center of pressure. As shown in Figure 10.3, the board has four load cells (i.e., strain gauges), one on each corner of the board. The

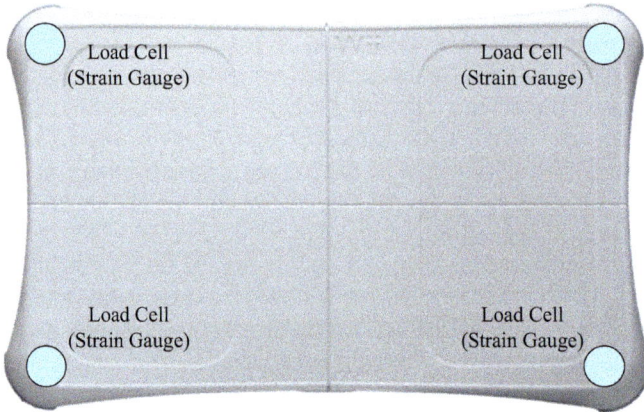

Figure 10.3   The Wii balance board can be used as a low-cost, wireless, and portable device to measure center of pressure

maximum load is 150 kg. The board communicates with the Wii game console via Bluetooth. The Wii balance board communication protocol has been fully decoded (together with that of Wiimote) (e.g., http://wiibrew.org/wiki/Wii_Balance_Board). This enables researchers to develop custom exergames that utilize the Wii balance board for research and clinical trials.

In [390], two custom-made balance games [391] were used in the study. Both games rely on the sensors embedded in the Wii balance board. The goal of the first game is to improve medio-lateral body movements. The game design is for the participant to control the basket at the ground to catch the apples falling from an apple tree by shifting his or her balance while standing on the Wii balance board. The goal of the second game is to improve both medio-lateral and anterior–posterior body movements. The game design is for the participant to control a sea-scape virtual avatar so that it could burst the rising bubbles that appear from the ground by shifting his or her balance while standing on the Wii balance board. The game virtual environments were developed using Virtools 4.0 software from Dassault Systems. Each of the two games has incremental difficulty levels that allow the participant to progress during the recovery. The custom games allow the collection of center of pressure position data during game plays. The study consists of a 5-week randomized trials with 76 healthy and fall-prone older adults. The participants were randomly assigned to either an experimental group or a control group. Each group consists of 17 fall-prone older adults and 21 healthy older adults. The control group would perform two 30-min sessions of balance training per week using the games. The control group did not participate any intervention but was required to keep a diary of daily physical activities. The assessment for the outcome of the intervention consists of the measures of balance confidence (i.e., Activities-Specific Balance Confidence Scale and the Falls Efficacy Scale [392]), balance control (i.e., Berg Balance Scale, custom static and dynamic balance tests), and multisensory function (i.e., the sound-induced flash illusion [393]). The experimental group exhibited improvements on balance and postural control. Furthermore, for fall-prone older adults, the improved functional balance was correlated with more efficient multisensory processing.

10.2.4 *Mobile apps*

A new trend is to provide user interface and guidance using a cloud-supported mobile apps running on smartphones or tablets. Some apps provide guidance based on the user's subjective input, such as the current pain level. Other apps use separate wearable devices embedded with IMUs to collect real-time motion data. A cloud-supported mobile app would collect and store data in a cloud service. The cloud service could also perform computational intensive algorithmic calculations.

In [394], a system consisting of a cloud-supported mobile app and two IMUs for orthopedic rehabilitation are described. To test the system, which is referred to as reflex, the 30 participants who have received anterior cruciate ligament (ACL) reconstruction surgery were recruited for a randomized 16-week trial. The goal is to increase the range of motion of the knee. The control group was treated in a clinical setting, and the experimental group received the same treatment at home via the re.flex system. While both group reached the full range of motion of the knee at the end of program, the experimental group recorded 31% higher adherence to the recovery program.

Kaia Health (https://kaiahealth.com/), Physitrack (https://www.physitrack.com/), BPMpathway (https://www.bpmpathway.com/), and SWORD Health (https://swordhealth.com/) are commercial platforms that facilitate remote rehabilitation using a mobile app. Kaia Health uses the smartphone's camera to track the user's motion. BPMpathway and SWORD rely on separate wearable IMU-based sensors for motion tracking.

Part III

Technology-facilitated rehabilitation

Chapter 11
Technology-facilitated physical rehabilitation

According to the National Institute of Health (https://clinicalcenter.nih.gov/rmd/pt/index.html), the goal of physical therapy is to restore or maintain sensory and motor abilities for patients who have functional impairments. Patients with a wide-spectrum of diseases and conditions can be benefited by engaging in physical rehabilitation prescribed by a certified physical therapist. Before a physical therapy can develop a plan of care, a thorough examination and evaluation of the patient's function levels are usually performed [395], which include the assessment of musculoskeletal functions, neuromuscular functions, cardiovascular and pulmonary metrics, integumentary status, communication ability, affect, and language levels, and cognitive ability.

Physical rehabilitation is one of the specializations in the larger scope of rehabilitation, as shown in Figure 11.1. In this part of the book, we will be covering occupational rehabilitation, pulmonary rehabilitation, cognitive rehabilitation, speech and language rehabilitation, and mental health rehabilitation. Among them, occupational rehabilitation has the broadest scope and it overlaps with physical and pulmonary rehabilitation (such as for the ability of walk and engage in activities of daily living), speech and language rehabilitation (such as for the ability to work and interact with others socially), cognitive rehabilitation (such as for the ability to engage in complex instrumental activities of daily living), and mental health rehabilitation (such as the ability to engage in normal social interaction). Cognitive rehabilitation and mental health rehabilitation also have overlap.

11.1 Framework for physical rehabilitation

Physical therapy is not limited to treatment of people with functional impairments [395]. The prevention of physical impairments and the promotion of wellness and healthy life styles are also within the scope of physical therapy. To facilitate a physical therapist to make the right clinical decision on the development of a plan of care, the international classification of functioning, disability, and health is used as a framework [395], as shown in Figure 11.2. This framework identifies factors important for reaching the desirable health outcomes, which usually means to attain a state of physical, mental, and social wellbeing beyond simply the absence of impairments. When an individual is not in the desirable health state, it is said that this individual

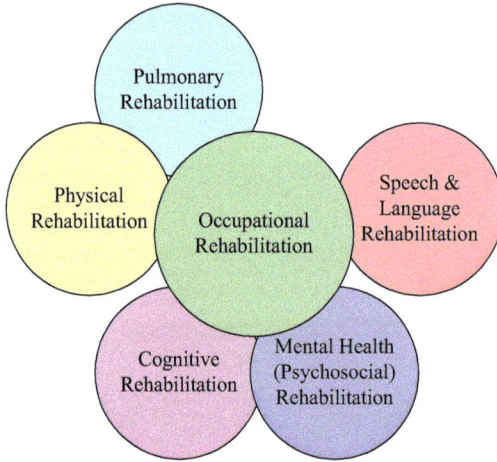

Figure 11.1   Different types of rehabilitations

has a health condition, which could mean that the individual is suffering from some disease, disorder, injury, trauma, or simply due to circumstances such as aging or stress.

The health condition is reflected as impairments in body structures and/or body functions [395]. Body structure refers to the anatomical parts of the human body such as organs and limbs. Body functions refer to physiological and psychological functions of the body. Besides the body structures and body functions, it is also useful to consider the contextual factors regarding the individual's past history and current living environment so that a more personalized plan of care can be developed.

The impairments could lead to changes to the body structures and body functions and could impact the individual's ability to carry out activities (here activity refers to the execution of a task) and to participate in social interactions. Activities of daily living (ADL) are the most common activities one engages in. Usually, ADL can be classified into basic ADL (BADL) and instrumental ADL (IADL). BADL are simpler and usually self-care activities such as eating, drinking, and dressing. IADL are more complex activities such as financial planning and online shopping.

Capacity refers to the individual's ability to perform activities and participate social interactions. Capacity qualifiers refer to the level of functioning while performing an activity considering the impairments the individual is suffering. The capacity qualifiers could be described by whether or not, and the kind of assistive devices or the environmental modifications needed to perform an activity. Performance describes what an individual does in the person's environment. Performance qualifiers refer to the extend of limitations in engaging in activities in the individual's environment.

Under the environmental factors, facilitators are positive factors (or assets) that would help an individual to perform activities, and barriers are negative factors (or disablement risk factors) that would hinder an individual to perform activities.

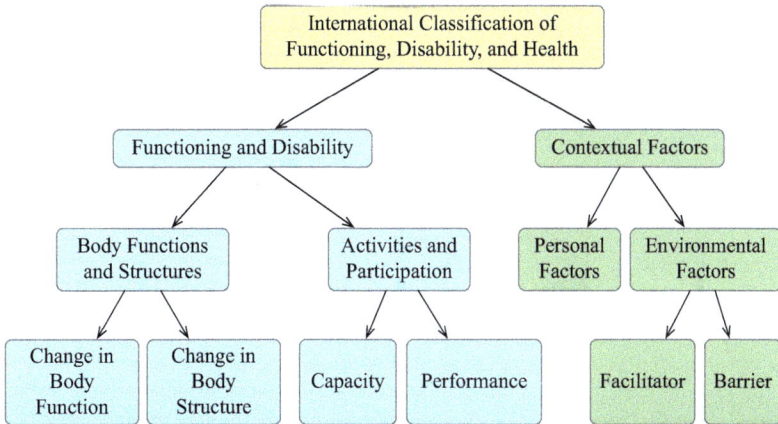

Figure 11.2   The international classification of functioning, disability, and health

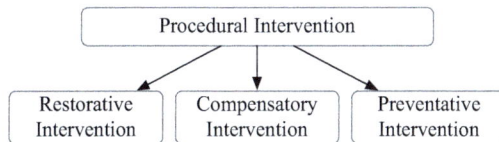

Figure 11.3   The three types of procedural interventions

Based on the tests and examinations of various functions and status of a patient, a physical therapist would proceed to providing a diagnosis and a prognosis. Diagnosis includes the determination of whether or not the patient health condition is addressable by physical therapy intervention (i.e., physical rehabilitation), and the identification of the levels of impairments with respect to the patient's movement system. Prognosis refers to the outcome in terms of the function improvements that can be attained by the plan of care and the time it takes to obtain the outcome. The primary component of the plan of care is intervention. The common intervention categories [395] include patient instruction, airway clearance techniques, the use of appropriate assistive technology, the use of biophysical agents such as electrical stimulation, light therapy, ultrasound, and iontophoresis, functional training for self-care and in everyday life, the application of integumentary repair and protection techniques, manual therapy techniques, motor function training, and therapeutic exercises.

Procedural interventions are typically used as part of the plan of care. Procedural interventions can be divided into three types of interventions as shown in Figure 11.3, i.e., restorative intervention, compensatory intervention, and preventative intervention. Restorative interventions are aimed at improving the motor function of the individual towards lifting the activity limitation and participation restrictions. Compensatory interventions are aimed at developing new motor patterns using remaining motor elements of the individual such that the individual could perform the desirable

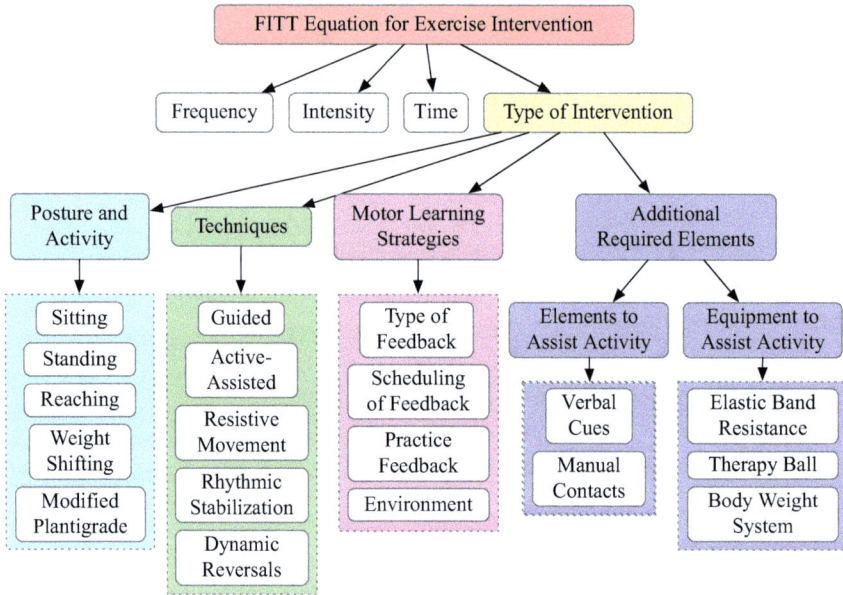

Figure 11.4   The frequency, intensity, time, and type of intervention equation for exercise intervention

function. Preventative interventions are aimed at maintaining good health and reduce the likelihood of further impairments.

The actual intervention for physical rehabilitation is often implemented according to some scheme. One of the most common schema is the frequency, intensity, time, and type of intervention (FITT) equation for exercise intervention, as shown in Figure 11.4. Here frequency means the number of treatment sessions per week. The intensity means the number of repetitions required for each prescribed exercise. Time means the total duration of the entire intervention program, and the length of each session. The type of intervention refers to the specific intervention strategies used in the plan of care. Regardless of the types of intervention used, typically four components are included: (1) what posture and activity are to be carried out during each treatment session; (2) what specific techniques would be used; (3) what learning strategies in terms of feedback types would be used; and (4) what additional elements are required, if any, during the treatment.

11.2   Motor control and motor learning

The principles of motor control and motor learning are the basis for making plan of care to improve the motor function of the patients. Figure 11.5 shows the key steps and elements in motor control. A prerequisite for human movements is the determination

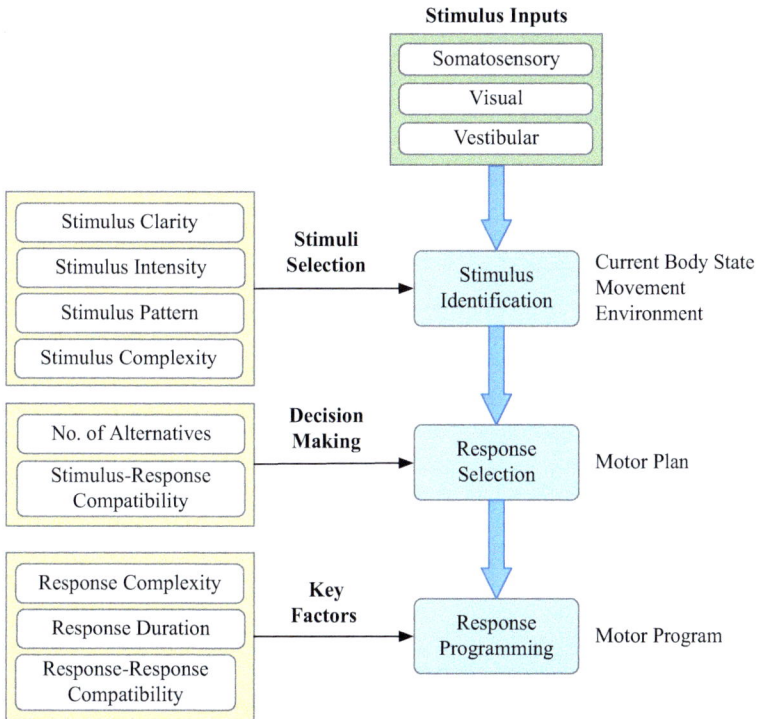

Figure 11.5 The main steps and elements in motor control

of current body state, current movement if any, and the environment [395]. This step is referred to as stimulus identification. In this step, various stimulus inputs, including somatosensory, visual, and vestibular inputs, are used to identify the current state by the central nervous system. This step is influenced by the clarity, intensity, pattern, and complexity of the stimuli. The next step in motor control is to determine a motor plan based on the selected stimuli and the current state of body. The central nervous system must consider the number of alternative motor plans and the stimulus–response compatibility and decides on the most optimal response. The final step in motor control is develop a motor program, which is referred to as response programming. In this step, the control nervous system would have to consider the response complexity, response duration, and the response–response compatibility.

Motor learning refers to how an individual learns a new motor skill. This is a complex process where the central nervous system would integrate and organize various spatial and temporal functions [395]. The closed-loop control theory [396] and the schema theory [397] are the two dominating theories of motor learning. According to the closed-loop control theory, the central nervous system would make continuous correction while the body is making a particular movement according to the sensory feedback received and the feedback received is compared with a perception trace of the intended movement in memory. The hypothesis of the scheme theory is that a set of rules is stored in memory for each movement and an individual learns motor skills by learning the corresponding rules about how to move the body. There are three stages of motor learning, namely, cognitive, associated, and autonomous. During the cognitive stage, an individual would learn what to do to complete a task. During the associated stage, an individual would learn how to complete the task by practicing movements for the task and refining the motor program for the task. During the autonomous stage, the individual would learn how to complete the task with high quality and with little cognitive efforts by making further practices of the movements and refinement of motor responses.

Motor skills can be categorized from different perspectives. As shown in Figure 11.6, the categorization of motor skills can be divided based on the muscle size, task boundary, environment, motor complexity, and multi-tasking [395]. Based on the muscle size, there are gross motor skills and fine motor skills. Gross motor skills involve large muscles and the movements are larger. In contrast, fine motor skills involve small muscles and the movements are smaller. Fine motor skills require more precision than gross motor skills. Based on the task boundary criteria, motor skills can be divided into discrete and continuous motor skills. Discrete motor skills refer to the skills needed to complete tasks that have well-defined beginning

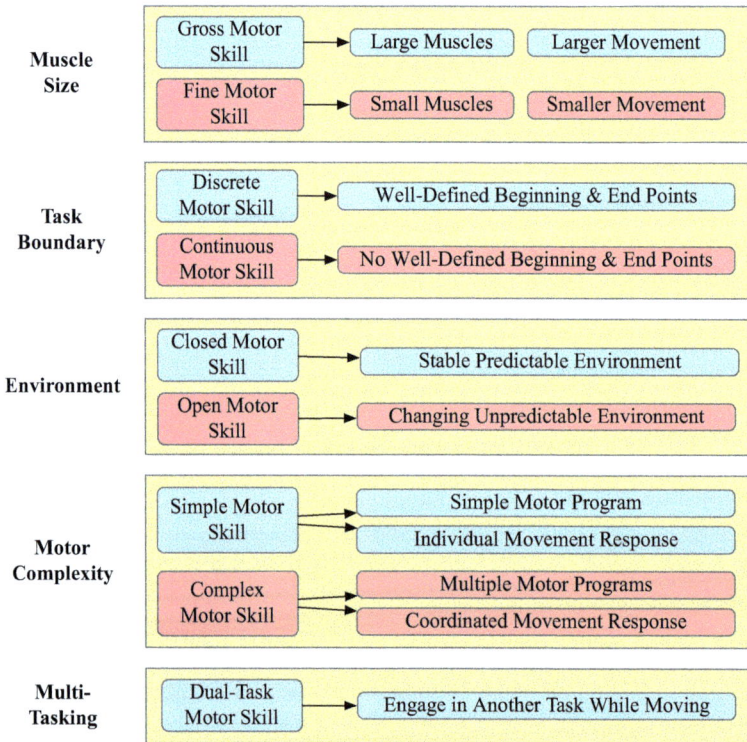

Figure 11.6   Common categories of motor skills

and end points. Continuous motor skills refer to the skills needed to complete tasks that do not have well-defined beginning and end points. Based on the environment where the tasks are performed, motor skills can be divided as closed and open motor skills. Closed motor skills are used to complete tasks in stable and predictable environment, whereas open motor skills are used to complete tasks in quickly changing and unpredictable environment. Based on the motor complexity, motor skills can be divided into simple and complex motor skills. Simple motor skills involve simple motor program containing individual movement response, whereas complex motor skills involve multiple motor programs containing coordinated movement response. All previously mentioned motor skills are assumed to be performed one task at a time. However, there are situations where it is desirable for an individual to engage in multiple tasks concurrently. Dual-task motor skills refer to the skills where an individual is capable of doing two tasks concurrently (such as holding/drinking a cup of water while walking).

11.3   Interventions for improve motor function

Motor control and motor learning theories form the foundation for designing interventions to improve motor function. Major types of interventions are summarized in Figure 11.7. The interventions may be geared towards three levels of rehabilitation targets. At the highest level is to promote recovery of motor function through restorative interventions. The primary means of intervention is task-oriented training. Task-oriented training should be guided by motor learning strategies, where practice and feedback are essential to achieve the rehabilitation goal. Task-oriented training aims to improve the patient's functional mobility skills, upper extremity skills, motor functions for activities of daily living, basic functional mobility skills, and gait and locomotion. In clinical-based intervention, it is important to provide a

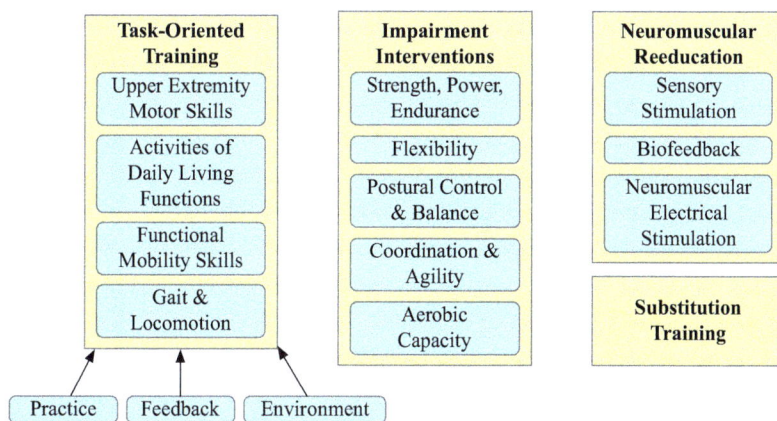

Figure 11.7   Different levels of interventions to improve motor function

supportive environment. Another level of rehabilitation target is to improve specific body functions and/or structures. At this level, there are impairment interventions and neuromuscular reeducation. Impairment interventions include strength training, flexibility training, postural control and balance training, coordination and agility training, and aerobic training. Neuromuscular reeducation is to stimulate muscles directly via electrical signal or biofeedbacks without via motor exercises. When it is unlikely for the impaired body parts to recover, then substitution intervention may be the only practical approach.

Common practice types, practice sequence, practice order, and practice strategies are summarized in Figure 11.8. In massed practice, practice sessions are conducted repeatedly with only a short rest period in between sessions. In distributed practice, practice sessions are conducted with a much larger rest in between sessions. During each practice session, how tasks are arranged are referred to as practice sequence or practice order. In blocked practice, an individual would practice one task repeatedly and then proceed to another task. In random practice, an individual would practice a given set of tasks in some random order. In blocked order, a set of tasks are completed in order where each task may be repeated a few times. In serial order, a group of tasks is practiced sequentially and then repeated as a unit. In random order, the set of tasks is completed in random order. There are four practice strategies. In mental practice, beyond the actual practice session, an individual is encouraged to image and visualize the actual tasks. It is also common for the individual to engage in some simple tasks (i.e., lead-up activities) as warm-up prior to the actual tasks. Part-whole practice is usually used for highly complex tasks, where an individual would practice parts of the tasks separately prior to the practice of the entire tasks. Finally, in transfer training, an individual would practice some tasks in order to enhance the skills needed for other tasks.

Feedback plays a critical role on the effectiveness of motor learning. This is important not only in traditional physical rehabilitation but also to guide the development of rehabilitation technology. Technology facilitates the creation various forms of feedback to the individual in physical rehabilitation. There are two types of feedback. One is intrinsic feedback, referring to the feedback naturally generated during the movement. Another is extrinsic feedback, referring to the external signals provided to help the individual to perform the task correctly. Extrinsic feedback is more often referred to as augmented feedback. Intrinsic feedback includes proprioceptive, visual, vestibular, and cutaneous signals. Extrinsic (augmented) feedback includes visual, auditory, and tactile cues.

Common types of feedback and feedback schedules are shown in Figure 11.9. Concurrent feedback is provided to the individual while the person is practicing a particular task as part of physical rehabilitation. Terminal feedback is provided after the individual has completed a practice. There are two forms of terminal feedback, namely, immediate feedback and delayed feedback. Immediate feedback is provided to the individual right after practice is over. Delayed feedback is provided after the practice with a short period of time to allow the individual to self-reflect. Feedback may be provided with less frequency. Summary feedback is provided after several practices have been completed. Summary feedback can be provided in two forms,

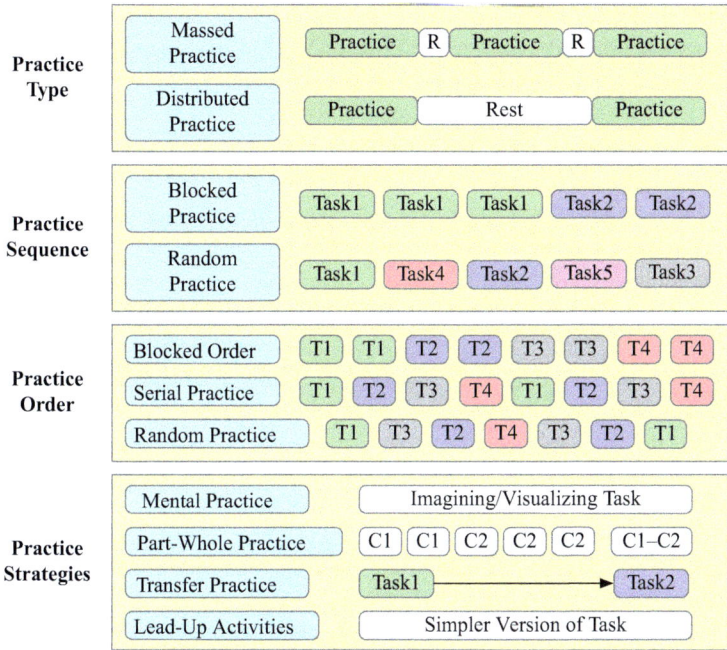

Figure 11.8   Common practice types, practice sequence, practice order, and practice strategies

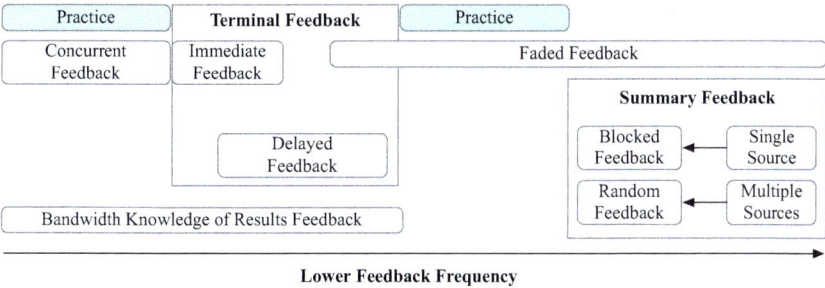

Figure 11.9   Different types of feedback and feedback schedules

blocked feedback and random (i.e., variable) feedback. Blocked feedback contains a single source of feedback, whereas random feedback contains multiple sources of feedback. Faded feedback is provided according to a predefined schedule. In faded feedback, feedback is provided immediately initially after every practice, then, feedback is provided with less and less frequency. It is also possible to provide feedback only when the practice is done wrong (i.e., deviated from some predefined boundary). Such feedback is referred to as bandwidth knowledge of results feedback.

11.4 Technology in physical rehabilitation

Technologies have been used pervasively for the testing and examination of various functions for physical therapy. Because physical rehabilitation focuses on the restoration of an individual's movement capability, objective measurement of the movements becomes a great opportunity for technology to play an essential role. In the previous ten chapters, we have elaborated: (1) various sensing modalities that have been used to measure motion, (2) various methods that have been developed to facilitate the motion measurements, to recognize the activities (including the assessment of the quality of the movements and the repetition counts), and (3) integrated systems in the form of exergames with virtual reality interfaces to not only automatically measure the movements, recognize the activity but also help make the physical rehabilitation tasks more engaging and entertaining. In addition, telerehabilitation, which can be used for all forms of rehabilitation, has also been used to conduct physical rehabilitation remotely [398]. The methods for conducting telerehabilitation range widely, from using short messaging, telephone calls, to using a Web interface, to videoconferencing, to using virtual reality systems [398], to using mobile apps with smartphones [399].

11.4.1 Augmented reality in physical rehabilitation

Beyond the technologies that have been discussed in previous chapters, augmented reality has been incorporated into physical rehabilitation in recent years [400]. Augmented reality is sometimes referred to as mixed reality. Augmented reality is a sophisticated technology that superimposes virtual objects into the physical real environment [401]. In contrast, virtual reality consists of entirely of virtual objects and environment. Augmented reality makes it possible to provide real-time visual cues to an individual either during game play, training [402], or various rehabilitation programs. There is a spectrum of augmented reality solution. At the higher end of the spectrum, 3D objects can be injected into the physical environment seamlessly on-demand in real-time. Such high-end augmented reality can be rendered by commercial products such as Google Glasses and Microsoft HoloLens. At the lower-end of the spectrum, the scene displayed to an individual is mostly virtual, but integrated with motion data that reflect the physical movements of the individual. This type of augmented reality can be rendered with commercial products such as Microsoft Kinect (the virtual scenes are usually designed by 3D game engines such as Unity). Exergames often belong to this type of low-end augmented reality applications and it is more commonly regarded as virtual reality applications. Some studies used custom-made solutions to achieve augmented reality towards the middle of the spectrum using depth-sensing devices such as Microsoft Kinect and a projector (to produce virtual objects) [403].

In [404], a low-end of augmented reality system was used in an exercise program to prevent muscle loss in the elderly. The system was based on Kinect v2 with a user interface composed by the color image frames of the actual physical environment with two subjects superimposed by the stick-figure skeletons detected by the Kinect software development kit. Additionally, the scene consists of some virtual objects such as grass, feedback information, and a menu. The application includes four workouts: (1) upper body resistance exercise, (2) lower body resistance exercise, (3) aerobic exercise, and (4) flexibility exercise. The system would measure the abduction and flexion angles of the shoulder, trunk, hip, knee. This randomized controlled study confirmed that the system is effective in achieving the goal of the study, particularly the exercise self-efficacy.

In [405], HoloLens were used to provide visual cue to patients with Parkinson's disease who suffer from the freezing of gait issue. The freezing of gait issue refers to lack of forward progression of the feet while walking [406], even when the individual's intention is to keep walking. This happens more commonly after the individual makes a turn. A hypothesis is that external cues could be provided for such individuals to alleviate the issue [407]. To test the hypothesis, a visual cue is continuously displayed via the HoloLens during each trial. More specifically, the cue consists of a sequence of small yellow spheres and a large yellow broken sphere resembling a mouth. As the subject walks forward, the mouth would eat the small yellow spheres one by one. The body of the subject is instrumented with a set of IMU sensors for post-trial analysis. The study showed that the proposed approach failed to improve the symptom of freezing of gait.

In [403], a custom-made mixed reality system was used in a study that aims to improve the arm, hand, and finger function in patients with chronic stroke. The system incorporates a Kinect sensor and a project facing downward to a flat desk surface where the virtual objects will be displayed. During the trial, the subject was asked to place his/her arm on the surface and interact with the virtual objects. The Kinect sensor was used to detect the movement based on depth sensing. The system facilitates a wide-range of exercises with planar arm and hand movements. These movements involve the flexion and extension of the elbow, the wrist, and the metacarpophalangeal joints. Such motor function is essential for activities of daily living. The study attributed the effectiveness of the system to several factors. First, the extrinsic visual, auditory, and tactile feedback provided by the augmented reality system facilitates motor learning. Second, the system increases the engagement level during the exercises due to the real-time feedback provided to the participants. Third, the system makes it possible to provide personalized exercise plan for each participant.

In [408], a low-end augmented reality system was used in a randomized controlled trial for improving balance and mobility in elderly. The system was based on Kinect sensor with a scene that integrates the color frames of the user and some virtual objects [409]. The system has three games. The balloon game focuses on hip flexion and extension while the user is reaching for the falling balloon. In the cave game, the user would need to flex and extend the knees while navigating in the cave. In the rhythm game, the user is expected to step on a specific point that is displayed on the virtual floor rhythmically. The rhythm game is designed to improve the one-leg standing ability. The result showed that the participants the experimental group with the augmented reality system attained better outcomes in terms of the fall risk index (i.e., better balance ability and lower limb function). This is attributed to several characteristics of using the augmented reality system: (1) the participants would engage in whole-body movements to maintain balance; (2) the visual feedback provided by the system encourages the participants to make the necessary adjustment towards the goal of the game; (3) the participants were more engaging during the exercises with the system.

In [410], a relatively high-end augmented reality device called visual–auditory walker was used in an at-home training program aiming to improve gait in patients with Parkinson's disease. The device consists of a head-mounted micro-display (i.e., augmented reality goggles), earphones, and a motion data collection unit. The goggles would display a virtual checkerboard-like floor over the physical world as the visual cue. Via the earphones, auditory cue is provided to the participant that reflects the rhythm of the steps. The visual cue also reflects the participant's movement and would move along with the participant. The participant is supposed to step on the tile in the virtual checkerboard, which promotes long strides while walking. The motion data collection unit consists of an accelerometer, a compass, and a microcontroller. No technical details are provided in the paper on how the motion data collected are analyzed. Presumably, the walking rhythm information is extracted from the unit and used to provide the visual and auditory cues. The study confirmed that the visual and auditory feedback from the augmented reality system improved gait velocity by 18–20%.

In [411], a custom-made augmented reality setup was used in an intervention designed to alleviate the phantom limb pain. The augmented reality setup consists of: (1) a conventional webcam and a computer monitor that displays the mirrored participant; (2) a virtual arm is superimposed on the actual mirrored frames displayed on the monitor. To make sure that the virtual arm is shown in the anatomically correct location of the missing arm and has the proper orientation, a fiducial marker is placed on the stump of the participant to indicate the exact location where the virtual arm should be placed, and the myoelectric pattern recognition is used to detect the muscle activation of the stump and guide the movement of the virtual arm. The goal of intervention is to reduce the phantom limb pain by promoting phantom motor execution. The tasks during each session of the intervention include a racing car game using phantom movements, and a posture matching task where the patient should control the virtual arm to match the target postures. The study found that the augmented reality-based intervention reduced phantom limb pain by 50%, and also reduced the negative impact of the phantom limb pain in activities of daily living and sleep by 50%.

In [412], an intervention program that incorporated augmented reality and functional electric stimulation with treadmill was used to improve the muscle strength, balance, and gait of stoke patients. The augmented reality is implemented via a head-mounted display. The display shows two views side-by-side. One view shows the actual movement of the participant while walking on the treadmill. The other view shows the target movement of a normal person. The participant can compare with his/her own gait with the ideal movement and makes the necessary adjustment. To apply electric stimulation, an electrode was taped into the proximal part of the tibia's anterior muscle and the distal part. The effectiveness of augmented reality and electric stimulation are examined by comparing the outcomes of three groups: (1) augmented reality with electric stimulation; (2) electric stimulation only (without augmented reality); and (3) treadmill only (without electric and stimulation and without electric stimulation). The study showed that the first group exhibited the most improvements on the muscle strength, balance, and gait speed. A follow-up study from the same research group improved the design by triggering electric stimulation using EMG signal instead of via a manual switch [413]. A further follow-up randomized controlled trial was reported in [414] using the same setup.

A large number of studies have incorporated Kinect with a low-end augmented reality interface for various physical rehabilitation and they have been thoroughly reviewed [169,415,416]. These studies have targeted patients with stroke [417–421], patients with Parkinson's disease [422,423], patients with cerebral palsy [424,425], patients with cystic fibrosis [426], patients with multiple sclerosis [427], patients with Leukodystrophy [428], patients with burn injury [429], children with ataxia [430], and the elderly [213,431,432].

11.4.2 *Smartphone use in physical rehabilitation*

In [433], an Android smartphone was used to measure the posture during a balance task in a study on the effectiveness of using biofeedback in balance training. The system consists of four components: (1) torso tilt measurement using the smartphone placed on the participant's waist (L2–L4 lumbar spine region); (2) data processing running at a PC; (3) a computer monitor displaying visual biofeedback; (4) a commercial device to provide kinesthetic biofeedback. Both the visual biofeedback and the haptic biofeedback are generated based on the data collected from the smartphone by the computer program. The torso tilt measured by the smartphone (via an Android app) is in terms of the mediolateral and anteroposterior angles, presumably computed based on the accelerometer and gyroscope motion data. The torso tilt reflects the degree of body sway during the stance and gait.

In [70], the accuracy of an iPhone app designed to measure the angular movement was reported. Instead of performing human subject tests with the app, the iPhone running the app was mounted to the movement arm of a commercial dynamometer device (Biodex isokinetic dynamometer system 4). The iPhone app uses the built-in 3-axial accelerometer to collect the motion data. During the test, the dynamometer simulates uniplanar knee flexion and extension motion. The dynamometer was used to establish the ground truth on the angular movement. The study confirmed the high accuracy and precision of the iPhone app.

In [434], the feasibility of using an Android smartphone app to assess the degree of chronic ankle instability was studied. The app uses the smartphone's built-in accelerometer to collect the acceleration data. The higher amount of motion detected, the stronger the instability. Fifteen human subjects were recruited for the feasibility study. Every participant has ankle instability in one leg and the other is absent of the issue. During the tests, the participant was required to stand in one leg for 20 s in eyes-open and eyes-closed conditions with each leg. The smartphone was attached to the middle of the shin of the leg being tested. The average acceleration for each leg and for each condition were calculated and the correlation was computed. The study found strong differences between the injuries leg and the healthy leg.

In [435], a smartphone app designed for functional mobility assessment was validated with a human subject trial of 32 healthy elderly people. More specifically, the app was used to measure the participant performance in two clinical tests, one is five-time sit-to-stand (FTSTS) [436], and the other is timed up-and-go [341]. The functional mobility level is an important factor related to the fall risk [437], quality of life [438], and mortality [439]. During the FTSTS test, an individual is asked to perform sit-to-stand movements consecutively for five times as quickly as possible. The time it takes to complete the FTSTS test is used as the metric of the motor function level. During the TUG test, an individual is asked to sit in a chair with arms placed on the armrests as the initial position, then stand up from the chair, walk forwards for three meters, turn back, walk towards the chair, and finally sit down. The individual is asked to complete these movements as quickly as possible, and the time it takes to complete the task is used as the metric for the motor function level. The mobile app uses the built-in inertial sensor to detection movement patterns and the initial and final orientation of the smartphone. During the test, the smartphone was affixed onto the participant's chest using Velcro straps. The mobile app has audio instructions for the participant when to start. For the timed up-and-go test, the mobile app stops taking the time when the phone returns to the original orientation. For the five-time sit-to-stand test, the app appears to be able to detect the sit-to-stand activity and the number of times the sit-to-stand is performed. The ground truth was established by placing a force sensor at the backrest of the chair used in the tests.

In [440], the effectiveness of a set of mobile games for upper limb rehabilitation in patients who have suffered from ischemic stroke was examined in a randomized controlled study. The motion data were collected via the built-in inertial sensors in the smartphone, and the graphical interface of the games was in the form of 2D cartoon-like virtual reality scenes displayed on a tablet. The smartphone was connected to the tablet via Bluetooth to transfer the motion data to the games running on the tablet. Four mobile games were developed and experimented in the study. The honey-pot-guard game was designed to induce the elbow flexion and extension. To collect the motion data, the smartphone was attached to the forearm during the honey-pot-guard game play. The protect-the-bunny game was designed to induce shoulder abduction and adduction. The put-out-fire game was designed to induce shoulder abduction and adduction, and shoulder flexion and extension. The lower-splash game was designed to induce shoulder adduction and adduction, elbow flexion and extension, and wrist pronation and supination. During the shoulder abduction and adduction game play, the smartphone was attached to the upper arm to collect motion data. During the elbow flexion and extension, the smartphone was attached to the forearm. During the shoulder flexion and extension, the participant was required to hold the smartphone. During the wrist pronation and supination movements, the participant was also asked to hold the smartphone. The outcome of the trial was measured using several clinical tests and scales, including manual muscle testing, the Fugl–Meyer Assessment of upper extremity [317], the modified Barthel index [441] (for activity limitation), part of the EuroQol scale [442] (for quality of life), and the Beck Depression Inventory [443] (for psychological state).

In [444], a mobile app-based system [445] was used to facilitate gait training in a randomized controlled study. Unlike other mobile apps, external wearable inertial sensors are used to collect motion data, and the mobile apps were used to receive the motion data from the inertial sensors, process the data, provide a simple user interface to the participant, and generate feedback and cues to the participant. The sensors transmit the motion data collected via a docking station using Bluetooth to the smartphone. Algorithms were developed to analyze the motion data to derive key gait parameters, including cadence, step length, gait speed, gait asymmetry, trunk flexion, and clearance, and to detect the freezing-of-gait problem. Two separate mobile apps were developed. One focused on gait detection (referred to as the ABF-gait app), and the other focused on the detection of the freezing-of-gait problem (referred to as the FOG-cue app). The ABF-gait app provides auditory biofeedback (i.e., ABF) to the participant. The FOG-cue app provides auditory cues to minimize episodes of freezing-of-gait (i.e., FOG). For gait training with the ABF-gait app, the two external inertial sensors were placed on the front side of the shoes. For gait training with the FOG-cue app, the two external sensors are placed above the ankles. The study found that the experimental group used the mobile apps achieved significantly larger improvement in gait speed compared with the control group. More technical details are reported for the FOG-cue app in [446].

In [447], a mobile app was used in wrist rehabilitation. The app uses the built-in accelerometer sensor to detect motion. During the wrist rehabilitation sessions, the participant would place a smartphone on the back of the hand via a removable splint. The mobile app has a graphical user interface that displays the range of motion in rotation angle, and the current exercise to be performed. Wrist rehabilitation exercises include the wrist flexion, extension, pronation, supination, radial deviation, and ulnar deviation. The study with 20 patients demonstrated that the range of motion for all wrist movements was improved.

In [448], a mobile app was developed and tested to facilitate the rehabilitation of knee proprioceptive function. The proprioceptive function refers to the ability of an individual to control a joint to reach a particular target angle. The app uses the built-in inertial sensors in a smartphone to detect the knee angle. The smartphone is connected to a vibrotactile buzzband via Bluetooth. During the exercise, the vibrotactile buzzband is wrapped around the leg and it two vibrators, where the front vibrator is on the tibia and the back vibrator is on the gastrocnemius muscle. The vibrotactile unit is controlled by the mobile app to provide vibrotactile biofeedback. The dead zone is defined to be 1 degree around the target knee angle. If the angle of the actual knee movement exceeded the target angle by the dead zone, the front vibrator is activated. If the angle difference is below the dead zone, the back vibrator is activated to provide biofeedback. A small human subject trial with 10 healthy participants was conducted, which confirmed the effectiveness of the vibrotactile biofeedback.

In [449], a mobile app was developed to detect the shoulder flexion angle using the built-in inertial sensors of the smartphone. The app provides biofeedback to the user. More specifically, the user is supposed to hold the smartphone while performing shoulder flexion. When the target angle is reached, the app would generate a short beep sound. Likewise, when the shoulder goes back to the initial position, the app would generate another beep sound. The effectiveness of the app was validated with 20 participants.

In [450], a randomized controlled study on the effectiveness of a smartphone-based system for trunk control training in stroke patients was reported. A smartphone-inserted balance board was used in the trial. The smartphone runs a set of mobile apps using the built-in gyroscope sensor to detect the posture and movements of the subject sitting on the board during trunk control training. More specifically, the mobile apps include CSMi center of pressure, CSMi limits of stability, CSMi weight bearing front-back, CSMi weight bearing left-right, CSMi weight shift, CSMi animal adventure apps. The screen is projected to a computer monitor to provide visual feedback to the participant. The study found that the experimental group trained with the smartphone-based system made significant improvements on step length and stride length.

Chapter 12
Technology-facilitated occupational rehabilitation

Occupational therapy is a discipline that studies how a human being engages in various activities in life [451], and how to help people to perform in these activities [452]. Trained occupational therapists are regarded as experts in doing [451]. Compared with other fields in rehabilitation, the scope of occupational therapy is extremely broad because it spans not only physical activities but also cognitive, psychological, and social aspects of human activities. Also because of this, intervention methods in occupational therapy vary significantly for different patients.

The common understanding of occupation is related to one's employment or means of making a living. In the discipline of occupational therapy, the term occupation is broadly defined, encompassing all forms of activities of daily life [451], which include those for self-care, for leisure, and for productivity. More specifically, an occupation is understood as an abstract concept, such as teaching, engineering, and singing. A list of typical occupations are provided in the fourth edition of occupational therapy practice framework: domain and process [453], as shown in Figure 12.1.

12.1 Framework for occupational therapy

Part of the framework for occupational therapy [451] is illustrated in Figure 12.2. An occupation is typically enacted by a group of activities that we do regularly, and these activities would have meaning either personally or from the social and cultural perspective. Through these activities, an individual could have the opportunity to participate in society. An activity in turn consists of a sequence of tasks. A task consists of a sequence of steps in the forms of actions and/or thoughts. Occupational therapy focuses on the observation and improvement of the performance at the granulation of occupation, activity, task. The term action appears to be not very precisely defined. On the one hand, action seems to be the smallest granularity in the composition of an occupation. On the other hand, action seems to be an umbrella term as used in [451] where occupation, activity, and task are regarded as three forms of action. In the following discussion, we follow the latter interpretation of action.

Actions have structures. The structures of actions are reflected as routines, habits, and rituals [451]. These three terms are often used interchangeably, referring to an established set of tasks, although there are subtle differences between them [451].

Activities of Daily Living

Instrumental Activities of
Daily Living

Health Management

Rest and Sleep

Education

Work

Play

Leisure

Social Participation

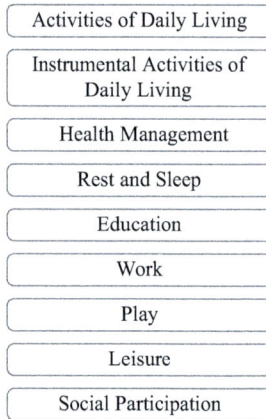

Figure 12.1   A list of typical occupations

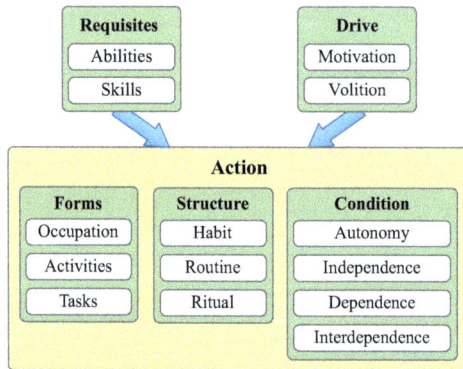

Requisites
Abilities
Skills

Drive
Motivation
Volition

Action

Forms	**Structure**	**Condition**
Occupation	Habit	Autonomy
Activities	Routine	Independence
Tasks	Ritual	Dependence
		Interdependence

Figure 12.2   Concept framework in occupational therapy

Habits refer to established patterns of a particular individual, which are typically
formed initially by making many repetitions day in and day out. A routine might be
a part of a habit. However, a routine could be what an individual does as part of a
job, such as a surgeon thoroughly cleaning his or her hand prior to a surgery. Once
a habit is formed, an individual rarely needs to exert a lot of attention in doing the
tasks as part of the habit. However, a routine as part of a job might require a great
amount of attention to ensure it is done properly. It is important for an individual to
allow certain flexibility in one's habit. Rituals refer to a sequence of actions that have
symbolic meanings from the culture, social, or spiritual perspectives [453].

How the actions of an individual are carried out can be inspected by the conditions
of the actions, including autonomy, independence, dependence, and interdependence.
Autonomy means that the individual has the free will of doing the action. Independence
means that the individual can perform the action alone without the help from anyone

or using any assistive device. Dependence means that the individual would need help from a caregiver or some assistive technology to perform the action. Interdependence captures the social function of some actions.

Whether or not an individual can perform an action, and the extend of the performance of the action rely on two major factors: (1) requisites for the action and (2) drive for performing the action. The requisites for an action refer to the abilities and skills of the individual. Sometimes the term "function" is used as well. Function is a direct reflection or measurement of one's ability. Skills refer to acquired abilities through learning and practices. The difference between abilities and skills can be understood in terms of the difference between activities of daily living (ADL) and instrumental activities of daily living (IADL) [454]. IADL refers to activities that require some levels of skills, such as using technology and doing financial planning. Another form of ADL is basic activities of daily living (BADL), such as putting on clothes and taking a shower. BADL can be performed without much effort once learned and usually without requiring any particular skill. Unless an individual suffers from mild cognitive impairment or dementia, the person could perform BADL easily without difficulty [454].

Other than the requisites for action, another factor for taking an action is the drive, which is referred to as energy source in [451]. Just having the requisites for an action might not lead an individual to take the action. The drive consists of two primary elements: motivation and volition. Motivation means that the individual wants to take the action. Volition means that the individual has the will power to complete the action. Volition is particularly important for demanding and sophisticated actions (e.g., running a marathon).

The requisites and the drive for action can be considered as the internal factors innate to an individual. An action may also be affected by the environment, which includes venue of the action, the culture, and the interaction with other individuals.

The basic abilities of an individual are evaluated in terms of a set of body functions, including mental functions, sensory functions, musculoskeletal functions, cardiovascular functions, hematological function, immune functions, respiratory functions, hearing and speech functions, digestive, metabolic, and endocrine functions, genitourinary and reproductive functions, and skin and related functions.

In [453], common skills considered in occupational therapy are provided and discussed in detail. We summarize these skills in Figure 12.3. There are roughly eight groups of skills, generally in the order of complexity levels.

- The first group is about controlling the body, including stabilizing one's body while performing tasks, aligning one's body while handling objects as a part of a task, positioning one's body with proper distance between the body and the objects.
- The second group is about organizing objects, including locating objects, gathering objects, organizing objects, restoring objects, and navigating between the objects.
- The third group is about handling objects, including reaching for objects, bending while placing objects, griping objects, manipulating objects, coordinating the

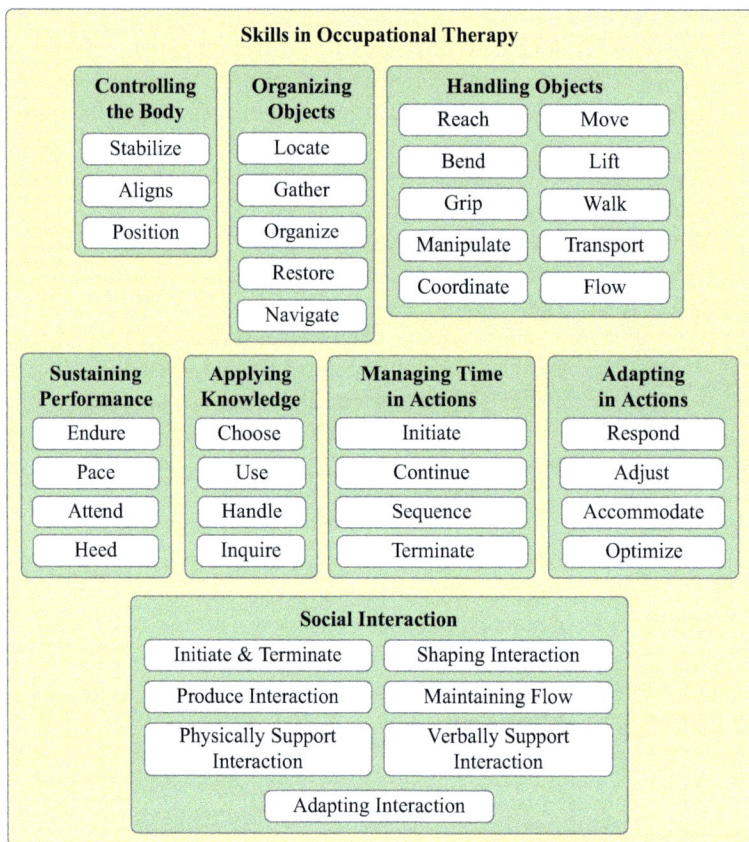

Figure 12.3   Common skills in occupational therapy

body parts while handling objects, moving objects, lifting objects, walking while handling objects, transporting objects, calibrating one's use of force and movement speed while handling objects, and finally, using smooth and fluid upper extremity movements while handling objects (i.e., flow).

- The fourth group is about sustaining performance while doing tasks, including keeping the right pace, attending to the tasks, and heeding to the original plan for actions.
- The fifth group is about the skills of applying knowledge while doing tasks, including choosing the right objects, using the right objects, handling objects properly, and making inquiries as needed.
- The sixth group is about managing time while doing tasks, including initiating and terminating tasks, and performing a set of tasks continuously in the right sequence.
- The seventh group is about making adjustment while doing tasks, including responding properly to various nonverbal cues, overcoming unexpected problems (i.e., adjusting) due to change of environment and/or work conditions, accommodating the demand of the tasks, and optimizing performance in doing tasks.

- The last group of skills is about social interaction, including initiating and terminating a social interaction, producing a social interaction, supporting an ongoing social interaction with proper body language and verbally, shaping the social interaction by raising questions, providing replies, showing appreciation, etc., maintaining the flow of the social interaction, and making appropriate adaptations during an ongoing social interaction.

12.2 Occupational therapy for return to work

In [455], a detailed occupational therapy plan was reported. The plan was designed to minimize sick days taken by workers. Hence, the plan is a form of return to work intervention. The plan consists of three main components: (1) mental training, (2) physical training, and (3) work-related problem solving, as shown in Figure 12.4. The

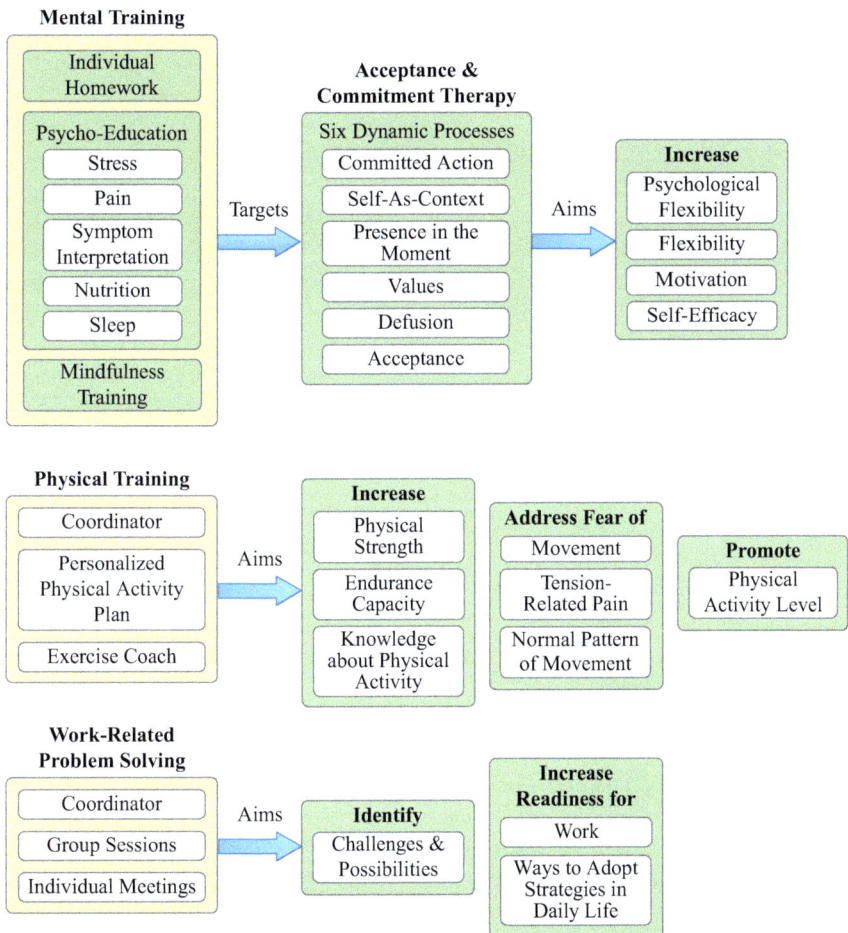

Figure 12.4   Occupational therapy plan for return to work

mental training is a major component and the training design is based on the acceptance and commitment therapy model [456] where six dynamic processes are targeted, including committed action, self-as-context, presence in the moment, values, delusion, and acceptance. The mental training aims to increase the trainee's psychological flexibility, flexibility, motivation, and self-efficacy.

The physical training is designed by the intervention coordinator and the exercise coach in conjunction with the trainee with a personalized physical activity plan. The physical training aims to increase the physical strength of the trainee, the endurance capability of the trainee, and the trainee's knowledge about physical activities. The training also addresses the trainee's fear of movement, tension-related pain, and normal pattern of movement. Ultimately, the physical training aims to promote the trainee's physical activity level.

The work-related problem-solving component is led by the intervention coordinator in individual meetings and group sessions. This component aims to identify the challenges and possibilities faced by the trainee, and increase the readiness of the trainee for work, and for ways to adopt the strategies developed during the intervention in daily life in the future.

12.3 Technology in occupational therapy

In occupational therapy, technology usually means assistive technology, which could help an individual to engage in various activities of daily living [457]. The use of assistive technology may be regarded as a way of the modifications to the environment where an individual with disabilities could performs activities [453]. Besides assistive technology, telerehabilitation has also been used heavily [458]. Although telerehabilitation could be done via traditional telephone or short messaging, nowadays teleconferencing is becoming the default platform for conducting telerehabilitation. It is also possible to use custom-made platform for telerehabilitation. Like other forms of rehabilitation, exergames have also been used in occupational therapy. Due to the gamification property, exergames are often considered more engaging than traditional therapy.

12.3.1 Assistive technology

In [459], Anson provided a systematic overview of the assistive technology, where assistive technologies are divided into three types, as shown in Figure 12.5.

The first type of assistive technology is to provide electronic aids to daily living. The simplest form of aids to facilitate the individual to control on and off of appliances remotely by pushing a switch. A more sophisticated assistive device could enable the individual to control features of electronic appliances beyond on and off, such as changing channels of the TV. Subsumed devices would include essential services, such has telephone, to make it much more convenient for the individual with disability to call a friend or answering a phone call.

**Electronic Aids to
Daily Living**

Levels of Control

Power Switching

Feature Control

Subsumed Devices

Appliances

**Augmentative and
Alternative
Communications**

Message Composition

Construct

Preview

Edit

Message Transmission

Print to Paper

Electronic Display

Speech Synthesizer

**Computer/
Smartphone as
Assistive Technology**

Take Notes

Social Networking

Screen Reader for
Blind Individual

Text-to-Speech

Augment Attention
& Thinking Skills

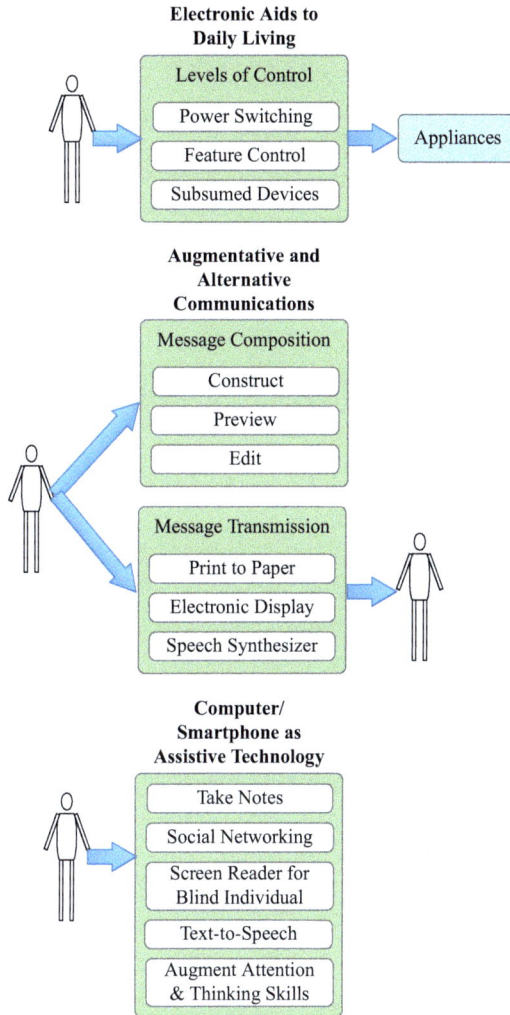

Figure 12.5   Common types of assistive technology in occupational therapy

The second type of assistive technology is to provide augmentative and alternative communications to the individual with disability. The technology consists of two components: (1) message composition and (2) message transmission. The message composition component would allow the individual to construct, previous, and edit a message via a number of ways that will be further discussed later. The message transmission can be done in several different forms, such as printing to paper, displaying the message on a computer monitor, and using a speech synthesizer enabled by the text-to-speech technology.

The third type of assistive technology is computer or smartphone/tablet. Although the use of computer by itself could be part of the activities of daily living, various computer programs have been developed to assist individuals with disability, for example, helping a student to take notes, helping an individual with limited mobility to engage in social networking, helping a blind individual to know what is displayed with a screen reader, helping an individual with speech impairment to communicate with another person using the text-to-speech program, and helping an individual with attention disorders to focus.

As shown in Figure 12.6, various types of input modalities have been used to accommodate individuals with different levels of disabilities. Physical keyboards may be modified to fit the disability of the individual, for example, some keyboards use the Dvorak hand handed, Dvorak one handed, and Chubon. Physical pointing devices include joystick, up/down keys, computer mouse, and track pad. Another approach is to use eye-tracking and human gesture tracking as an input method. For individuals with limited range of motion and lack of fine motor control, switch-encoding input could be useful. A switch-encoding input consists of a small set of switches. Different functionalities could be controlled by identifying which switch is on and for how long. Morse code is an example encoding method for the input. The Tongue Touch Keypad is a special switch-encoding input device. The maturity of speech recognition enables an individual to use speech as a form of input to control various devices. Scanning input is useful for individuals with very limited cognition and motor control. A scanning input device facilities an individual to make specific choices with very little effort. Integrated controls are devices that offer multiple modalities to control, which could be necessary for individuals with severe disabilities such as cerebral palsy.

The input rate can be enhanced with several mechanisms. Even if the input modality is spelling, pictograms or icons could be used to represent whole words, and a two-level method could be used where an individual would first select the category and then a specific word under the category. It is also possible to make prediction on

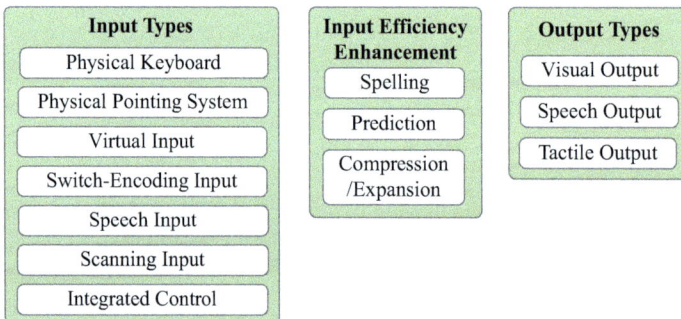

Input Types	Input Efficiency Enhancement	Output Types
Physical Keyboard	Spelling	Visual Output
Physical Pointing System	Prediction	Speech Output
Virtual Input	Compression /Expansion	Tactile Output
Switch-Encoding Input		
Speech Input		
Scanning Input		
Integrated Control		

Figure 12.6 Input and output in assistive technology, as well as mechanisms to enhance the input efficiency

the next word based on the previous words. Using compressed expressions such as acronyms could improve the input rate significantly.

There are three major output modalities: (1) visual, (2) speech, and (3) tactile. Visual output may need to consider colors, contrast, and image/font size to improve accessibility. The speech out is enabled by the text-to-speech technology, which is rather mature nowadays. Braille output is one of the most well-known tactile output modalities. According to [459], only a small fraction of the blind population use Braille.

12.3.2 Telerehabilitation

Another form of technology in the context of occupational therapy is telerehabilitation. Telerehabilitation can be considered a form of delivery method in occupational therapy where the therapy can be conducted remotely online via telephone and/or video conferencing. Telerehabilitation has also been conducted via more advanced technology such as a mobile app [460]. A comprehensive review on telerehabilitation in occupational therapy is provided in [458]. The target population and therapy aims are summarized in Figure 12.7.

In [461], telerehabilitation is used to facilitate cognitive rehabilitation in patients with acquired brain injury. In [462], telerehabilitation is used to address executive dysfunction in patients with traumatic brain injury. The difference between acquired and traumatic brain injuries is that an acquired brain injury is caused by internal factors such as lack of oxygen, and a traumatic brain injury is caused by external force.

In [463], telerehabilitation is used to conduct disability prevention education for breast cancer patients who are undergoing chemotherapy. In [464], telerehabilitation is used to conduct home safety education for community-living older adults. In [465–467], telerehabilitation is used to improve hand function of children/adolescences with cerebral palsy. In [468], telerehabilitation is used to improve handwriting skills in students with fine-motor or visual–motor impairment. In [469], telerehabilitation is used to conduct oral care education in patients with tetraplegia.

Several studies used telerehabilitation to facilitate various occupational therapy in post-stroke patients, including improving quality of life and depression via home-exercise [470], conduct training on energy conservation with fatigue management [471], perform electrical stimulation [472], and improving motor recovery using a mobile app [460].

12.3.3 Exergames

In [473], several exergames were designed to mitigate cognitive impairment in MS patients. Microsoft Kinect sensor is used in these games to track human movements. These games are designed to practice a number of perspectives related to cognitive functions: (1) hand–eye coordination during arm movements, (2) working memory, (3) visual memory, (4) reaction time, (5) processing speed, (6) inhibition and selective attention, (7) weight shifting, and (8) postural balance and correction reactions. These games typically have 4–5 difficulty levels, and have cartoon-like graphical user

Patients with Acquired Brain Injury	Address Cognitive Dysfunction
Patients with Traumatic Brain Injury	Address Executive Dysfunction
Breast Cancer Patients Undergoing Chemotherapy	Disability Prevention
Students with Fine-Motor and/or Visual-Motor Impairment	Improve Handwriting Skills
Children with Cerebral Palsy	Improve Hand Function
Community-Living Older Adults	Home Safety Education
Patients with Tetraplegia	Oral Care Education
Post-Stroke Patients	Improve Quality of Life & Depression via Home-Based Exercise / Energy Conservation (Fatigue Management) / Electrical Stimulation / Improve Motor Function

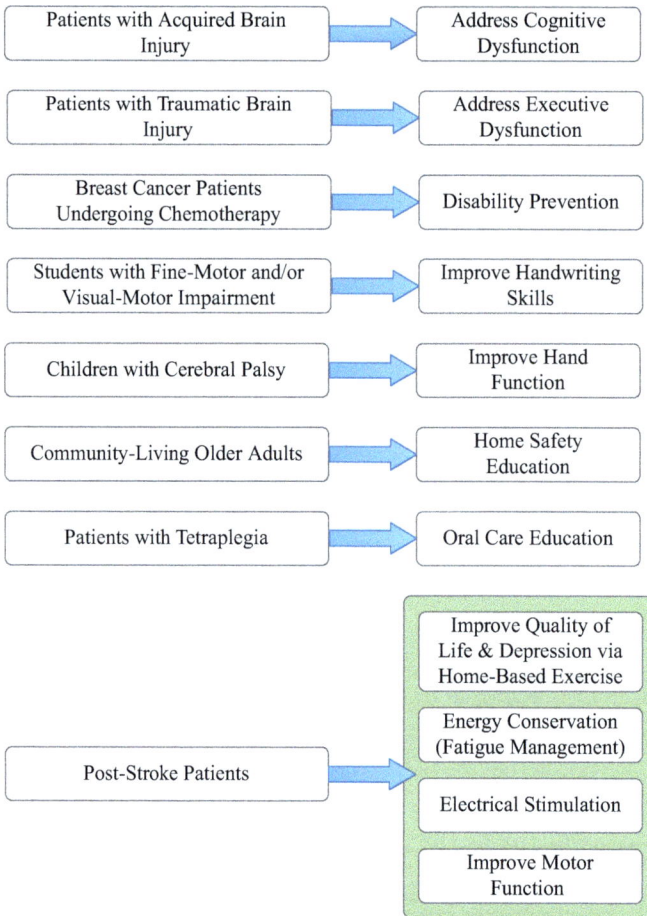

Figure 12.7   Target population and occupation therapy aims using telerehabilitation

interface. For each exergame, a number of objective metrics are used to determine the cognitive level.

12.4   Tracking of activities of daily living

One main goal of occupational therapy is to facilitate a patient to regain the energy and function to engage in normal activities of daily living (ADL). Although currently rarely incorporated into occupational therapy practice, monitoring the activities of daily living could increase the effectiveness of the therapy because the functional and energy level of the patient can be more accurately determined in the nature setting prior to, and during, the therapy program.

That said, it could be expensive and intrusive to fully track ADL. Traditionally, a dedicated smart home fully instrumented with various sensors for tracking [454]. Common sensors include inertial sensors, pressure sensors, proximity sensors, passive infrared sensors (PIR), and recently Bluetooth sensors. These sensors will need to be deployed in the entire home and some may need to be attached to appliances (such as the door of the refrigerator) or household items (such as water cup) in some way. It is also possible to use a set of programmable cameras, such as Microsoft Kinect, to follow and track an individual's activities [230,402,474–476]. However, this approach may be perceived as too intrusive.

Recently, we explored using a much less intrusive method to partially track ADL. A key hypothesis is that the pattern and the change of the pattern established from partial observation of ADL (particularly IADL) can still effectively associated with the individual's functional status and energy levels. The mobile app tracks three elements in ADL: (1) physical activity levels using the built-in inertial sensors, such as the number of steps taken in walking, jogging, and running; (2) indoor dwelling pattern tracking using Bluetooth beacons deployed in key locations such as kitchen, bedroom, restroom, and living; (3) pattern of IADL in terms of usage statistics of common mobile apps [477]. By applying conventional machine-learning algorithms (such as KNN), we were able to establish patterns of ADL reliably.

Chapter 13
Technology-facilitated speech rehabilitation

Oral communication is the dominating means of social interaction between humans. Hence, speech impairments could significantly impact an individual's career, mental health, and quality of life. Speech rehabilitation is a field that aims to address various speech impairments. Traditionally, speech rehabilitation is conducted by a trained clinician. In the United States, such clinicians are referred to as speech–language pathologists. To reduce the financial cost and to increase the convenience for the patients, various technologies have been adopted in speech rehabilitation, particularly mobile apps, which could enable a patient to practice alone at home. The main form of technology-facilitated speech rehabilitation is computer-based speech therapy.

As shown in Figure 13.1, speech therapy typically adopts a variety of speech exercises to improve one or more specific target areas that could improve the patient's ability to properly speech as well as to understand the other person's speech, ranging from loudness, phonation, articulation, resonance, prosody, to oral muscles, social communication, and cognitive communication. There are a large number of disorders that could lead to speech impairments, including aphasia, dysarthria, apraxia of

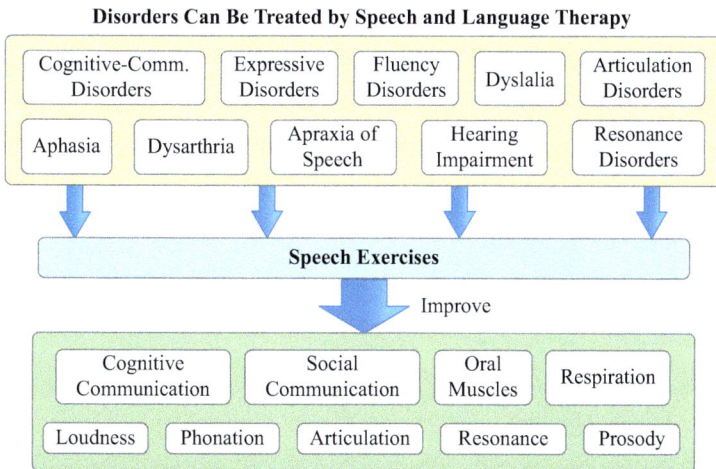

Figure 13.1   Common disorders that can be addressed by speech rehabilitation and common therapy methods

speech, dyslalia, hearing impairment, resonance disorders, fluency disorders, articulation disorders, expressive disorders, and cognitive–communication disorders. Due to the heterogeneity of the disorders, speech exercises could differ significantly.

13.1 Common speech-related disorders

13.1.1 Aphasia

Aphasia is a disorder caused by damages or injuries to specific areas in the brain that are responsible for language. Patients who suffer from the aphasia usually have impaired comprehension and all forms of communication. The most common cause of aphasia is stroke [478].

13.1.2 Dysarthria

Dysarthria refers to a set of speech disorders due to impairments in the neuromuscular control of speech [479], and there are six types of dysarthria: (1) flaccid dysarthria, (2) spastic dysarthria, (3) ataxic dysarthria, (4) hypokinetic dysarthria, (5) hyperkinetic dysarthria, and (6) unilateral upper motor neuron dysarthria. Hypokinetic dysarthria is often caused by Parkinson's disease [480]. Dysarthria is often present in patients who have suffered from acute ischemic stroke [481]. The distinctive characteristics of dypokinetic dysarthria are reflected in three types [480]: (1) Voices: reduced loudness, monotone, and breathy; (2) Articulation: imprecise consonants and vowel centralization; and (3) Rate: increased, decreased, or variable. Apraxia of speech is primarily a childhood neurological disorder that negatively impacts a child's ability to speak with accuracy and correct rhythm [482].

13.1.3 Apraxia of speech

Apraxia of speech is due to the impairment in the pathways of the brain that are involved in planning and coordinating the movements for speech. There are two types of apraxia of speech. Childhood apraxia of speech is due to genetics and is usually diagnosed during childhood. Children with the autism spectrum disorder, epilepsy, or cerebral palsy could have childhood apraxia of speech. Acquired apraxia of speech usually occurs in adults who have suffered from brain damage, such as stroke, traumatic brain injuries, or brain tumor.

13.1.4 Dyslalia

Dyslalia is a speech disorder where the patient is unable or has difficulty to articulate words properly, particularly some specific phonemes and syllables. Dyslalia is predominately among children. There are several types of dyslalia. The most common type of dyslalia is functional dyslalia. Functional dyslalia is caused by abnormality of the peripheral organs of speech. Audiogenic dyslalia is caused by hearing loss. Physiological dyslalia is due to immaturity in the organs of speech. Educational dyslalia is due to improper education while a child is learning to speak. Organic dyslalia is due to defects in the organs of speech, such as in the lips, tongue, teeth, palate. Organic dyslalia is also referred to as dysglossia.

13.1.5 Hearing impairment

People who suffer from hearing impairment could have difficulty in communicating with other orally due to the inability of understanding other people properly.

13.1.6 Resonance disorders

Resonance disorders are caused by too much or too little nasal and/or oral sound energy in the speech sound [483]. Symptoms of resonance disorders may have hypernasality (i.e., too much energy in the nasal cavity), hyponasality (i.e., too little energy in the nasal cavity), cul-de-sac resonance (i.e., when the resonance is trapped in the cavity due to obstruction), and mixed resonance. Velopharyngeal dysfunction is a resonance disorder with typical symptoms of hyper- and/or hyponasality, and nasal air emission [484]. Such symptoms are a result of improper closure and/or opening between the nasal and oral cavities (i.e., the velopharyngeal valve) during speech tasks [484].

13.1.7 Cognitive communication disorders

Cognitive communication disorders refer to communication difficulties caused by cognitive impairment. Cognitive communication could be a result of traumatic brain injury [485,486], stroke [487], mild cognitive impairment [488], and dementia [489].

13.1.8 Expressive disorders

Expressive disorders include expressive language disorder [490] and expressive phonology disorders [491]. Individuals with expressive disorders have difficulty expressing themselves verbally. This could be due to the difficulty in finding the right words (i.e., world retrieval problems), putting the words together to form a sentence (sentence formulation problems), grammatical impairments, and receptive language impairments [492]. Fluency disorders primary refer to stuttering. Articulation disorders refer to the difficulty in producing certain word sounds.

13.1.9 Fluency disorders

Fluency disorders refer to the symptoms where a person has chronic problems with making continuous speech. The most common type of fluency disorder is stuttering. The other type is cluttering. The primary behaviors of stuttering include repetitions of sounds, syllabus, or whole words. The primary behaviors of cluttering is merging words or cutting off parts of words.

13.1.10 Articulation disorders

Articulation refers to the process that produces sounds, syllables, and words. People who suffer from articulation disorders are usually children when they have impaired

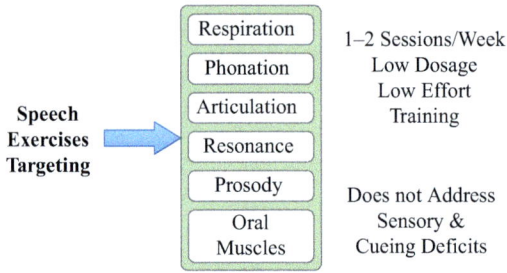

Figure 13.2   Standard speech and language therapy

oral–motor functions required to make some speech sounds and syllables. Specific symptoms are articulation disorders include: (1) adding sounds or syllables to words, (2) distorting a sound, (3) omitting certain sounds out of a word, and (4) substituting one sound for another in a word.

13.2   Standard speech and language therapy

Standard speech and language therapy is usually administered by certified speech–language pathologists via a set of speech exercises targeting multiple domains, including respiration (i.e., breathing), phonation, articulation, resonance, prosody, facial/oral muscles, as shown in Figure 13.3. Standard speed and language therapy would typically have one to two sessions per week with low dosage and low effort training, and there is no consideration on addressing sensory and cueing deficits [480].

The respiration, phonation, resonance, and articulation are the four most important processes of speech production [493]. Respiration provides energy for the sound production through breathing (i.e., inhalation and exhalation). When a person is speaking, exhalation takes longer time than inhalation. Phonation is the process of creation of sounds via rapid vibration of the vocal folds when the air (from the lung) passes between them. The movements of the vocal folds determine the pitch (i.e., the frequency of the vibration), loudness (i.e., the volume or the intensity of the sound; louder sound requires greater air pressure and larger amplitude of the vibration), and quality (determined by the movement patterns of the vocal folds) of the sound produced. Resonance is the process of enhancing the loudness and the quality of the sound produced by the phonation process via the air-filled cavities while the air is coming out of the mouth. The cavities include pharynx, nasopharynx, nasal cavity, and oral cavity. Articulation is the process of shaping of phoned voice into comprehensive words with a number of articulators, including lips, teeth, alveolar ridge, tongue, hard palate, soft palate (i.e., velum), etc.

Prosody conveys information beyond individual words in a sentence via rhythm, intonation, stress patterns, pausing, and loudness variations. Patients with some speech disorders, such as childhood apraxia of speech, may produce speech that is

LSVT LOUD　　　　　**LSVT ARTIC**

Increased Movement Amplitude Directed to Respiratory-Laryngeal Systems	**Increased Movement Amplitude** Directed to Orofacial-Articulatory Systems

High-Dosage
High-Effort

Daily Exercises

Hierarchy Exercises

Homework Exercises

4 Sessions/Week
Over 4 Weeks

Integrate

Address Sensory
& Cueing Deficits

Self-Monitoring

Self-Generate

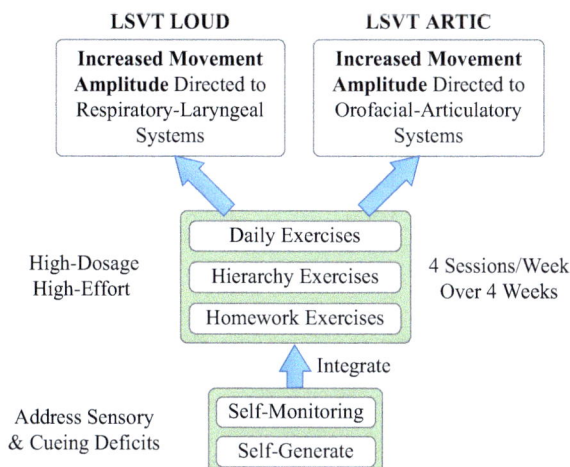

Figure 13.3　Lee Silverman Voice Treatment

aprosodic (lack of prosody, i.e., monotone), or dysprosodic (using improper prosody instead of the expected prosody pattern).

Oral–motor skills are important to make proper sound. That is why oral muscles exercises are often included in speech therapy.

The actual speech exercises used in a therapy program are rarely defined in the academic literature, probably because they vary significantly and they are customized for each individual patient. In the webpage from Cleveland Clinic (https://my.clevelandclinic.org/health/treatments/22366-speech-therapy), several speech exercises are provided, including tongue exercises, smiling, puckering lips, reading out load, and playing word games.

13.3　Lee Silverman Voice Treatment

The Lee Silverman Voice Treatment (LSVT LOUD) was developed by Dr Lorraine Ramig to treat the speech disorder in patients with Parkinson's disease [494]. Patients with Parkinson's disease often develop a speech disorder called hypokinetic dysarthria, which leads to significant reductions in vocal loudness. The lack of loudness in speech would lead to speech intelligibility because it is difficult for other people to hear. The LSVT LOUD program aimed to increase the movement amplitude towards respiratory-laryngeal systems. Later another LSVT-based program called LSVT ARTIC was introduced by Dr Ramig [480]. LSVT ARTIC is short for LSVT Articulation, and it was designed to increase the movement amplitude directed to orofacial-articulatory systems.

The main characteristics of LSVT LOUD and LSVT ARTIC are summarized in Figure 13.3. Both LSVT programs are high-dosage and require high effort from the patients. The programs consist of four 1-h sessions per week and last for 4 weeks.

Each training session consists of daily exercises, hierarchy exercises, and homework exercises. Each type of exercises is fully specified in [480]. The daily exercises consist of three sets of tasks and last for 30 min. The first set of tasks was designed to achieve maximum sustained movements during the first 12 min. The second set of tasks was designed to achieve directional movements during next 11 min. The third set of tasks was designed to make functional movements during the next 7 min. The hierarchy exercises last for 25 min and were designed to train the use of rescaled vocal loudness (for LSVT LOUD) or enunciation (for LSVT ARTIC) achieved in daily exercises in context-specific speaking activities. The final 5 min in each session are spent for the therapist to assign homework exercises to be done outside the therapy room. During the treatment days, the homework is to practice a subset of the daily exercises and hierarchy exercises for 10 min per day. During the non-treatment days, the homework is to practice a subset of the daily exercises and hierarchy exercises for 15 min per day. Furthermore, there is a conversational carryover assignment where the patient is encouraged to engage in a real-world conversation using louder voice (for LSVT LOUD) or more enunciated speech (for LSVT ARTIC).

The LSVT programs were also designed to address sensory and cueing deficits with self-monitoring and self-generate techniques. The training for loudness (for LSVT LOUD) or enunciation (for LSVT ARTIC) is through the use of modeling or tactile and visual cues, and while practicing, the patient is encouraged to focus attention on how it feels and sounds while talking with louder voices or increased enunciation (for LSVT ARTIC) (i.e., self-monitoring) and to make effort to increase the loudness (for LSVT LOUD) or enunciation (for LSVT ARTIC) internally (i.e., self-generate).

Another variation of LSVT LOUD program is the Pitch Limiting Voice Treatment (PLVT) [495]. This treatment aimed to address an issue with LSVT LOUD where the extra effort made by the patients could raise their pitch together with their voice loudness. PLVT addresses this issue by promoting loud (voice) and low (pitch). PLVT requires accurate feedback to the patients regarding the loudness and pitch level.

13.4   Computer-based speech therapy

Technology has been used to help diagnose some speech disorders [496], to facilitate therapy, and to enable at-home self-monitored rehabilitation. This line of research has been reviewed in [497,498]. The user requirements and user acceptability issues have been studied in [499]. Major benefits of using technology include reduced cost of therapy [500], more accurate and reproducible control during rehabilitation [501,502], more convenience to patients, and more accessibility. Many patients could enjoy low-cost speech therapy facilitated by technology, such as mobile apps, who otherwise would not choose or have the opportunity to do so.

Technology has been integrated into speech therapy in different degrees. Many computer-based systems of different complexity and functionalities have been developed to provide more controllable stimuli to patients and richer information to teach patients how to speak correctly during speech therapy (such as visualize the speech [503]), and to make the speech therapy less labor-intensive, more customizable, and more accessible to patients. Some highly repetitive tasks have been automated by either pre-recording words and sentences to be played during a speech therapy session, or using text-to-speech technology. With the maturity of the speech recognition technology, not only normal speech can be recognized with high accuracy but speech spoken by patients with speech impairments with customized models. There is also attempt to automate the assessment of patient progress and speech exercise recommendations using fuzzy logic [504].

It is common for recent computer-based speech therapy systems to incorporate virtual reality scenes and avatar user interfaces (this type of systems have been referred to as 3D virtual head or talking head in [497]). Some systems incorporated gamification mechanics to make the systems more attractive and engaging to patients. These systems often are customizable and can collect data regarding therapy frequency and durations, as well as performance.

In [503], a visual speech computer program was used in a study to train perception and production of speech for individuals with hearing loss. The visual speech computer program (called Baldi) was used as an assistive technology in the training. Using such an assistive technology has several benefits: (1) it reduces the workload of the clinician; (2) it helps tailor the training session based on the specific needs of the individual, and most importantly, (3) it teaches the individual how to read the speaker's face and the accompanying visual information towards better perception and understanding of the speech. The incorporation of visual speech information can compensate the degraded auditory speech due to hearing loss. The computer program relies on text-to-speech technology, and incorporates visual information while generating pre-programed speech via a virtual head, including vocal cord vibration for voiced segments, and the turbulent airflow emanating from the mouth for voiceless segments. The virtual head has transparent skin so that inside articulators are visible to the trainee.

In [505], a rather simple computer program was used in an auditory training for adults with hearing impairment. The program is used in the word training task and the sentence training task. Various test materials were pre-recorded. During the word training task, two words were displayed on the computer monitor screen. Then, the target word was played acoustically. The patient was required to select one of the words corresponding to the target word played. The patient was then provided correctness feedback on the screen. If the selection was correct, the training proceeded to the next trial. During the sentence training task, a pre-recorded sentence was played while the computer monitor screen was displaying six words. The patient was asked to select three of the six words that appeared in the sentence played.

In [506], the design and implementation of a mobile speech therapy game called SpokeIt was introduced. The game was developed to demonstrate correct pronunciation with the corresponding lip animations. The game uses an offline speech recognition engine called Pocketsphinx. The game has a cartoon-like game interface and incorporates a storybook game style with a set of mini-games.

In [501], the design and implementation of a speech therapy system was reported. The system was designed to improve word naming ability of patients with aphasia. The system provides a user interface where an avatar is functioning as a virtual therapist.

The system runs a custom speech recognition engine running in a cloud server. The system provides one interface for the patient, and another interface of the clinician. The clinician interface allows the clinician to define the exercises to be conducted by the patient. The system also collect data on patient practice and performance so that the clinician could keep track of the progress made by the patient.

In [502], the design, implementation, and evaluation of a speech therapy system called AphasiaScripts was presented. The system offers computerized script treatment to aphasia patients. The system's user interface consists of an avatar serving as a virtual therapist, prompts for the patient what to say, and provides a visual speech trajectory. The avatar also demonstrates oral–motor movements for words the patient is supposed to say. This is an example of visual speech (or visible speech). The visual speech trajectory was based on a codebook. The foundation for designing the visual speech trajectory include: (1) individual sounds differ by phonetic context and the manner of speaking; and (2) the temporal characteristics of the sounds are based on physiological and acoustic constraints on the movement of oral–motor articulators during speech. Another impressive feature of the system is the integration of cueing to the patient. Three forms of cues are provided to the patient: (1) auditory cues from the virtual therapist, (2) visual cues from the mouth movements displayed by the avatar, and (3) written cues. There are two levels of cue treatment. At the high cue level, the patient would receive the auditory and visual cues during all stages of the treatment. At the low cue level, only the written cues are available during sentence practice.

In [507], the effects of a computer-delivered speech and language treatment were compared against clinician-delivered treatment program. The treatment program is called treatment of underlying forms (TUF). It is a linguistically based therapy program for individuals with agrammatic aphasia to improve agrammatic sentence deficits. The computer program is referred to as Sentactics. The primary user interface of Sentactics consists of an avatar serving as a virtual therapist with a name Sabrina, and spaces to show two cartoon-like pictures of the patient to make a selection. This interface is to help the patient to practice sentence production via the sentence–picture matching task. Sentactics would play the pre-recorded sentences during the sentence–picture matching task. The patient performance is assessed using the sentence production priming test. The actual assessment on the patient performance is not automatically evaluated by Sentactics. Rather, Sentactics could record the patient's responses for offline analysis. The study found that the effects of the computer-delivered treatment program are on par with the clinician-delivered program.

In [484], a custom-developed computer system was used in a speech therapy program for patients with velopharyngeal dysfunction. The system uses a microphone to record the voice of the participant, and an accelerometer attached to the nasal skin to measure nasal skin vibration. The system provides a cartoon-like game interface for the participant. In each round of game play, the participant is prompted to repeat three times of one of the nasal and non-nasal alternating consonant–vowel–consonants sequence (e.g., man, bag, mean, bead, nine, guide, nun, bug, mom, dog, noon, and dude). After each round, the system calculates a score. Initially, no feedback is provided to the participant and the data collected are used to calculate a baseline score for the participant. The baseline score is used to set nasal and non-nasal targets such that the participant would be judged as meeting the targets 70% of the time. This design is to encourage the participant to continue while making the game challenging. The feedback for each repetition of a word is in terms of a smiley face for meeting the target or a frown face for failing to meet the target.

In [508], the design and the implementation of a custom-developed computer game system were introduced. The system integrated with a motion tracking system referred to as the Wave system. The Wave system uses several electromagnetic sensors attached on the tongue, jaw, lips of the patient to track their movements relative to the head. The motion tracking system was claimed to have up to 0.5 mm accuracy in the target 300 mm^3 field space. The motion data are collected and used to visualize the patient's speech via a 3D science developed using the Unity 3D game engine. This system was used in a clinical study [509] to assess its efficacy in treating patients with Parkinson's disease. This study inferred that movement reduction due to the articulatory deficit in the patients could be the reason for the reduced intelligibility, imprecise articulation, and elevated speaking rate.

In [510], a Web-based computer program was introduced to help patients with dysarthria to engage in speech training at home. Patients would download the audio examples and follow the instructions of the program. The main task is imitate each of the audio examples. The patients would then compare with their own speech with the audio example. To facilitate this comparison, the program provides visual feedback to the patients. This program has been used as a part of the PLVT training, where the visual feedback include the intensity of the sound (reflecting the loudness) the patient made, and the fundamental frequency (reflecting the pitch).

In [511], a case study that used a mobile speech therapy app was presented. The app would provide visual-acoustic biofeedback for a patient who suffers from residual rhotic errors to learn how to speak properly. The feedback was generated by analyzing the speech signal using linear predictive coding spectrum where the horizontal axis displays the frequency and the vertical axis display the amplitude of the speech signal. Rhotic can be distinguished from the sonorants using the low height of the third formant of the spectrum.

In [512], the efficacy of the Web-based computer program [510] and a game-based system for treating patients with dysarthria was evaluated. The game is called treasure hunters. The game is a two-player cooperative game where the task is to navigate a map to find a treasure chest buried in the ground together. One player would play a digger and attempt to find the treasure chest. However, this player needs the help from the other player, who plays a diver when water is present to look for the key to the treasure chest. The two players are connected remotely and use voice to communicate. During the game play, the system collects the voice data, performs real-time analysis on the audio, and provides visual feedback regarding the voice loudness and pitch based on the PLVT guideline. The loudness feedback is in the form of a loudness meter where the meter would turn red if the voice is not loud enough (using 60 db as the threshold). The feedback on pitch is in the form of a notification generated when the pitch is too high (using 170 Hz as the threshold).

Practice Conditions	
Amount	→ Large
Distribution	→ Distributed
Variability	→ Variable
Schedule	→ Random
Attentional Focus	→ External
Target Complexity	→ Complex

Feedback Conditions	
Type	→ Knowledge of Results (KR)
Frequency	→ Low (with Summary KR)
Timing	→ Delayed (e.g., 5 seconds)

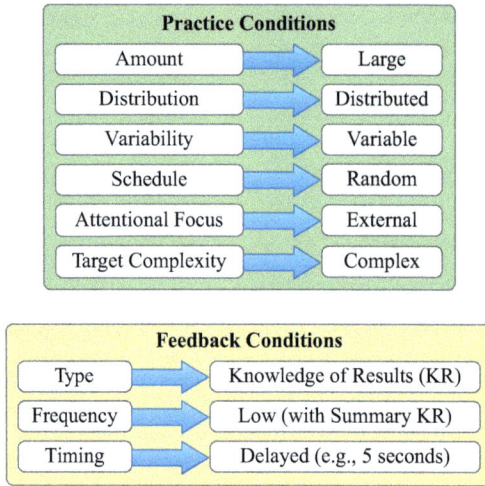

Figure 13.4   Principles for speech therapy practice conditions and feedback conditions

In [513], the design, implementation, and evaluation of a speech therapy computer game was reported. The game was designed for patients with Apraxia and has a traditional 2D cartoon-like user interface with animal avatars. The system assumes the availability of a microphone for the patient to record the voice and has a speaker. The design followed several best practices in speech therapy. First, all utterance attempts are counted towards the session goal. Second, using speech to control the game play is made as a secondary input method of the patient to perform tasks in order to collect in-game assets. Third, to encourage the patient to speak during the game play, an energy bar is displayed on the game interface indicating the remaining energy left in the avatar. If the energy drops to zero, the avatar would move very slowly and hard to control. Fourth, the system provides two forms of feedback to the patient: (1) to inform the patient the correctness of the response (this is referred to as the knowledge of response), and (2) to inform the patient what is the correct response (this is referred to as the knowledge of correct response). In the latter case, the patient is expected to judge the own response [514]. To enable the former, the system has implemented automatic speech recognition functionality. The speech exercise design followed the principles of motor learning [515,516].

In the context of speech therapy, as shown in Figure 13.4, the principles are divided into two conditions: (1) practice conditions and (2) feedback conditions. The authors of [515] performed a comprehensive literature review and found the best practices in the practice conditions in terms of practice amount, practice distribution, practice variability, practice schedule, attentional focus, and target complexity. The feedback conditions are in terms of feedback type, feedback frequency, and feedback time. These principles obviously could be applicable to a much broader inventions for human behaviors.

Chapter 14
Technology-facilitated pulmonary rehabilitation

Many respiratory and lung diseases are chronic and some of them are incurable. Respiratory diseases impact hundreds of millions of people worldwide and incur significant medical costs and loss of productivity. One of the most well-studied respiratory diseases is the chronic obstructive pulmonary disease (COPD), which is currently incurable and is projected to be the third most frequent cause of death worldwide [517].

Pulmonary rehabilitation has gone beyond the stage of academic research and is now a part of standard treatment protocol for patients with chronic respiratory and lung diseases [518]. Pulmonary rehabilitation provides a comprehensive intervention consisting of a spectrum of modules and is supported by a multidisciplinary team including not only chest physicians and nurses but also physical therapist, occupational therapist, nutritional expert, physiologist, and social workers. The goal of pulmonary rehabilitation is to improve the patients' physical as well as psychological conditions and ultimately to change the patients' behavior so that they will live a more active and therefore health-enhancing lifestyle [518].

Exercise training is a center piece for pulmonary rehabilitation because it is essential to break the vicious circle that could aggravate the patient's health. On the one hand, engaging in physical activities might induce breathlessness due to air trapping and increased hyperinflation in the lungs, which could in turn trigger anxiety or even panic. On the other hand, avoiding physical activities could weaken the muscles, and ultimately reduce the capacity to engage in physical activities. Physical inactivity has been confirmed as a predictor of mortality in patients with COPD [519].

The key elements in pulmonary rehabilitation are illustrated in Figure 14.1. Almost all chronic respiratory and lung diseases are suitable for pulmonary rehabilitation. In this chapter, we cover selected studies on COPD and COVID-19. Traditionally, COPD has been the focus of pulmonary rehabilitation. Since the COVID-19 pandemic, pulmonary rehabilitation for COVID-19 patients have received much attention, usually adapting the rehabilitation protocols originally developed for COPD. There are much fewer published studies on pulmonary rehabilitation for other diseases than COPD and COVID-19.

The pulmonary rehabilitation targets a large number of symptoms or issues caused by the chronic respiratory and lung disease [520]. The symptoms and issues can be roughly categorized into groups. Each group of symptoms and issues is addressed by

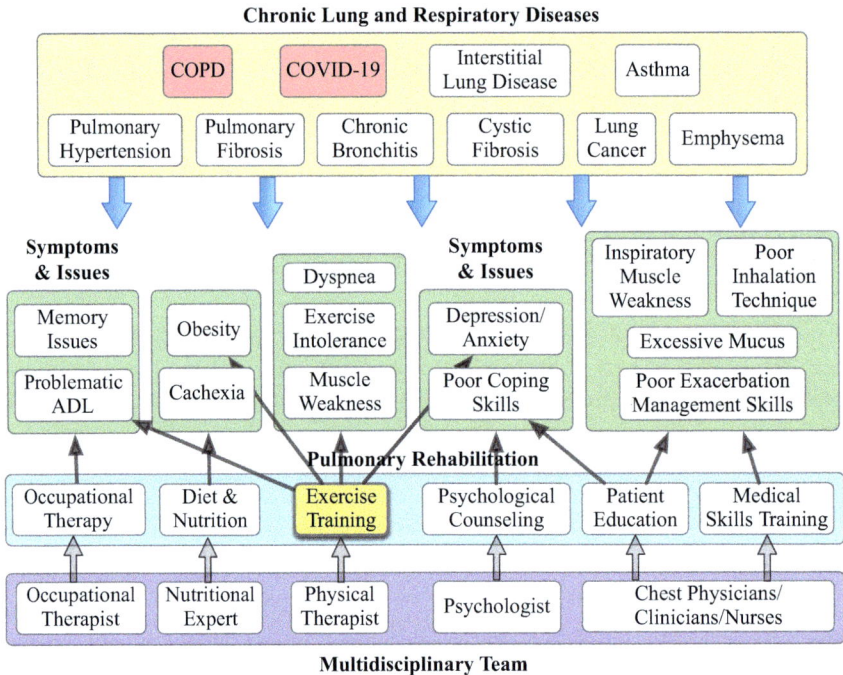

Figure 14.1   Key elements in pulmonary rehabilitation

one or more pulmonary rehabilitation modules. The pulmonary rehabilitation modules are supported by a multidisciplinary team of professionals.

Perhaps the most common respiratory symptom is dyspnea, which means short of breath. While learning proper inhalation technique could help alleviate dyspnea to some extend, exercise training has proven to improve the functional capability that improve dyspnea significantly. Muscle weakness and exercise intolerance could be caused by prolonged sedentary lifestyle, which has also been shown to improve with exercise training. Exercise training has also been shown to reduce obesity, improve activities of daily living, and reduce depression and/or anxiety levels. Besides exercise training, patient education and medical skills training are also essential elements of pulmonary rehabilitation. Medical skills training could help patients improve inhalation technique, improve exacerbation management skills, overcome respiratory muscle weakness and the presence of excessive mucus. Medical skills training could be regarded as a form of patient education. Patient education also help patients to learn and engage in self-management, such as enhance one's self-efficacy, perform goal setting, collaborating with peers. Cachexia (i.e., unintentional weight loss) could also happen in patients with respiratory and lung diseases, and it may be addressed by diet and nutrition training offered by nutritional experts. Occupational issues such as memory issues and problems in doing activities of daily living could be addressed with occupational therapy offered by occupational therapists. Psychological issues such as depression, anxiety, and poor coping skills can be addressed with psychological counseling offered by psychologists. Social workers are also mentioned in the literature as a member of the multidiscipline team [521].

14.1   Clinical scales and tests in pulmonary rehabilitation

14.1.1   The Borg Rating of Perceived Exertion

The Borg Rating of Perceived Exertion (RPE) scale measures the physical activity intensity based on the individual's perception [522]. The lowest rating level is 6, indicating no exertion. The highest rating level is 20, indicating the maximum exertion. A rating level between 12 and 14 is considered as moderate exercise level. A greater exertion intensity will be reflected from increased heart rate, increased breathing rate, increased sweating, and increased muscle fatigue [523]. The reason for the 6–20 range was because there is a strong correlation between the perceived exertion rating level and one's heart rate, i.e., given a Borg rating level of n, the heart rate is likely to be $n \times 10$ [524]. The Borg RPE is often used to prescribe and regulate the exercise intensity [525]. Some studies [526,527] used a modified version of the Borg scale that was designed for cycling strength test and has a range of 0-10 [528].

14.1.2   Dyspnea ratings

The dyspnea ratings scale was developed originally for prescribing exercise intensity in patients with COPD [525]. The scale has a ratings range of 10, with the smallest value being 0.5 indicating very slight dyspnea, and 10 being very severe dyspnea. This scale was cited in the official statement on pulmonary rehabilitation as a tool for exercise intensity regulation [518].

14.1.3   The Wisconsin Upper Respiratory Symptom Survey

The Wisconsin Upper Respiratory Symptom Survey (WURSS) [529] is designed to assess the negative impact of acute upper respiratory infection (originally for common cold). The survey has several versions with different number of questions (WURSS-44, WURSS-24, WURSS-21, and WURSS-11) (https://www.fammed. wisc.edu/wurss/). The scale was used in [530] to evaluate the symptoms of post-acute COVID-19 patients.

14.1.4   Numeric rating scale as a measure of dyspnea

The numeric rating scale as a measure of dyspnea has a range of 0–10, where 0 means there is no shortness of breath and 10 means shortness of breath as bad as can be [531]. This scale was used in [527] to evaluate the outcome of the pulmonary rehabilitation.

14.1.5 Medical Research Council dyspnea scale

The Medical Research Council (MRC) dyspnea scale was designed to evaluate the degree of breathlessness on daily activities (original for COPD patients) [322]. The scale has a range of 0–4, where 0 means feeling breathlessness only when engaging in strenuous exercise, and 4 means feeling breathless when doing slight activity such as dressing. This scale was used in [527] to evaluate the outcome of the pulmonary rehabilitation. The is also a modified version of the MRC scale (mMRC) [532]. It was used in [533] as a measure for dyspnea severity.

14.1.6 Functional independence measure

The functional independence measure (FIM) was designed to provide a comprehensive assessment of disability [383,534]. FIM has questions on motor and cognitive capabilities. Questions on motor capability include the activities of daily living on self-care, sphincter control, mobility, and locomotion. Questions on cognitive capability include communication (such as comprehension) and social cognition (such as problem solving and memory). The scale has a range of 1–7. A score of 1 means that total assistance is needed, and a score of 7 means complete independence. The scale was used in several pulmonary rehabilitation studies [535,536].

14.1.7 Cumulative illness rating scale

The cumulative illness rating scale (CIRS) was developed to assess multi-morbidity where all medical problems are considered [537]. The original scale had evaluation of 13 main systems with a range 0–4 for each system. The scale was later modified to include 14 systems [538]. This scale was used in [536] to evaluate the outcomes of pulmonary rehabilitation.

14.1.8 St. Georg's respiratory questionnaire

The St. Georg's respiratory questionnaire (SGRQ) is a popular scale to measure the health-related quality of life for respiratory patients [539]. SGRQ contains 50 items and they are divided into three domains: (1) symptoms, (2) activity limitations, and (3) the psychosocial impact. The final score ranges between 0 and 100, where 0 is the best status and 100 is the poorest health status. This scale was used in [533] to measure the quality of life of long-COVID-19 patients, and in [540] to measure the quality of life post-intervention in patients with COPD.

14.1.9 Feeling thermometer

The feeling thermometer (FT) is a visual analog scale presented in the form of a thermometer with a range of 0–100. FT is widely used in political science to gauge the popularity of the candidate prior to an election where a lower value means cold towards the candidate, and a higher value means hot towards the candidate. In health science, FT has been used to gauge the perceived quality of life of the patient [541–543]. FT was used in several studies on pulmonary rehabilitation [536].

14.1.10 The six-minute walk test

The six-minute walk test (6MWT) is used to measure the patient's functional capacity. During this test, the patient aims to walk as far as possible in 6 min. A common variable used is the distance the patient has walked during the test [544]. This test is often used in pulmonary rehabilitation studies [545].

14.1.11 Short physical performance battery

The short physical performance battery (SPPB) was developed to assess the lower extremity function [546]. The battery consists of the time it takes (1) to walk 8 feet, (2) to stand up and seat down in a chair five times, respectively, and stand with the feet side-by-side, in semi-tandem, and in full-tandem positions up to 10 s. This test was used in [547] to measure one of the outcomes.

14.1.12 Functional ambulation category

The functional ambulation category (FAC) determines six levels of walking ability regarding the amount of manual assistance required for ambulation based on the measurement of 9-m walk [548]. This scale was used to measure one of the outcomes in [547].

14.2 Exercise training

As shown in Figure 14.2, exercise training in pulmonary rehabilitation may consists of a variety of exercise types with different intensity to accommodate individual's exercise capacity and to target specifically the muscles in the respiratory system [518]. The goal of exercise training is to strengthen the muscles needed for walking without any kind of assistance (i.e., ambulation) and to improve cardio-respiratory fitness with reduced breathlessness and fatigue towards a more physically active lifestyle. According to the pulmonary rehabilitation standard [518], the exercise training plan must follow three principles: (1) the plan must reflect each individual's specific requirements; (2) the plan must exceed the loads experienced when the individual is doing activities of daily living; and (3) the plan must increase the exercise intensity gradually as the individual is making progress in the physical conditions.

14.2.1 Endurance training

Endurance training is the most intensive form of exercise training. The pulmonary rehabilitation standard [518] recommends that endurance training be conducted three to five times per week, and in each training session, the exercise intensity level should be maintained at or above the 60% maximal work rate for 20–60 min continuously. The exercise intensity can also be measured using the Borg dyspnea or fatigue score (a target range of 4–6), or using the rating of perceived exertion (a target range of 12–14).

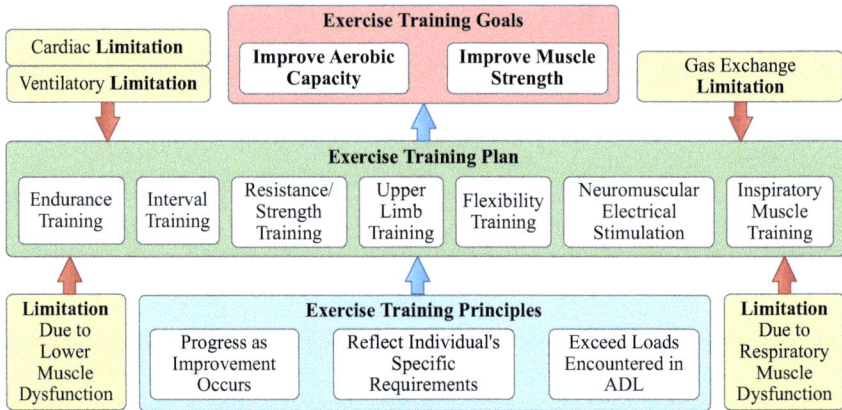

Figure 14.2   Exercise training principles, exercise training plan, and exercise training goals in pulmonary rehabilitation

14.2.2   Interval training

Interval training is designed to accommodate patients who cannot sustain the exercise intensity for the recommended period of 20–60 min for endurance training by reducing the duration of each episode of intensive exercise [518]. The episodes of intensive exercises are often distributed over time to give the patient enough time to recover with rest or lower intensity exercise in between.

14.2.3   Resistance/strength training

Resistance training is also referred to as strength training. Strength training focuses on strengthening local muscle groups by doing repetitive lifting of relatively heavy loads [518]. Strength training is appropriate for patients who have reduced muscle mass and strength, such as those who have COPD.

14.2.4   Upper limb training

Patients with chronic respiratory disease may have difficulty doing some activities of daily living, such as dressing and bathing due to lack of strength in upper limb muscles. Upper limb training could address this issue by aerobic upper extremity exercise, such as using arm cycle ergometer, and strength training of upper extremity muscles, such as biceps, triceps, and deltoids [518].

14.2.5   Flexibility training

Postural impairments are associated with pulmonary function decline, increased work of breathing, back pain, and decreased quality of life. Flexibility training aims to correct postural impairments in patients with chronic respiratory disease [518]. Flexibility training consists of the stretching of muscle groups in upper and lower body (such as calves, hamstrings, quadriceps, and biceps), and the range of motion exercises for the neck, shoulder, and trunk. The recommended target training frequency is two to three times per week.

14.2.6 Neuromuscular electrical stimulation

Transcutaneous neuromuscular electrical stimulation (TNES) is a method to elicit muscle contraction by delivering electrical stimulation on selected muscles [518]. TNES could control the amplitude, frequency, and duration of the electrical stimulus to elicit the designed muscle response. A particular advantage of TNES is that it does not trigger dyspnea and it poses very little cardio-circulatory demand. Furthermore, because TNES does not require active participation from the patient, the treatment can overcome the lack of motivation, or the inability due to severe physical impairment, to participate in conventional forms of exercise training.

14.2.7 Inspiratory muscle training

Inspiratory muscle training is used to improve the pressure-generating capacity of the inspiratory pump muscles [518]. Patients with severe respiratory diseases, such as COPD, have compromised capacity in inspiratory pump muscles. Conventional forms of exercise training, such as endurance or interval exercise training, are ineffective in improving the capacity of inspiratory pump muscles. Special devices are used for inspiratory muscle training. Such devices could apply a resistive load on the muscles to strength them over time [549].

14.3 Pulmonary rehabilitation for COPD

As a progressive chronic lung disease, the most prominent characteristic of COPD is the obstruction in airflow. Direct consequences of this obstruction include persistent and progressive dyspnea, coughing, fatigue and repeated chest infection [517]. Longer-term impacts of COPD include muscle wasting, osteopaenia, cardiovascular disease, and even depression [550]. COPD is incurable and is associated with significant financial costs [551]. Pulmonary rehabilitation was initially developed to help COPD patients to manage their symptoms and reduce the likelihood of hospital readmission. Pulmonary rehabilitation for COPD has been very comprehensively reviewed recently [517,552–555]. To avoid redundancy, we focus on studies that used technology in some ways [556,557]. In Figure 14.3, we illustrate key technologies that have been incorporated into pulmonary rehabilitation. At the highest levels, we have telehealth/tele-rehabilitation, exercise training, and self-management. Telehealth is facilitated by computing hardware, computer networking, and cloud computing. Exercise training is typically facilitated by exergames (either commercial off-the-shelf health/sports games, or custom made games) where various sensors play an important

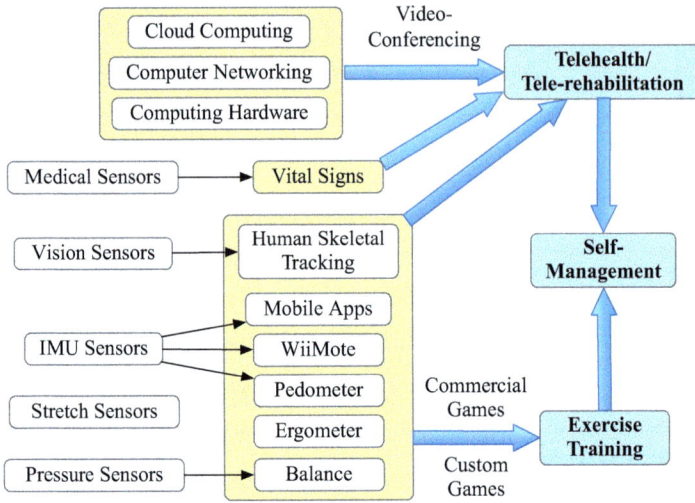

Figure 14.3   Technology in pulmonary rehabilitation

role in collecting motion data so that the user performance can be evaluated automatically and feedback can be provided. Telehealth is often relying on medical equipment that measures vital signs of the user. Tele-rehabilitation often integrated exercise training. Self-management is often monitored via telehealth and has a component in exercise training or exercise self-monitoring.

14.3.1   Functional testing and measurement of physiological parameters

Pulmonary rehabilitation requires the monitoring of the patient's symptoms closely so that the rehabilitation programs can be adjusted accordingly. Various medical equipment are used to accurately gauge the severity of the symptoms. For example, many devices are available for pulmonary function testing, such as body plethysmograph, pulmonary gas analyzer, blood gas analyzer, Silverman pneumotachometer, and pulse oximeter.

Vital signs are commonly measured during recalibration. For example, heart rate monitors have often been used to measure the exercise intensity and also to make sure that the heart rate is in a safe range [526]. Some physical capacity tests, such as the six-minute walk test, can now be measured using a mobile app [558].

14.3.2   Telehealth

The basic form of telehealth is to provide medical advises and rehabilitation guidance remotely, typically over video conferencing [559,560]. More advanced form integrate with home-based systems and/or mobile apps to collect information such as vital signs, and to help guide the patient to perform the required actions, such as rehabilitation exercises. Telehealth has becoming increasingly popular during the COVID-19 pandemic [558,561,562].

14.3.3 Technology-facilitated exercise training

There have been different levels of technology integration into exercise training for COPD. At a primitive level, motion sensors have also been used to collect data regarding the intensity and the amount of exercises done by the patient. For example, pedometer was used in several studies to measure the number of steps taken by the patient [563–565]. Some studies used mobile apps running on smartphones to measure the steps taken based on the built-in sensors [558].

At a more advanced level, technology-facilitated exercise training systems have been integrated with pulmonary rehabilitation [556]. A common strategy is to utilize commercial off-the-shelf game consoles, such as Wii/Wii-U (with Wii/Wii-U remote) and Microsoft Xbox (with Kinect sensor), and exercise-oriented commercial games, such a Wii Fit and Kinect Sports. Custom-developed exergames have also been used. More details can be found in Chapter 10.

14.3.3.1 Microsoft Xbox with Kinect

In [566], a pilot study using four min-games from the Shape-Up commercial Kinect game for pulmonary rehabilitation in COPD patients was reported. In [567,568], a pulmonary rehabilitation program for COPD patients used the set of mini-games (rafting, cross-country running, hitting a ball, and roller-coaster ride) in the Kinect Adventures game.

14.3.3.2 Wii

In [569], the Wii EA Sports Active game was used in a pulmonary rehabilitation program in patients with COPD. More specifically, four activities in the game were used: marching, dancing, and two different air-punching were used in the experimental group. The control group engaged in walking on a treadmill. For both groups, the exercise intensity was regulated using the Borg exertion scale with a target range of 14–16. The study found that the Wii game elicit a similar exercise response to the treadmill.

In [570,571], the Wii Fit game was used in a 6-week pulmonary rehabilitation program with two Wii exercise session per week. The study found significant improvements in exercise tolerance, reduced dyspnea symptoms, and improved quality of life in patients with COPD.

In [540], the Wii Fit Plus game was used as part of a pulmonary rehabilitation program (about 3 months long) for COPD patients. More specifically, yoga, jogging, and twisting and squat activities were used. The primary outcome was measured using the 6-min walk test and it showed significant improvement in exercise tolerance capability among the patients.

In [572], three Wii games, Wii Fit Plus, Wii Sports, and Wii Resorts, were evaluated in a pilot study for their feasibility and acceptability in pulmonary rehabilitation. More specifically, basic run, basic steps, obstacle course, island cycling, and rhyme parade games from Wii Fit Plus, the boxing game from Wii Sports, and the canoeing game from Wii Resorts are used for lower body and upper body work out. The study concluded that moderate exercise using these games was safe, feasible, and enjoyable for pulmonary rehabilitation.

14.3.3.3   Cycling exercise machine

In addition to commercial games with game consoles, commercial off-the-shelf cycling exercise equipment that integrates technology for exercise intensity control, visual display, and the measurement of the pedaling speed has also been used for pulmonary rehabilitation studies. In [573], a custom-made group exercise system called virtual exercise environment was developed where a commercial off-the-shelf cycling machine (Bremshey BE5i home trainer) was a key component. The system incorporated a heart rate belt (polar T31) to measure the heart rate while a subject is doing cycling exercise. The virtual reality interface of the system was provided by a cycling mobile app from WebAthletics and displayed on a 22-in monitor in front of the cycling machine. The controller component of the system communicates the data collected from the cycling machine and the heart rate belt to an online portal. However, even though the system was developed for COPD patients to perform pulmonary rehabilitation, only healthy subjects were recruited to test the feasibility of the system.

In [574], a virtual park system was developed for pulmonary rehabilitation. The custom system is integrated with a commercial cycle-ergometer (COSMED 100k) and a pulse-oximeter (NONIN 3150 wrist-worn, measuring both heart rate and the level of oxygen saturation), and provided virtual reality interface displayed on a wide projected screen. The system displays the cadence of the cycling machine, the distance traveled, and the current time on the patient-side interface. The system also provided a clinician-side interface. The usability and acceptability of the system was evaluated by nine patients with COPD or bronchial asthma.

14.3.3.4   Custom-developed exergames

In [575], a virtual reality headset (from Pico) with a custom-made app was used in an 8-week pulmonary rehabilitation program to provide virtual reality environment for the participants. A small probe Nonin 3150 was used to measure heart rate and oxygen saturation levels during the exercise. The virtual reality app consists of eight modules, and it transmitted data collected in real time to a server. The usability of the system was evaluated with 10 older adults with COPD.

In [576], three computer games (the balloon game, the eating game, and the penguin game) were developed to facilitate breathing training (pursed-lip breathing and holding breath and coughing). The breathing movements during the exercise training were captured using stretch sensors (from Thought Technology) attached to the clothes. In [577], a mobile app called flappy breath was developed to help doing breathing training where the breathing movements were captured using both the acoustic sensor in the smartphone and a commercial breath belt (from Advanpro).

14.3.4 Technology-facilitated self-management

Although self-management is listed as one of the modules in pulmonary rehabilitation, it is typically offered in the form of education and the duration of the program is often short ranging between a few weeks to a few months. It is well known that the positive effects of pulmonary rehabilitation are usually short term. Ideally, patients with COPD (and other chronic respiratory diseases) should adopt a more physically active lifestyle and manage their symptoms proactively to enjoy a higher quality of life. Technology could facilitate long-term self-management with little burden to the patients and incur minimum financial cost. Such long-term self-management programs are usually enabled by a Web-based app or a cloud-enabled mobile app, and ideally some equipment for the patients to: (1) measure parameters related to their symptoms and (2) measure the amount of physical activities (such as a pedometer to the number of steps taken, or a ergometer for cycling) [552]. The Web app or the mobile app would facilitate communication between the clinicians and the patients, offer coaching either virtually (such as chatbot and videos) or remotely by a clinician/nurse, and collect data about the patients' activity and status.

In [578], a randomized trial on the efficacy of a technology solution for self-management of COPD was reported. The study formed two experimental groups and one control group. Each participant in either experimental group was given a cloud-enabled health kit and was asked to record a set of information: (1) oximetry and blood pressure; (2) optionally temperature and weight; and (3) results for the COPD assessment test and the modified medical research council scale. The control group received standard care with routine in-person appointments and access to a certified respiratory educator. The cloud-enabled health kit has the following hardware components: (1) a tablet computer; (2) a wrist cuff blood pressure measurement device; (3) an oximeter; (4) a weighing scale; (5) a thermometer. The devices were connected to the tablet via Bluetooth so that the reading can be recorded automatically (presumably). The tablet also had the digital version of the COPD assessment test and the modified medical research council scale so that the participants could complete the assessments on their own. One experimental group received remote monitoring. The other experimental group used a self-monitoring program. The primary outcome of the trial was evaluated using the partners in health scale [579]. The secondary outcomes were the COPD severity and knowledge evaluated by a number of scales. The study found that there was no statistically significance between the three groups.

In [580], three mobile apps designed to help patients with COPD to engage in more frequent regular walking were evaluated by a group of patients with COPD. The development of the mobile apps followed persuasive design principles [581]. There are four persuasive design principles: (1) primary task support, meaning helping users to carry out the designated task; (2) dialogue support, meaning motivating users via feedback and interaction; (3) social support, meaning facilitating social influence; and (4) credibility support, meaning the app/system itself should appear credible to users. This study developed three mobile apps, with each one followed a different persuasive design principle (the first three principles), respectively. Interviews and questionnaires were used to collect participants' feedback and preferences. It is unclear how long the participants have used the apps prior to providing feedback, and to what extend the apps were used by the participants. The study the mobile app that followed the dialogue support principle received the most positive feedback.

In [582], a Web-based app was used to patients with COPD for self-management. The app provided online self-management resources, including video instructions for seven strength and balance exercises that can be performed at home and link to related websites, were also included. Furthermore, the app provided tailored feedback and prompts for doing home exercises and other related information. The outcome of the program was evaluated after 6 months.

In [583], a 9-week home pulmonary rehabilitation program was conducted to access the acceptability of conducting rehabilitation online with 10 participants in two groups. The primary technology is a compact computer with a webcam and Internet connection using TV as the monitor. Online communication was done via videoconferencing. The study consisted a component of supervised exercise sessions two times a week conducted online. Each participant was given a pedometer to motivate walking (and a pulse oximeter as well because its readings were required). Each participant was required to keep a digital diary with a list of specific questions to answer with step counts and pulse oximetry values. The usability of the technology was assessed via interviews and the system usability scale questionnaire. The feedback from the participants was positive.

14.4   Pulmonary rehabilitation for COVID-19

The coronavirus disease 2019 (COVID-19) has vastly heterogeneous severity on the population [584]:

- About 40% of COVID-19 patients have mild symptoms.
- About 40% of patients have moderate symptoms.
- About 15% of patients have severe symptoms.
- About 5% of patients develop critical symptoms.

A portion of the COVID patients suffer from various aftereffects (i.e., sequelae), such as fatigue, breathlessness, and neuropsychological impairments. Usually, these patients are referred to as long-COVID-19 patients [585]. For these patients, pulmonary rehabilitation would be necessary to help them manage and improve their symptoms.

In [530], the effects of an exercise training program were studied in a randomized controlled trial (with 30 participants who had mild or moderate COVID-19). The experimental group engaged in a 2-week (three sessions per week) aerobic exercise program. Both the experimental group and the control group received standard medications according to the Turkish Ministry of Health. Each exercise session consists of 30 min of moderate intensity walking/running on a treadmill or bicycling on a stationary bicycle. The exercise intensity is regulated using the Borg RPE rating level between 12 and 14. The Wisconsin upper respiratory symptom survey was used to evaluate the negative effect of acute upper respiratory infection. The study found that the Wisconsin scale score was significantly reduced in the experimental group compared with the control group after the exercise intervention.

In [586], the outcome of an 8-week unsupervised pulmonary rehabilitation program for 20 previously hospitalized COVID-19 patients was reported. The pulmonary rehabilitation program consists of an exercise program and nutrition recommendation. The compliance to the program was checked using two phone calls per week. The participants were required to engage in three sessions per week. Each session lasted for about 100 min, 50 min of which were aerobic exercise with walking. The participants were assessed with a number of tests. The study saw significant improvements in dyspnea during the 6-min walk test, sleep quality, oxygen saturation, hemodynamic parameters, plasma antioxidant capacity, and body composition.

In [545], the outcome of a 12-week pulmonary rehabilitation program for 98 previously hospitalized COVID-19 patients was reported. Those who have completed at least five sessions of pulmonary rehabilitation (27 of the participants) formed the experimental group. Those who have not participated any rehabilitation session formed the control group. The pulmonary rehabilitation program consists of: (1) inspiratory muscle training, (2) pursed-lip breathing and active cycle of breathing technique, (3) maximal voluntary diaphragmatic contractions, (4) interval training with two 4-min bicycle or treadmill high intensity exercise (4-min low intensity exercise in between). The interval training was offered 3–5 days per week. The exercise intensity of the interval training was controlled via Borg RPE rating level of 13–16, or 40–80% of the maximum heart rate (i.e., defined as 220 minus the age per minute). The experimental group showed significantly longer distance of the 6-min walk at the end of the program then the control group.

In [562], the effects of a virtual physical therapy intervention program on improving functional capacity in COVID-19 patients who were previously hospitalized were evaluated. The 105 participants were divided into four group: (1) virtual physical therapy group (44 participants), (2) home physical therapy group (25 participants), (3) independent exercise group (17 participants), and (4) no exercise group (20 participants). The program duration was approximately two and half months (April 24, 2020–July 13, 2020). The virtual physical therapy took place one to two times per week. The exercise program consists of eight exercises: (1) diaphragmatic breathing, (2) incentive spirometry, (3) sit to stand, (4) standing marching, (5) shoulder scaption, (6) standing heel raises, (7) sidestepping, and (8) wall push-ups. Each exercise session lasted for about 30–60 min. During the exercise sessions, the patient's rate of perceived exertion was recorded and monitored. The participants' lower body strength was measured using the 30-s sit-to-stand test, the cardiopulmonary endurance was measured using the 2-min step test. The study found that the virtual physical therapy group had more significant improvements compared with the other three groups.

In [526], the outcomes of a pulmonary rehabilitation program for patients were reported. The program included aerobic durance training and resistance training in each session, and the training took place twice per week. The exercise intensity was regulated using a modified version of Borg scale in the range of 4–6. The endurance training consists of two sessions of cycling in continuous mode and two sessions of cycling in interval mode using a computer-guided cycle-ergometer. During the continuous mode, the training intensity was at 20–30% of the maximum peak work rate. During the interval mode, the maximum intensity (for 4 min) was at 50% of the maximum peak work rate. The outcome of the intervention was measured using the 6-min walk distance. Nine out of 12 patients who completed the training exhibited significant improvement in the 6-min walk distance.

In [527], the outcomes of a comprehensive pulmonary rehabilitation program for 108 patients who have overcome acute COVID-19 but still symptomatic were reported. The pulmonary rehabilitation program has the following components: physical exercise training, respiratory physiotherapy, general physiotherapy, patient education about COVID-19, medical supervision, psychosocial support, nutritional counseling, occupational therapy. Not all patients participate in all the components. The patients were divided into three groups based on the initial symptom severity levels (severe acute, severe after interval, mild after interval). The physical exercise training consisted of endurance training and strength training, which all participants would participate, and two particular trainings designed for specific patients. Whole-body vibration training was offered to patients who satisfied certain requirements (i.e., no thromboembolic complication and no elevated D-dimer level). Inspiratory muscle training was offered to patients who had inspiratory muscle weakness. Endurance training was done in the form of cycling and it was offered 3–5 sessions per week and lasted 30–60 min in each session. The training intensity was controlled by the modified Borg scale with a target range of 4–6. During the training, the heart rate and the oxygen saturation in the blood (via a pulse oximeter) were measured. The strength training took place 2–3 times per week and each session lasted 45–60 min. The study found significant improvements in exertional dyspnea, physical capacity (as measured by the 6-min walk test), quality of life, fatigue, and depression in the participants.

In [535], a controlled study on the effectiveness of pulmonary rehabilitation in severe and critically ill COVID-19 patients was reported. The study formed two groups. One group consists of 51 COVID-19 patients, and the other group consists of 51 patients with common pneumonia. The COVID-19 group has two subgroups, one with severely ill patients, and the other subgroup with critically ill patients. The rehabilitation plan followed the standard recommendation [518]. The study found significant improvements in both groups according to a number of metrics, including 6-minute walk test, functional independence measure [383], and the chronic respiratory questionnaire [587].

In [588], a study on the effectiveness of pulmonary rehabilitation in post-ICU COVID-19 patients was reported. The study formed two groups. One group consisted 21 post-ICU COVID-19 patients. The other group consisted 21 post-ICU non-COVID-19 patients with respiratory failure. Both groups participated pulmonary rehabilitation according to the standard recommendation [518]. Both groups showed improvements according to the 6-min walk test. The COVID-19 group made more significant improvements over the non-COVID-19 group.

In [536], a prospective study on the effects of pulmonary rehabilitation for COVID-19 patients compared with patients with other lung diseases was reported. The pulmonary rehabilitation program was 3-week long and had a total of 25–30 therapy sessions on 5–6 weekdays. The study followed the standard pulmonary rehabilitation recommendations [518]. The pulmonary rehabilitation consisted individualized endurance exercise and strength training customized based on the severity of the patients and their functional physical limitations. The outcomes of the program was assessed using FIM, 6-min walk test, CIRS, and FT. The study found that the COVID-19 group made significant improvements after pulmonary rehabilitation as measured by FIM, 6-min of walk test, and FT.

In [584], the effects of a prospective, comprehensive 3-week pulmonary rehabilitation program for COVID-19 patients were reported (24 mild/moderate and 26 sever/critical). The pulmonary rehabilitation program followed the standard recommendation [518]. The study found that participants made significant improvements as measured by 6-min walk test and a lung function test.

In [547], the effects of a therapeutic exercise program for 33 post-acute COVID-19 patients were reported. The program included a daily 30-min exercise session for several days (the mean duration was 8.2 days). The exercise session consisted of resistance training, endurance training, and balance training. The exercise intensity was regulated using the modified Borg scale with a target range of 3–5. In addition, breathing exercises and manual therapy were provided to patients who needed them. The outcomes were evaluated using the short physical performance battery, the Barthel index, functional ambulation category, and the single leg stance test. A portion of the participants performed the 6-min walk test. The study found that all participants made significant improvements over all outcomes.

In [558], the outcomes of a one-week telerehabilitation program for COVID-19 patients were reported. The experimental group (with 18 participants) engaged in daily resistance and strength training with a set of 10 exercises. The number of repetitions for the exercises depended on the intensity level measured by the Borg scale. The control group (with 18 participants) did not receive any exercise training. The outcomes of the program were evaluated using the 6-min walk test (the distance was measured by a mobile app), 30 sit-to-stand test, and the Borg scale measured at the end of the 30 sit-to-stand test. The experimental group exhibited statistically significant more improvements according to all three measurements than the control group.

In [533], the outcomes of a pulmonary rehabilitation program for 68 long-COVID-19 patients were reported. The intervention program were delivered remotely via Zoom, and it consisted of 18 40-min sessions at three times per week. The pulmonary rehabilitation program consisted of sanitary education, respiratory related skills education, aerobic exercise training, and other specifically targeted training. The study found that the participants made significant improvements in activities of daily living (measured by the exertion made during activities of daily living using the modified Borg scale), dyspnea severity (measured by the modified MRC dyspnea scale), and quality of life (measured by the St. Georg's respiratory questionnaire [539]).

Chapter 15
Technology-facilitated cognitive rehabilitation

Although the terms "cognition" and "cognitive" have been pervasively used, they are rarely defined clearly. In [589, p. XV], cognitive science is defined as "the study of mental representations and computations and of the physical systems that support those processes." Presumably, "those processes" refer to the processes of mental computations. The Wikipedia page used the dictionary definition for cognition. Nevertheless, it appears that cognition is an umbrella term referring to anything with mind and brain. Hence, the scope of cognition is very broad, and there are many theories under the scope. The Wikipedia defined the scope of cognition as "all aspects of intellectual functions and processes such as: perception, attention, thought, intelligence, the formation of knowledge, memory and working memory, judgment and evaluation, reasoning and computation, problem solving and decision making, comprehension and production of language" without citing any references (https://en.wikipedia.org/wiki/Cognition). The scope of cognition can be implied from the chapters included in [589], which have the following: coordinate transformations in the genesis of directed action, attention, categorization, reasoning, cognitive development, the brain basis of syntactic processes, cognitive neuroscience, and emotion. In [590], 20 conscious brain events and 20 unconscious brain events are provided. Some of the listed conscious brain events include immediate memory, declarative memory, episodic memory, automatic memory, intentional learning, attended information, effortful tasks, strategic control, explicit inferences. Cognitive rehabilitation often aimed to improve these functions in the patients.

In [591], the mutual beneficial relationship between regular physical exercises and cognition are highlighted. Although it is well recognized that regular physical exercises would have physiological benefits, they could have cognitive benefits as well. More specifically, regular physical exercises could improve cognitive performance [592]. The most well-recognized cognitive attributes that benefit from physical exercises are executive functions [593–598] and effortful control [591], although other attributes have also been observed, such as memory [599], attention [600], spatial learning, and psychomotor functions [601]. On the other hand, an individual with high capacity in effortful control and executive functions could help adherence to physical activity programs. This cyclic relationship between regular physical exercises and cognition is illustrated in Figure 15.1.

Executive functions consist of three components: (1) the management of working memory; (2) the ability of doing intentional inhibition of harmful impulses, negative

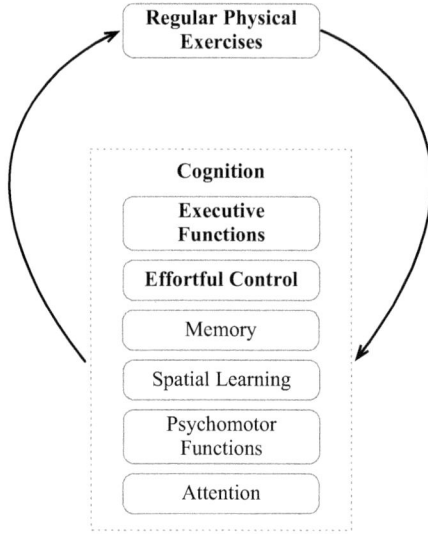

Figure 15.1 The cyclic relationship between regular physical exercises and cognition

thoughts and emotions; and (3) cognitive flexibility, which is the ability to adapt one's behavior towards the goals when the environment changes. Effortful control (also called self-regulation) [602] refers to the ability to regulate one's attention and to control one's inhibit and activate behavior as needed [603]. According to [591], executive functions are prerequisite of self-regulation.

There are several hypotheses on the underlying mechanisms for the cognitive benefits of regular exercises [604]. In [605], brain-derived neurotrophic factor (BDNF) was believed to play a critical role in brain plasticity because BDNF could mediate changes in cortical thickness and synaptic density in response to physical activity and environmental enrichment. This hypothesis is supported by [606]. In [607], the introduction of angiogenesis and increased blood flow in brain were observed in rats after vigorous physical activity and motor skill training. In [608] the authors proposed a hypothesis that the hippocampal neurogenesis, brain angiogenesis, and the synthesis of monoamines are stimulated by the neurotrophic molecules released while exercising. According to this hypothesis, the improvement on cognition is a reflection of stronger brain integrity and brain functioning induced by the stimulation caused by regular physical activities. In [609], the reduction of pro-inflammatory cytokines and improved BDNF levels were observed at the end of a 16-week multimodal physical exercise intervention program. At a higher level, the proposed mechanisms include increased cerebral blood flow [610], better cardiovascular and cerebrovascular fitness [611], and the stimulation of neuronal growth and survival [612]. The cardiovascular fitness has been widely recognized as a good predictor for positive impact on cognition level [601,613,614].

Physical exercises have the strongest effect on the cognition of older adults and children [591] for good reasons. Children are developing their cognitive function, and the cognitive functions for older adults are declining. By doing regular appropriate physical exercises, children could develop more balanced cognitive functions, and older adults could slow the decline cognitive functions or even recovery some lost cognitive functions.

15.1 The impact of physical activities on cognition for children and young adults

The interest in incorporating regular physical activity into the curriculum of school children is primarily due to its positive effect on improved academic performance [615]. It has also been explored as a way to reduce overweight and obesity among children [616,617]. The mediating factors for the improvement in academic performance are hypothesized to be improved cognitive levels, particularly the executive function. Some studied hypothesized that there is direct correlation between physical fitness (particularly the cardiorespiratory fitness) and the academic performance [618]. In addition, other mediating factors were also recognized, such as increased arousals, reduced boredom, and better self-esteem [619]. The increased arousal and reduced boredom could lead to increased attention span and concentration, and the increased self-esteem may lead to improved classroom behavior, which are all contributing to better academic performance [619]. The mediating factors due to regular physical activities towards better academic performance are summarized in Figure 15.2.

For children, a common practice is to use academic performance as a metric for the efficacy of physical activities. Academic performance can be considered to reflect the cognitive levels of the students. In [620], college students were surveyed regarding their levels of physical activities and the results are correlated with their academic

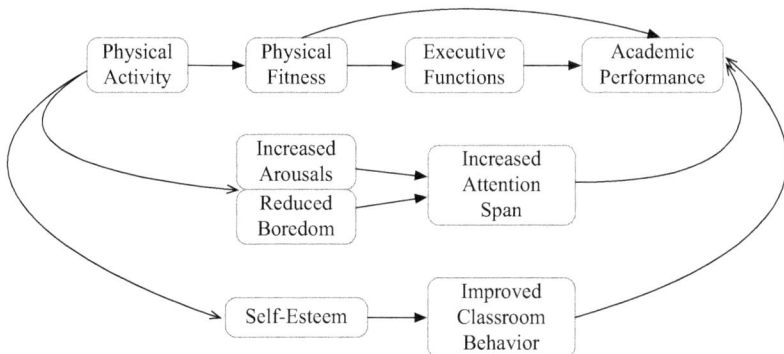

Figure 15.2   The mediating factors due to regular physical activities towards better academic performance

performance. The study found that doing physical activities for two-to-three hours per week can be positively associated with higher academic performance.

In [621], a randomized controlled trial of 3-year long was conducted to evaluate the impact of physical activities to the elementary students' academic performance and obesity issue. The experimental group incorporated a physical activity across the curriculum program to promote 90-min per week of moderate to vigorous physical activities as part of the curriculum. The control group did not incorporate the intervention program. Significantly better improvements in academic achievement were observed for the experimental group compared with the control group in all subjects, including composite, reading, math, and spelling.

In [622], the correlation between two types of physical activities and the academic achievement of elementary students was studied using a large database (Early Childhood Longitudinal Study-Kindergarten) and questionnaires to parents and school administrators. The school administrators were surveyed regarding the school-based physical education, and the parents were surveyed regarding the additional physical activities outside school. The study found a strong correlation between physical activities and the mathematical and reading achievement.

In [623], a randomized trial was conducted to evaluate the effectiveness of a developmental movement program on the first-grade children. The study consisted of four groups: an experimental group that incorporated the intervention program, a control group, a free-play group, and an educational toys group. The study showed that the experimental group made the most significant improvement in spatial development, reading and mathematical skills compared with the other three groups.

In [619], the correlation between the engagement of moderate to vigorous physical activity and the academic performance was studied in elementary students. Students were surveyed regarding the physical activities they performed on the previous Sunday, Monday, and Tuesday on every Wednesday (the method is referred to as 3D physical activity recall). The academic performance was assessed using the grades in four core courses and a standard test score. The assessment was done for each participating student at the beginning, middle, and end of the school year. The study found that students who performed some or met a target for moderate to vigorous activity had significantly better academic performance than those who performed no vigorous activity. The target for the moderate to vigorous activity is defined as doing moderate activity are 30 min per day for at least 5 days per week, or doing vigorous activity for 20 min per day for at least 3 days per week.

In [624], the efficacy of a 2-year health-related school physical education program in a school district was evaluated regarding their impact on the academic performance of the students. Two experimental groups and one control group were used in the study. In the first experimental group, specialists taught the sports, play, and active recreation for kids curriculum. In the second experimental group, the same program was taught by trained classroom teachers. In the control group, the usual programs were used. Overall, the two experimental groups exhibited better academic achievement.

However, some studies have reported ineffectiveness of physical activities on improving the cognitive levels of students. For example, in [625], a 16-month randomized controlled trial was conducted to access the impact of school-based physical activity intervention on academic performance in elementary children. The experimental group incorporated school-based physical activity intervention. The control group used usual practice. Although the physical activity in the experimental group increased by 47 min per year, the academic performance between the two groups showed no significant difference.

In additional to studying the effectiveness of intervention programs, some studies examined the relationship between fitness and cognitive capabilities. In [616], the study focused on finding the relationship between specific physical fitness components and various cognitive functions of 100 elementary children with overweight and obesity. The physical fitness components include muscular strength, speed-agility, and cardiorespiratory fitness. The fitness components were evaluated using the ALPHA battery [626], and physical activity and sedentary time (measured by accelerometry). Three specific executive functions were considered, including cognitive flexibility (measured by the Design Fluency Test and Trail Making Test [627]), inhibition (measured by the Stroop test [627]), and planning ability (measured by the Zoo Map Test [628]). The study found that cognitive flexibility was robustly associated with all fitness components, planning ability was positively related to muscular strength (handgrip strength), and inhibition was positively related to speed agility. The study also found that the metrics of physical activity and sedentary time have no correlation with executive function performance.

In [599], the impact of aerobic exercise on affect and specific cognitive functions was studied with 28 students (ranging 17–29 years of old). The experimental group was asked to engage in three 30-min running sessions per week for 6 weeks. The control group proceeded with their normal activities. The affect was measured using the German version of the Positive and Negative Affect Schedule (PANAS) [629]. The Visual and Verbal Memory Test (VVM) was used to assess short-term maintenance of visuospatial and verbal material. The d2 Test of Attention was used to assess an individual's attention and concentration. The experimental group exhibited significant improvement in positive affect, and significant improvement in visual spatial memory performance. However, concentration performance and verbal memory were not impacted.

In [618], the correlation between physical activity, physical fitness, cognition, and academic performance was studied using data collected from 186 elementary students (9–11 years old). The cardiorespiratory fitness was used as the physical fitness metric. Three executive functions were examined, including inhibition, cognitive flexibility, and working memory. The study found that (1) vigorous physical activities were positively correlated to cardiorespiratory fitness, (2) cardiorespiratory fitness was positively correlated to academic performance, (3) inhibition and cognitive flexibility were positively correlated to academic performance, and (4) only inhibition was significantly correlated to vigorous physical activities, probably it is more sensitive to physical activities.

15.2 Technology-facilitated detection of mild cognition impairment and dementia

As of 2013, close to 15% of seniors (70 years old and older) in the United States suffer dementia, and the annual financial cost of dementia ranges between $41,689 and \$56,290 per person, and between \$157 and \$215 billion total cost nationwide every year [630]. Alarmingly, there are about 10 million new cases of dementia each year [631]. To improve the quality of life of the seniors and to mitigate the cost of care, early diagnosis dementia is essential, which requires accurate detection of symptoms and incidents in the pre-dementia stage of mild cognitive impairment (MCI).

Clinical assessment of MCI is usually done via several clinical scales, such as the neuropsychiatric inventory [632], Diagnostic and Statistical Manual of Mental Disorders criteria [633], and Geriatric Depression Scale [634]. These scales aim to capture neuropsychiatric symptoms that are associated with heightened risk of dementia [635]. Although these scales are useful, they depend on self-reporting or informant reporting. Inevitably, they do not consistently capture important clinical changes that could be predictors of early dementia, particularly in areas including sleep duration, lability of mood and changes in attention with inability to sustain activities, changes in meal times and appetite and circadian changes in functional capabilities. These factors are often difficult to notice because they require the caregivers or patients to have acute awareness to these changes.

MCI was initially characterized by the Peterson criteria [636], where the patient has memory complaints and abnormal memory for age, but could engage in normal activities of daily living and has normal general cognitive function. Further research [637] has demonstrated the need to expand the scope of MCI to include amnestic and nonamnestic MCI, or single-domain or multi-domain, and categorize activities of daily living into basic activities of daily living (BADL), such as eating, drinking, and taking clothes on and off, and instrumental activities of daily living (IADL), which refers to more complex activities such as financial planning and operating everyday technology at home (e.g., smart phone). IADL of MCI patients could show signs of deficits.

Many clinical studies have been conducted characterizing IADL of MCI patients. Two types of instruments have been employed in these studies: (1) self-report or informant-report rating instruments via interviews and surveys [638]; and (2) performance-based instruments where a patient is asked to conduct a number of designated activities in a controlled environment where the person is observed directly by a trained clinician [635,637]. A standardized everyday technology use questionnaire (ETUQ) was proposed and has been widely used [639]. Most of these studies are randomized controlled studies using an experimental group with patients who have MCI, and a control group with healthy subjects. Both groups are required to perform a set of predefined IADL tasks. The performance of these tasks is evaluated using a number of metrics, such as the time it takes to complete each task, and the number of errors made while performing the tasks. Instead of directly observing a subject to perform the designated tasks, it is also possible to video-record the actual movements of the subject for online observation [640] or offline analysis [641,642]. The assessment is more ecologically valid due to the absence of raters as seen by the participant. For offline analysis, the video enables a rater to accurately measure the time it takes to complete each IADL task and the completion quality. Performance-based assessment is generally regarded to be more ecologically valid than self-reporting and informant-reporting because it avoids the issues of under-estimate or over-estimate

Clinical Scales			
Performance-Based Instruments	Tracking of Everyday Technology Use	Fully-Instrumented ADL Tracking	
Self Reporting	Informant Reporting	Video-Based Tracking	Minimally-Instrumented ADL Tracking

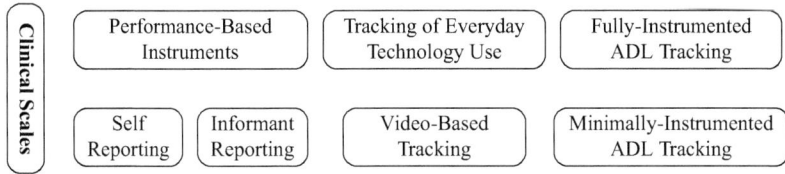

Figure 15.3   MCI detection methods

of functional abilities of the patients [643]. These studies reported consistent results on the validity of using IADL as a means to differentiate patients who have MCI from healthy subjects [644–648].

The common methods for MCI detection are summarized in Figure 15.3. For all methods, the clinical scales are used in some ways, either directly by clinicians for subjective determination of MCI symptoms, or translated into way that can be matched by classification based on the data collected. As we have mentioned earlier, traditionally, clinically studies predominately used self-reporting, informant-reporting, or performance-based instruments to collect data, which would require subjective observation and determination. Facilitated by the development of sensing technologies [649–651] and machine learning algorithms [652,653], objective detection of MCI symptoms has been investigated extensively. The methods include video-based tracking of ADL, tracking of ADL in a fully-instrumented smart home, and tracking of ADL in a minimally instrumented smart home.

15.2.1   Video-based detection of MCI

Several studies have reported the capability of automatically characterizing the ADL performed by a subject [654,655] using the computer vision technology [656–658]. To facilitate the recognition of ADL, an event-driven context model was proposed [657]. This model aims to capture the dynamic hierarchical structure of human actions, human–object interactions, and the environment information. This model consists of three layers. The bottom layer uses a variety of sensors to detect the presence of a human and to collect the state information of household appliances and objects (e.g., TV, range, refrigerator, and tables). The middle layer uses the computer vision technology to determine the human actions such as lying, sitting, standing, walking, and the states of environment objects, such as whether or not an appliance is on, or the fridge door is open. The top layer recognizes human activities using some ontology predefined for ADL, and the human actions and the states of the environment objects detected in the middle layer.

In [654], automatic activity recognition was accomplished with four modules: (1) human subject detection using background subtraction; (2) human subject tracking; (3) gait analysis; and (4) event recognition based on an ADL model, which was extended from the event-driven context model [657]. The model is based on a set of predefined physical objects (e.g., a human subject or stationary objects) and a set of sub-events. The study demonstrated a relately high recognition rate in ADL (e.g., 93% for the prepare-medication task). The study also showed a fairly high precision (83.67%) in classifying healthy subjects, patients with MCI, and patients with dementia. In the study, each subject performed two sets of activities. The first set are guided activities consisting walking, counting backwards, doing both walking and counting backwards at the same time, repeating a sentence, and articulating controls. The second set are semi-guided IADL consisting watering plant, preparing tea, preparing medication, managing finance, watching TV, making and receiving phone calls, and reading article and answering to questions. The classification was based on the following metrics: (1) single task duration; (2) single task gap duration; (3) single task standard deviation steps; (4) dual task gap duration; (5) dual task maximum steps taken; (6) person using medicine box frequency; and (7) person using medicine box duration.

15.2.2   *MCI-detection via fully-instrumented smart home*

Performance-based assessment and the video-based MCI detection use a carefully crafted set of activities for the subjects to complete. A concern for such a design is that the activities are not performed in their natural context. Hence, the ecological validity of the acquired data could be questionable.

A residential home could be instrumented with a variety of sensors to fully monitor the ADL of its residents. This type of homes are referred to as smart homes. The smart home could facilitate the observation of a human subject living in a near realistic environment where the subject could sleep and carry out the daily activities as usual. The sensors used in a smart home could be roughly divided into three types [659–662]: (1) human motion sensors; (2) contextual sensors; and (3) wearable sensors. These sensors combined could make it possible to track fine-grained movements of a participant, which could be used to detect subtle abnormal behaviors.

Human motion sensors are used to collect data on human movements. Common human motion sensors include passive inferred (PIR) motion sensors, presence sensors, proximity sensors that trigger when a human subject moves from one room to another, and moves about within a room. It is also possible to install depth cameras such as Microsoft Kinect [169,663] around the household to detect not only the presence of a human subject but also the posture of the subject. Sleep sensors have also been used (usually placed underneath the mattress) to collect data about the subject's sleep patterns.

Contextual sensors are used to establish the context of the ADL of the human subject and to facilitate the detection of human–object interactions. Contextual sensors include IMU-based sensors, plug sensors, temperature sensors, light sensors, humidity sensors, smoke sensors, and gas sensors [664]. IMU-based sensors are usually attached to furniture, appliance, and other key areas/objects, such as chairs and refrigerator door, microwave oven, stove, kitchen sink, and medicine box. Plug sensors are attached to electronic devices to collect power consumption data. Temperature sensors are placed near the stove to detect cooking activities.

Wearable sensors include smartwatches and wristband health trackers worn by the human subject to collect motion data and possibly heart rate data.

A particular challenge in this line of research is how to meaningfully process the data collected from a large number of sensors in different modalities. In [660], defeasible logics [665] was used to help semantic interpretation of the data collected from many sensors that are very heterogeneous and noisy. Ontology was first constructed to represent activity-related information. Then, rules are used to define the temporal relationship and contextual conditions that are essential to recognize complex activities to aggregate activities.

In [659], a hybrid technique that combines a knowledge-based inference engine and statistical reasoning was used to detect fine-grained abnormal behaviors. Statistical reasoning was used to recognize activities based on temporal features of the data collected. Then, the knowledge-based inference engine was used to detect anomalies based on the models defined by clinicians. In [666], a recurrent neural network and an auto-encoder-based method were used for abnormality detection in MCI patients.

In [667], the activity pattern was established by observing the number of events in three different periods: (1) 24-h day (midnight to midnight), (2) day time (8 am–8 pm), and (3) nighttime (midnight–6 am). The dimension of the data was reduced using the principal component analysis. Then, fuzzy pattern tree and support vector machine were used to detect abnormality in ADL.

In [668], an unsupervised learning method was used to recognize the activity patterns based on four sets of time-series data: (1) the presence of the subject in each room; (2) types of activities of the subject; (3) activity level changes throughout the day; (4) mean heart rate of the subject. The unsupervised learning method consists of three steps: (1) preprocessing, discretization using k-means, and temporal aggregation; (2) feature mining by finding the most relevant subsequences; and (3) clustering based on the features obtained in step 2.

In [669], the Hidden Markov Model was used to detect individual activities and a clustering algorithm was used to recognize activity patterns. In [664], activity abnormality was detected using the temporal relationship of low-level events. The low-level events are grouped pairwise with an interval, e.g., the lamp was turned on and then off. This approach enables the use of temporal logic to associate different event intervals with temporal relationship, which helps recognize abnormality.

In [659], several commission anomalies in IADL were detected using a fully-instrumented smart home, including substitution with a wrong object (e.g., using a spoon instead of a knife to spread the butter on the break, doing some wrong (e.g., putting a milk container to a cabinet instead of the refrigerator), doing an activity repetitively (e.g., eating breakfast twice). In [662], a fully-instrumented smart home was used to identify deficiencies across multiple domains such as sleep, ADL, mood, and social interaction of patients with MCI.

15.2.3 *MCI-detection via minimally instrumented smart home*

A different strategy in MCI-detection is to collect just enough data to establish patterns of ADL with the lowest cost and minimum intrusiveness to the human subject being monitored [670]. This approach has greater chance of being adopted for MCI-detection (and other behavioral changes) using ecologic valid data.

In a pioneering study [670], the circadian activity patterns and deviation from the norm were recognized. Present sensors were deployed in different rooms in a smart home so that the occupancy rate in each room per hour can be established. The expectation is that the distribution of the occupancy rate follows normal distribution, and the expected distribution is used to quantify deviations from the normal pattern. In [671], activity patterns are established by analyzing spatiotemporal data collected from motion sensors. In particular, staying in the same location for too long would imply a sedentary life style. Furthermore, a decreased level of ADL would imply a level of fatigue.

In [672], the speed of walking was used as a potential predictor of future cognitive function. A smart home was instrumented with multiple PIR detectors. A hidden-semi-Markov model was used to fuse the data from multiple sensors and derive the speed of walking. In a follow-up paper [673], the response time from hearing an event to the action was used as a predictor in conjunction with the walking speed for predicting future cognitive function.

In [674], PIR sensors were also used in a smart home to detect the occupancy rate in each of the four rooms, including the main bedroom, the main bathroom, the kitchen, and the living room. The occupancy probability was calculated per 30-min (instead of 1-h as in [670]). The model was referred to as a generalized linear model (GLM). The Kullback–Leibler divergence [675] was used as a measure of difference between the corresponding GLMs for healthy seniors and patients with MCI. The same research group employed the k-means clustering algorithm to establish mobility patterns in an earlier study [676].

In [477], Bluetooth beacons instead of PIR sensors were used to track indoor mobility patterns of one or multiple subjects in a home with several room for human behavior study, although it was not specifically designed to track patients with MCI.

15.2.4   *MCI-detection via non-mobility IADL tracking*

Several studies have focused on detecting MCI based on voice analysis and use patterns of everyday technology, particularly smartphones. In [677], patients with MCI and patients with dementia were differentiated via voice analysis using support vector machine. The voice data were collected via the subjects were carrying out four speech tasks. In [678], the computer-use patterns were used to differentiate patients with MCI from healthy seniors. The computer used in the study was pre-configured to track the computer usage, including the number of days used, mean daily use amount, and coefficient of variation of the computer use. The study found that patients with MCI have significantly reduced computer usage. In [679], the topic of using mobile apps to assess cognitive levels and to detect MCI was reviewed. The review concluded that the usage patterns of mobile apps can be used to detect sensitive declines in cognitive functions. In [477], we also investigated how to capture the usage patterns of mobile apps.

15.3 Cognitive rehabilitation for older adults

In [605], the impact of physical activity on brain-derived neurotrophic factor (BDNF) in older healthy adults was studied. Participants were asked to engage in three activities: (1) physical exercise using EA Sports Active 2 with Microsoft Xbox 360, (2) cognitive training, and (3) mindfulness practice. Each activity has the same duration of 35 min. BDNF analysis was done via blood samples. The study found that a single bout of physical exercise had significantly larger impact on improving the serum BDNF levels than cognitive training or mindfulness practice in the same person.

In [680], a meta-analysis on the efficacy of physical activities on cognition function in patients with dementia was reported. Physical activities used in the intervention studies are typically divided into three types: (1) aerobic, (2) non-aerobic, and (3) the combination of aerobic and non-aerobic. The meta-analysis found that aerobic exercises have positive impacts on improving the cognitive functions in patients with dementia.

In [597], a 10-month randomized trial was conducted with 57 older adults to study the relationship between exercises, fitness, and cognitive functions. One group engaged in aerobic exercise, and the other group engaged in strength-and-flexibility exercise. Several cognitive tasks were used to evaluate the cognitive functions of the participants. The study found that the group engaged in aerobic exercise did not exhibit superior performance in tasks that required little executive control. Improved performance was observed in the aerobic exercise group in the Stroop Word-Color task. The study also reported that the aerobic fitness of the participants was not related to the cognitive tasks performance, which is different from the mainstream hypothesis that the cognitive improvement is due to improved fitness as a result of engaging in regular physical exercises.

In [681], the roles of video games as a means for improving cognitive functions, slowing the decline of, or even reverse the decline of, cognitive functions are examined. Studies have shown that playing video games appeared to improve reaction time and visa-motor coordination, spatial skills, and visual attention. However, the authors of [681] cautioned that many studies lacked adequate controls for arousal, engagement, and motivation. As a result, it is unclear whether non-video-game-based cognitively engaging tasks such as word puzzles, playing golf or chess could lead to similar benefits.

In [682], a Kinect-based system was developed to track psychomotor exercises, which could facilitate home-based cognitive rehabilitation at home and objective assessment of cognitive levels [683]. The accuracy of the system was tested with 15 human subjects where 10 were healthy, 2 had frontal lobe injury, and 3 had mild dementia using 14 psychomotor exercises (touch left/right eye with left/right hand, touch left/right ear with left/right hand, touch head with left/right hand, raise left/right/hand).

In [684], a 4-month four-condition randomized controlled trial was conducted to study the impact of mild aerobic exercise and cognitive training. The first group engaged in only cognitive training. The second group engaged in only mild aerobic training. The third group engaged in both cognitive training and mild aerobic training. The fourth group engaged in only book-reading activity. The study found that cognitive training played a critical role in explaining the improved cognitive functions (improved hand–eye coordination, and global visual memory, speed of information processing, visual scanning, and naming) of the participants. The study result is not consistent with the mainstream finding that engaging in regular aerobic exercises would improve cognitive functions.

In [685], a 24-week study on the feasibility and efficacy of using Wii video gaming as an intervention (90-min per week) to improve the cognitive level of older adults with MCI. Due to the small size of participants (20), only medium effect sizes for cognitive and physical functioning were observed.

In [377], the efficacy of using interactive video game (Kinect Sports) as an intervention method (control group) was compared with conventional multimodal supervised physical exercise (control group) in older adults in a 12-week randomized trial. The study found significant advantage in both physical functional ability and cognitive function improvement in the experimental group. The cognitive function was assessed using the Stroop task. The experimental group exhibited a significant improvement in the total number of correct responses on the Stroop task, shorter average reaction time of correct color-words. The physical functional ability was measured by several clinical tests, including 6-min walk, dynamic balance, timed up and go, and functional reach. This suggests that the entertaining and engaging elements of serious games are contributing to the cognitive function improvement in addition to physical exercises.

Chapter 16
Technology-facilitated mental health rehabilitation

Mental health is a serious issue globally. According to [686], direct spending on mental health illnesses exceeded $225 Billion US Dollars in the United States in 2019, which is about 5.5% of all health spending in the country. According to Diagnostic and statistical manual of mental disorders: DSM-5 [687], there are a long list of mental health disorders. The primary categories are shown in Figure 16.1, including neurodevelopmental disorders (autism spectrum disorder and attention-deficit/hyperactivity disorder belong to this category), trauma- and stressor-related disorders (post-traumatic stress disorder is a well-known example in this category), neurocognitive disorders (major and mild neurocognitive disorders due to Alzheimer's diseases are well-known examples), substance-related and addictive disorders, depressive disorders, anxiety disorders, schizophrenia spectrum and other psychotic disorders, bipolar and related disorders, obsessive–compulsive and related disorders, dissociative disorders, somatic symptom and related disorders, feeding and eating disorders, elimination disorders, sleep–wake disorders, sexual dysfunctions, gender dysphoria, disruptive, impulse-control, and conduct disorders, personality disorders, and paraphilic disorders.

Mental health rehabilitation is much more challenging than rehabilitation of physical illnesses or cognitive rehabilitation because mental health issues may have drastically different causes, and typically a multi-faceted approach is required for the interventions to be effective, as shown in Figure 16.2. The most common mental health care includes pharmaceutical intervention with psychiatric drugs and psychotherapy. Psychotherapy is also called talk therapy. Via talking to, and conversing with, people with mental and emotional issues, a therapist would aim to help them eliminate or control the symptoms so that they can heal and function better.

There are many types of psychotherapy as shown below and the psychotherapy can take place in an individual, family couple, or group setting:

- Cognitive behavior therapy aims to identify the thinking and behavior patterns that are harmful or ineffective, and ultimately replace them with more productive patterns.
- Dialectical behavior therapy is a special type of cognitive behavior therapy that focuses on emotions.
- Interpersonal therapy aims to help patients address troubling interpersonal issues.

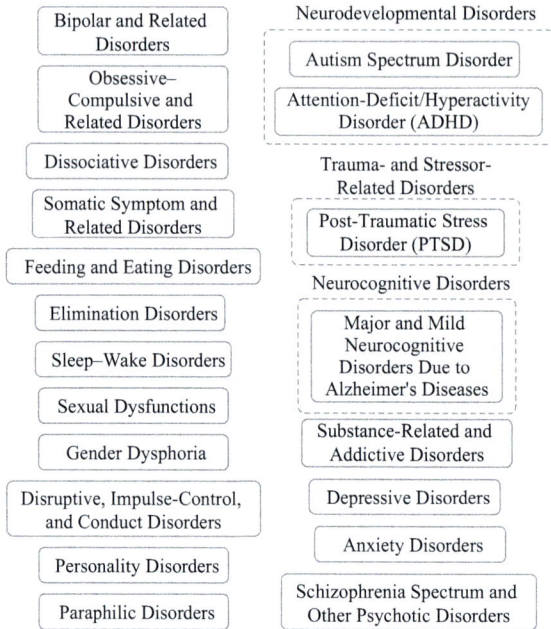

Figure 16.1   Common mental illnesses

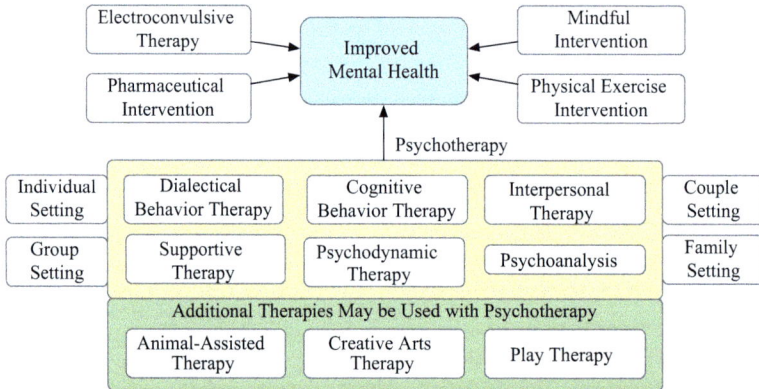

Figure 16.2   Common interventions for mental health rehabilitation

- Psychodynamic therapy aims to help patients identify unaware negative thoughts and unpleasant childhood experiences so that the patients could improve self-awareness and change their old thought and behavioral patterns.
- Psychoanalysis is a more intensive form of psychodynamic therapy that has more frequency therapy sessions.

- Supportive therapy aims to help patients improve their self-esteem, reduce anxiety, enhance coping mechanisms so that they can function better in their life via guidance and encouragement.
- Additional therapies may be used in conjunction with psychotherapy include animal-assisted therapy where trained animals are used to help the patient to feel comfortable and cope with trauma, creative arts therapy where art, dance, music, etc. to help the patient, and play therapy where game play is used to help children.

Electroconvulsive therapy (ECT) is an established psychiatric treatment since 1930s [688]. This therapy uses electrical charge delivered to the brain via scalp electrodes to induce a generalized seizure for about 20–60 s [689]. This therapy has been primarily used to treat depression. In addition, mindful intervention has also been proposed to treat mental health issues [690,691]. Mindfulness is typically understood as paying attention to the present moment in a non-judgmental way [692]. The other aspect of mindfulness include being open-minded while rooting in the present [693]. Quite interestingly, studies have found that regular aerobic exercise increases mindfulness [694]. Similar to cognitive rehabilitation, regular physical exercises have also been recognized to be instrumental to mental health rehabilitation. It was hypothesized that the positive impact of regular physical exercises are through psychological, physiological, and inflammatory mechanisms [695,696]. In the previous chapter, we have shown that regular physical exercises could enhance cognitive functions of the patients.

Mental health assessment and rehabilitation have been predominately relying on self-reporting or expert observation on one's mood and experiences via questionnaires and interviews [697]. During the last decade, computerized assessment has been increasingly used to increase the ecological validity of the data collected. There is also active research on measuring the mood state and stress level objectively via facial expression [698], voice analysis [699], heart rate variability [700], and physiological signals [701].

Immersive virtual reality, typically via a head-mounted goggle, appears to be a sound technology tool to enable effective intervention for mental health rehabilitation, with several randomized clinical trials reported [702].

Due to the large number of mental disorders and the vast literature on mental health interventions, in this chapter we focus on the general discussion on the relationship between regular physical exercise and mental health, and three well-researched mental disorders, namely, autism spectrum disorder, major depressive disorder, and post-traumatic stress disorder. Furthermore, patients with severe and/or chronic physical diseases could trigger symptoms of anxiety, depression, and stress. A plenty of studies have focused on the reducing anxiety, depression, and stress for these patients [703–706]. We will not cover these studies in this chapter.

16.1 Regular physical exercises and mental health

The positive effects of regular physical exercises on improving mood, reducing depression, anxiety, and stress have long been recognized [695,707–709]. Various

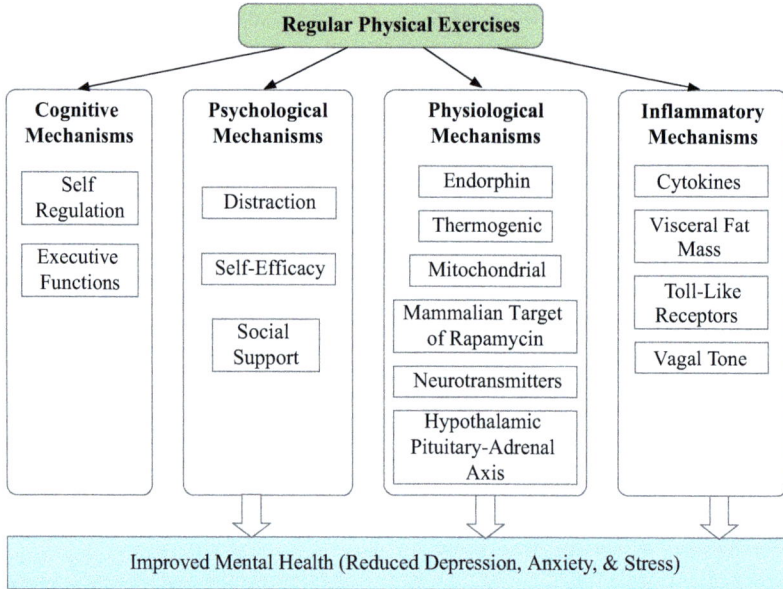

Figure 16.3   Hypothesized mechanisms of regular physical exercises towards improved mental health

hypothesized mechanisms have been proposed regarding how regular physical exercises could improve mental health, including psychological, physiological, and inflammatory mechanisms [695,696]. In addition, numerous studies have shown that regular physical exercises could improve cognitive functions, which could be another mechanism towards improved mental health because many mental health issues are caused by lack of self-regulation. These mechanisms are summarized in Figure 16.3.

16.1.1   Psychological mechanisms

A number of psychological factors have been hypothesized as the reason for the improved mood. *Endorphin* is one of the first recognized factors. Endorphins are generated by the body during physical exercises (and other pleasurable and painful experiences). After the exercises, an individual usually experiences feelings of euphoria, which helps improve the mood. Higher levels of endorphins after exercises have been detected [710]. In [711], a randomized controlled clinical trial was conducted to assess the efficacy of a physical exercise-based intervention program on treating patients with depression (with 60 participants). Both groups took antidepressant drugs. The experimental group engaged in additional physical activities. The endorphin levels were tested before and after the intervention. The study observed higher levels of endorphin and reduced depression levels in the experimental group. In [712], a 6-week randomized controlled trial was conducted with 30 moderately depressed males. The participants were randomly assigned to one of three groups with varying levels of excise intensity (low, moderate, and high). Each group engaged in three 1-h sessions per week. The study found that the depression levels were reduced in the moderate and high intensity groups, but not in the low-intensity group. The study also measured the endorphin levels, but the result was inconclusive due to potential insensitivity of the assays used.

Another popular hypothesized mechanism is referred to as the thermogenic factor. The hypothesis is that the body temperature will be elevated while doing exercises, which could lead to better mood and reduced anxiety level [713,714]. However, existing studies failed to provided conclusive evidence to support this hypothesis [715–717].

The mitochondrial factor is another popular hypothesized mechanism. The mitochondrial function is crucial in human health [718]. More specifically, mitochondrial biogenesis takes place at a higher rate during neuronal development and repair. Hence, mitochondrial dysfunction could lead to the inability for the nervous system to take part in neuroplasticity, neurogenesis, and neuronal development [718,719]. For example, depression and anxiety have been attributed to mitochondrial dysfunction [720,721]. It is also evidence that regular physical exercises could improve mitochondrial function [722]. However, the does not appear to have randomized controlled trials that prove the mitochondrial factor as a key mechanism in improving mental health.

The mammalian target of rapamycin (mTOR) regulates cell growth, protein synthesis, and metabolism [723]. mTOR is believed to play an important role in long-term memory, and perturbation of mTOR signaling cascade is regarded as a common pathophysiological feature of human neurological disorders [724]. Some recent study has shown evidence that regular physical exercise could increase mTOR signaling in brain regions that are associated with cognition and emotional behaviors [725].

Neurotransmitter is another hypothesized mediating factor between regular physical exercises and improved mental health. A neurotransmitter is produced by a neuron as a signaling substance to communicate with another cell across a synapse. Neurotransmitters have a large variety and more than 100 of them have been identified. They are believed to play an essential role in the function of complex neural systems. Neurotransmitters that are directly related to one's mental health include dopamine, serotonin, and adrenaline. Both dopamine and serotonin help regulate one's emotion, mood, reward, and arousal. Adrenaline regulates one's arousal and stress. Regular physical exercises are found to increase the levels of serotonin and adrenergic levels, which resulted in reduced levels of depression [726].

The hypothalamic—pituitary—adrenal axis (HPA axis) refers to a neuroendocrine system consisting of hypothalamic, pituitary, and adrenal [727]. These three components interact with each other and together as a system it controls the responses to stress and regulate mood and emotions, sexuality, and other essential functions. HP axis imbalances are believed to be the pathophysiology of depression [728,729]. A common hypothesis is that regular physical exercises may reduce depressive levels due to exercise-induced attenuation of the HPA axis response to stress [730]. However, more clinical studies are needed to confirm the hypothesis [731].

16.1.2 Inflammatory mechanisms

The immune system is found to play an important role in the pathogenesis of mental health disorders [732]. There are evidences that regular physical exercises could reduce inflammation, and reduce inflammation-related depression levels [732]. There are four hypothesized mechanism for the anti-inflammatory effects of regular physical exercises [733]. Regular physical exercises were found to lead to (1) increase release of anti-inflammatory cytokines, (2) reduced visceral fat mass, (3) decreased toll-like receptor expression, and (4) increased level of vagal tone.

In [734], a randomized controlled trial was conducted to assess the immunological effects of regular physical exercises in patients with major depression. Both the experimental group and the control group received cognitive-behavior therapy (which is a standard treatment for depression). The experimental group additionally engaged in four 40-min homework exercise sessions (at moderate level of activities such as walking, jogging, swimming, and gyms) per week. The control group only included additional euthymic activities. The experimental group was found to have higher levels of anti-inflammatory cytokines (interleukin-10) and reduced levels of C-reactive protein than those of the control group. However, the depressive levels were reduced after the trial for both the experimental and the control groups, and they were not significantly different between the two groups.

16.1.3 Psychological mechanisms

The distraction hypothesis is the earliest explanation for the positive effects of exercises on mental health [735]. It is based on the hypothesis that a diversion from a negative stimulus could lead to an improved mood state [736]. Exercise provides a good means of diversion from stress, anxiety, and depression.

The self-efficacy hypothesis is another popular explanation for the effects of exercises. The hypothesis is that when an individual could gain a greater sense of mastery and control through engaging in regular physical activities, and this could help the individual better cope with stress and depression while doing other tasks [735]. This hypothesis was supported by a randomized controlled study [737] and a cross-sectional study [738].

Social support is a recent hypothesis on why exercises could improve mental health [735]. It is common for individuals who are keen in doing exercises to join some exercise clubs. This provides the individuals the opportunity to establish new personal relationship and gain a sense of belonging in a community. Studies have shown that older adults could particularly benefit from social support due to joining an exercise club [739,740], which helped reduce depressive symptoms.

16.2 Rehabilitation for patients with autism spectrum disorder

Autism spectrum disorder (ASD) is a neurodevelopmental disorder with symptoms in three domains: (1) stereotypic behavior (i.e., restrictive and repetitive behaviors),

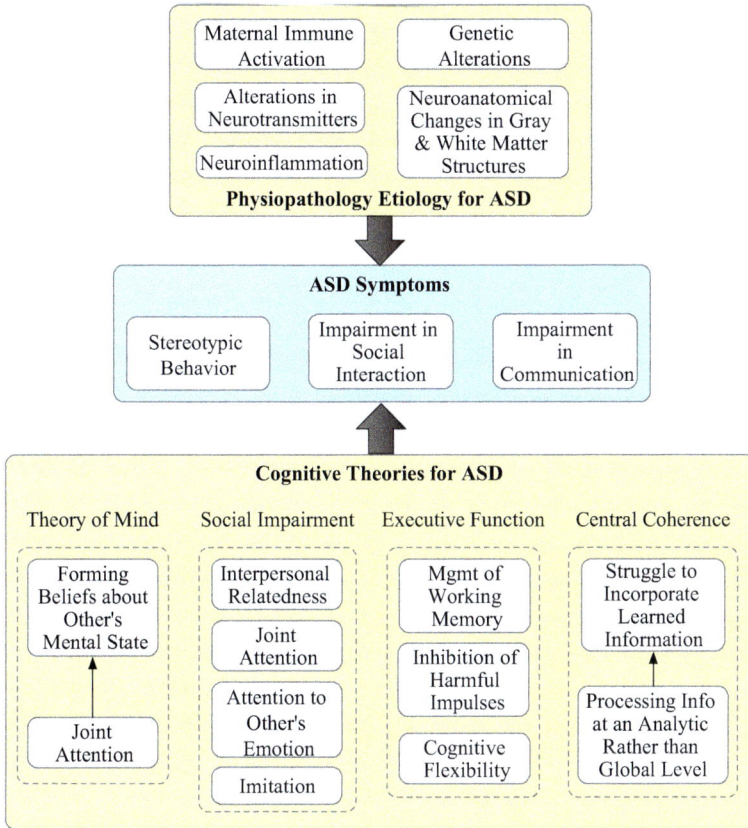

Figure 16.4   ASD symptoms and the hypothesized root causes

and (2) impairment in social communication, and (3) impairment in social interaction [687]. There are many hypotheses and theories regarding the root causes for the ASD symptoms, as shown in Figure 16.4, one approach is to determine the physiopathology etiology for ASD, and the other approach is to explain causes of the symptoms at the cognitive level with various cognitive theories.

Some of the most dominating cognitive theories have been reviewed in [741], which is shown in 16.4. These theories hypothesized the root causes from different angles with some degree of overlaps. The theory of mind [742] hypothesized that individuals with ASD have impairment in forming beliefs about other people's mental state, probably due to impairment in joint attention. The social impairment theory attributed to symptoms to the impairment/deficiency in several factors, interpersonal relatedness, joint attention, attention to other's emotion, and imitation. A closely related theory called social attention impairments [743], attributed the ASD symptoms to the impairment in social orienting (i.e., orient to social stimuli as they occur), joint attention (i.e., coordinate attention to objects and events in an interactive session with another individual), and attention to other's distress, which can be roughly regarded as a subset of the social impairment theory. The executive function impairment theory [744] hypothesized that the root cause for the ASD symptoms, particularly the stereotypic behavior, is the impairment in executive function. As we have introduced in Chapter 15, the executive function consists of management of working memory, the inhibition of harmful impulses, and cognitive flexibility. The central coherence theory [745] hypothesized that individuals with ASD often struggle to incorporate learned information and derive meanings due to attention to parts/piecemeal (i.e., at analytic level) instead of at the global level.

The study on the physiopathology etiology for ASD is still in its early stage [746]. Currently, the hypotheses include neuroinflammation, alternations in the neurotransmitters system, maternal immune activation, genetic alterations, and neuroanatomical changes in gray and white matter structures [746].

Over 1% of the population in the United States (about 3 million people) suffer from the autism spectrum disorders (ASD). Similar to other mental illnesses, the actual damages caused by ASD are far beyond financial costs. The lack of social skills for patients with ASD could significantly impact their personal lives and career opportunities negatively, and could put undue burden on their caregivers.

The recent technological development is giving hope for early diagnosis of ASD, and more effective and cost-conscious treatment. Technology could help produce controlled stimuli and focus on a single phenomenon [747,748]. This has two benefits: (1) help collect objective evidence on the behavior that is phenotype of patients with ASD, which would help early and more accurate diagnosis of children with ASD; and (2) deliver the stimuli according to the treatment plan, which would result in more consistent treatments. The most common methods in autism research are serious games and robotics. Serious games sometimes are referred to exergames, as we have covered in detail in Chapter 10. However, in the context of rehabilitation of patients with ASD, the primary purpose of serious games is not to get the patients to engage in physical exercises, but rather, to learn, practice, and improve social skills. In the serious games for ASD rehabilitation, virtual avatars constitute the main virtuality reality interface [749].

3D immersive virtual reality can be realized via sophisticated surrounding displays such as the CAVE system from Visbox, or head-mounted displays. Some studies used this type of 3D virtual reality systems [750–753]. Most other studies used conventional computer monitors to display virtual reality scenes [754–764]. The impacts of the level of immersion in virtual reality systems on ASD treatment have been reviewed in [765].

16.2.1   *Clinical scales in ASD studies*

A number of clinical scales have been used to select participants with ASD and to ensure the control group have compatible intelligent quotient. Scales for related disorders have also been used when those factors were considered, such as attention-deficit/hyperactivity disorder (ADHD) and anxiety.

- The autism diagnostic observation schedule (ADOS) [766] is a semi-structured standardized assessment scale for diagnostic evaluations of individuals for ASD in several areas, including communication, social interaction, play, and imagination. It was regarded as the gold standard in clinical ASD diagnosis according to [767] and it was used in [758,767,768].
- The social communication questionnaire (SCQ) (formerly known as the autism screening questionnaire) is a 40-item scale typically to be completed by parents of children who are being assessed for possible autism [769]. This scale was used in [750,758,767,768].
- Autism spectrum screening questionnaire (ASSQ) is a 27-item checklist to be completed by untrained caregivers [770]. This scale is used in [750].
- The autism diagnostic interview-revised (ADI-R) is a 93-item interview questions regarding the functional domains of language & communication, reciprocal social interactions, and restricted, repetitive, & stereotyped behaviors & interests [771]. This scale was used in [758].
- The social responsiveness scale (SRS) is a 65-item scale that measures deficits in social behavior of individuals with ASD [772]. This scale was used in [750,758, 767,773].
- The Gilliam autism rating scale (2nd edition) (GARS-2) assesses the characteristics of individuals between 3 and 22 years of old with ASD [774]. It has 42 items divided into three sub scales: (1) stereotyped behaviors, (2) communication, and (3) social interaction. This scale was used in [775–777].
- Childhood autism rating scale (CARS) is a 15-item behavioral rating scale used for the diagnosis of ASD and the determination of the severity of ASD [778]. This scale was used in [779,780].
- The childhood joint attention rating scale (C-JARS) consists of 55 questions to determine the level of joint attention skill [781]. This scale was used in [782].
- The Vineland adaptive behavior scales (VABS) measure the adaptive behavior of individuals, especial children with mental disorders [783]. The scales are often used as supplementary measures for individuals with ASD [768,780,784–786].
- The sensory profile [787] is a scale with 125 questions to determine the levels of social functioning in nine subscales, including sensory seeking, emotionally reactive, low endurance, oral sensory sensitivity, inattention, poor registration, sensory sensitivity, sedentary, and fine motor. This scale is used in [773].
- The aberrant behavior checklist [788] is a 58-item scale for assessing treatment effects on mentally retarded individuals. The items determine the following factors, including irritability/agitation, lethargy/Social Withdrawal, Stereotypic Behavior, Hyperactivity, and Inappropriate Speech.
- The Bruininks–Oseretsky test of motor proficiency assesses both the gross and fine motor skills with 46 items [789]. This scale has been used by evaluate motor skills for individuals with ASD [785].
- Conners-3 is a scale that measures ADHD symptoms [790]. This scale was used in [750].
- The multidimensional anxiety scale (MASC) for children is a 39-item self-report instrument that assesses the anxiety in individuals between 8 and 19 years of old [791]. This scale was used in [750].

- The Wechsler abbreviated scale of intelligence (WASI) consists of subtests on vocabulary, block design, similarities, and matrix reasoning [792]. This scale was used in [750,767].
- The peabody picture vocabulary test (PPVT) is a measure of single-word receptive vocabulary, which is used to assess cognitive function [793]. This scale was used in [758].

16.2.2   Social attention

ASD can be considered as an attention-related disorder and recent attention-related studies on patients with ASD have been reviewed in [794]. Social attention is a complex cognitive process because it is rooted at perception, memory, and decision-making while interacting with another person [795].

In [750], an experiment was conducted to test the social attention model of ASD using a public speaking serious game delivered via a head-mounted display. Each participant was asked to perform a task in the virtual reality game. Via eye-tracking as part of the virtuality system, the participant's degree of attention to each of the nine avatars in the 3D virtual classroom while answering questions regarding the participant. The study recruited 37 students with high functioning children with ASD, and 54 peers without ASD. The study confirmed the social attention model of ASD where social attention disturbance can be used as a school-aged phenotype of autism. More specifically, the study has three research hypotheses: (1) children with ASD would pay less attention to social avatars compared with peers without ASD; (2) there exist strong heterogeneity in children with ASD due to their differences in a number of factors such as age, IQ, ADHD symptoms, and social anxiety; (3) impairments among children with ASD would be appear more severe in a public speaking task than in a nonsocial task; and (4) social attention would be related to learning problems. The study used a variety of measures for autism symptoms, intelligence quotient, ADHD, learning problems, social anxiety, social attention (using the serious game and the virtual reality system), nonsocial attention, academic achievement and learning. The social attention measurement consists of a measure of orienting and a measure of fixation length, both are based on data collected by the virtual reality system, including the start and the end times, each look event in the virtual classroom. The measure of orienting means the total number of looks to each individual avatar. The measure of fixation length refers to the average duration of fixations to any of the avatar.

In [758], a study was presented on how to objectively capture the performance of the participant during a social communication task in a virtual reality serious game via eye gaze tracking, and how to provide specific feedback to the participants. The study showed that the system with feedback improves the social engagement performance during the social communication task for children with ASD (8 adolescents with ASD participated the study). The objective data collected include gaze pattern, pupil dilation, and blink rate during the task. Based on the data collected, the mean pupil diameter, the mean blink rate, and the average fixation duration at each region of interest were derived. The main criteria on providing feedback is the percentage of time the participant looked at the avatar (the social communication task is between the participant and the avatar). The desirable range is between 70% and 90%.

In [782], the study focused on how children with ASD react to different joint attention cues via a virtual reality interface displayed on a touchscreen monitor. The hypothesis is that the difficulty for children with ASD in engaging in social communication is due to the deficiency in picking up joint attention cues. More broadly speaking, the joint attention skill consists of response to joint attention and initiating joint attention [796]. This study focused on how to train children with ASD to improve on how to properly respond to joint attention. For example, during a two-person interaction session, when one person is pointing her figure to some object, the other person is supposed to follow the cue and look towards that object. The authors argued that using a virtual reality interface is more advantageous than robot-assisted techniques in that it incurs much lower cost and offers higher flexibility, and would cost far less than manual training by a therapist. The study presented four different types of joint attention cues delivered by an avatar in the virtual reality interface:

- Gaze-based cue.
- Gaze-based cue plus head turn.
- Gaze-based cue plus head turn plus finger-pointing.
- Gaze-based cue plus head turn plus finger-pointing plus visual prompt at the target cue object.

The participant was asked to do a screen tap on the cued object within 10 seconds after the cue was presented. The system would provide a feedback if the participant did a screen tap. The system also collected physiological data during the training sessions, including PPG and electrodermal activity (EDA), via Biopic MP150. Based on the response time and the percentage of screen taps on the set of predefined region of interests, the performance of each participant was computed. PPG was used to derive the mean pulse rate, and EDA was used to compute the tonic mean. The changes of mean pulse rate and tonic mean with respect to the base line during the training sessions were computed.

In [797], the gaze pattern of children with ASD while learning pronunciation using a virtual reality tutor system was studied. This study showed that a 3D avatar received more attention from than a real human face, probably because the 3D avatar could display the movement of external and internal articulators and airflow changes in a more prominent manner. Around the mouth area was defined as the area of interest (AOI). Data for three parameters were collected for this AOI. The first parameter is the entry time, which is the response time defined as the time difference between the start of the trial and the first hit on the AOI. The better attention of the participant, the shorter of the entry time. The second parameter is the fixation count, which is the number of fixations on the AOI. If the gaze is within the AOI for more than 100 ms, then it is counted as one fixation. The third parameter is the proportion of fixation duration, which is the ratio of fixation duration inside the AOI to that of the other area. The better the attention of the participant, the higher proportion of fixation duration.

In [768], an intervention program based on a set of serious games was used in a clinical trial to test their efficacy in improving the attention (and imitation) skills of young children with ASD. The set of games is referred to as GOLIAH, short for Gaming Open Library for Intervention in Autism at Home [798]. GOLIAH has four attention games, including follow the therapists pointing, cooperative drawing, bake a cake, and receptive communication. The study found that the time to perform joint attention games was shortened. The intervention program improved the participants' ASD symptoms as measured by ADOS [766], Vineland adaptive behavior scales [783], and SCQ [769].

16.2.3 Imitation

Lack of skills in imitation and joint attention is said to be the primary reason for individuals with ASD having difficulty in engaging proper social interactions [798]. Imitation refers to the process of doing something with the intention of repeating an action carried out by another individual [799]. Imitation involves a learning step where the imitator must learn how the other individual is doing the action. In a social interaction scenario, imitation could also be a way of communication.

In the same study, we mentioned earlier [768], the intervention program used a set of seven imitation games (as part of GOLIAH [798]) to help improve the imitation skills of children with ASD. The six imitation games include imitate free drawing, imitate step-by-step drawing, imitate speech, imitate sounds, imitate actions with balls, imitate actions and build with cubes, and guess the instrument. The study showed that the imitation scores were improved for the participating children with ASD.

16.2.4 Cognitive load

In [800], the cognitive load of adolescents with ASD was characterized while learning driving using a virtual reality driving simulation system. Cognitive load refers to the working load imposed on an individual's cognitive system while doing a learning task. The study collected gaze data via eye tracking, electroencephalography (EEG) (via Emotive EPOC neuro-headset), electrocardiogram (ECG), electromyography (EMG), respiration (RSP), skin temperature (SKT), pulse plethysmograph (PPG), and galvanic skin response (GSR) (via Biopac MP150). The data were used to classify whether or not the cognitive load is zero or one. Several supervised learning algorithms were experimented for classification, including SVM, KNN, decision tree, artificial neural networks, and linear discriminant analysis.

16.2.5 Facial expression and emotion recognition

In [767], a virtual reality-based system was designed to help understand how adolescents with ASD react to various facial expressions using objective data. The study collected gaze data (via eye tracking) and physiological data, including ECG, PPG, SKT, and GSR. The primary hypothesis of the study is that the participant with ASD would express less stress if the individual correctly cognized the facial expression than if the recognition was wrong. However, the current study only used unsupervised learning to highlight the differences in patterns due to the lack of labeled data. Ten high functioning adolescents with ASD and ten typically developing adolescents were recruited in the study.

In [801], the result of an intervention program designed to teach children with ASD was reported. The intervention is based on a serious game with virtual reality scenes and avatars, as well as a game controller that has tactile actuators. The avatars would show facial expressions that indicate a variety of emotions, including neutral, anger, disgust, fear, sadness, pain, surprise, funny, and happy. Unlike some studies that only showed the facial expressions via an avatar, this study incorporated context in the virtual reality scenes. The participants were asked to identify the emotion of the avatar in different scenes. Furthermore, some avatars only showed faces while some others showed body gestures in addition. For comparison, real person's faces and real-life social scenes were also used in some emotion recognition tasks. The game is first used to train the participants on recognition, and then it was used to test the participant's emotion recognition skills. The study showed that the game can be used to effectively improve the emotion recognition skills of children with ASD.

16.2.6 Physical exercise-based intervention

The impact of exercise-based intervention for children and youth with ASD has been reviewed from several different aspects, including the positive impact on reducing neuroinflammation [746], positive impact on behavioral outcomes [802], improving stereotypic behavior [803], improving executive function [804], and improving social learning [805]. The benefits of exercise-based intervention on individuals with ASD also include the improvement on non-primary symptoms/factors, such as improvement on emotion regulation [806].

The types of exercises used in the intervention programs include walking/jogging/running, horseback riding, martial arts, yoga and dance, and swimming [802],

16.2.6.1 Walking/jogging/running

In [807], the effects of physical exercises on stereotypic behaviors in three individuals with ASD were evaluated. The study was experimented with a moderate exercise program, which has 15-min of walking, and a vigorous exercise program, which has 15-min of jogging. The study evaluated the frequency of stereotypic behavior of the participants prior to the exercise, right after the exercise, and 90 min after the exercise. The study found that only the vigorous exercise program caused significant reduction in stereotypic behaviors of the participants. Furthermore, the positive effect was temporary.

In [808], the effects of a jogging-based intervention on academic performance in four children with ASD were assessed. The intervention program runs for 2 weeks and each week the participants were asked to engage in 12-min of jogging followed by a 5-min cooldown period with walking and stretching. The study found that the program resulted in significantly improved academic engagement time.

In [809], the effects of an exercise-based intervention on several variables in seven children 3–6 years of old with ASD (an additional participant with intellectual disability and one more participant with developmental delay) were studied. The intervention consists of 15-min of running and jogging. The control group does not engage in the exercise intervention. Both the experiment group (with exercise intervention) and the control group will engage in a classroom task. The evaluated dependent variables include the number of stereotypic behaviors, percentage of on-task behavior (i.e., seated and producing academic responses as required by the teacher in the classroom), and the number of correct responses (i.e., respond correctly to a directive provided by the teacher). The study found that only the correct responding improved significantly in the experimental group. No significant differences in on-task time and stereotypic behavior was observed after the exercise intervention in the study.

In [810], the effects of single-exercise-session-per-day and multiple-exercise-session-per-day on stereotypic behaviors in children with ASD were evaluated. A single exercise session consists of 10-min of walk/jog. Stereotypic behaviors of the participants were observed prior to and right after each exercise session. Although a significant reduction in stereotypic behaviors was observed after a single exercise session, the reduction was often short-lived. The study found that using multiple sessions over the same day had more substantial reductions in stereotypic behaviors.

16.2.6.2   Horseback riding

Therapeutic horseback riding has been used as a treatment to improve posture, balance, and mobility while developing a bond between the patient and horse [773]. In [773], the effects of a 12-week horseback-riding program on social functions in 19 children with autism were studied. The study showed that participating children ASD had significant improvements in sensory seeking, sensory sensitivity, social motivation, inattention, distractibility, and sedentary behaviors.

In [785], the effects of a 10-week therapeutic horseback-riding intervention program on self-regulation behaviors in 42 children and adolescents with ASD were evaluated. The study observed significant improvements in self-regulation behaviors (irritability, lethargy, stereotypic behavior, and hyperactivity), as well as the motor skills.

In [779], the effects of a therapeutic horseback-riding intervention program in children with ASD were evaluated. The evaluation was done at four points: (1) 3–6 month prior to the start of the program, (2) right before the start of the program, (3) after participating in the program for 3 months, and (4) after participating in the program for 6 months (20 participants remaining at this point). The study observed a significant reduction in the severity of ASD symptoms at 3 months and 6 months of horseback-riding as measured by the CARS scale [778].

In [775], the effects of a therapeutic horseback-riding intervention program on several variables in 21 children with ASD were evaluated. The study showed that the participants significantly improved their social interaction and sensory processing, and reduced the severity of ASD symptoms following the intervention program. Not surprisingly, the improvements were not maintained after two 6-week breaks.

16.2.6.3   Martial arts

In [776], a form of karate (called Kata) was used as the means of intervention for 30 children with ASD. The intervention program runs for 14 weeks. Stereotypic behaviors were evaluated prior to the intervention, week-14, and one month after the intervention. Significant reduction in stereotypic behaviors was observed at the end of the intervention and one month after the intervention. In a follow-up study from the same research group, the same intervention program designed was used also for 30 children with ASD [777]. The study showed that the program significantly improved social interaction. For both studies, GARS-2 [774] was used to measure the ASD symptoms.

16.2.6.4 Yoga and dance

In [811], the efficacy of an intervention program in 24 children with ASD was evaluated. The intervention program combined yoga, dance, and music therapy with the aim of eliciting relaxation response and ran for 8 weeks. The study found a significant improvement in behavioral and cognitive symptoms.

16.2.6.5 Swimming

Swimming is a relatively popular intervention method for children with ASD. According to a recent review on the subject [812], 23 studies reported the effects of swimming-based intervention programs (most of the studies focused on teaching swimming skills and/or improving motor skills).

In [813], the effects of a 10-week swimming intervention program on social behaviors of 16 children with ASD were evaluated. The study found significant improvements in social behaviors of the participants as measured by the School Social Behavior Scale (2nd edition) [814].

In [786], the efficacy of a 12-week swimming intervention program on motor skills and social skills in three adolescents with ASD was evaluated. The outcome was assessed using VABS [783] and it showed significant improvements in both motor skills and social skills.

In [780], the effectiveness of a 10-month long swimming-based intervention program on functional symptoms in children with ASD was studied. The study found significant improvements in various functional factors, including functional adaption skills according to VABS, emotional response and adaptation to changes according to CARS [778].

16.3 Exercise-based intervention for patients with major depressive disorder

The criteria given by the diagnostic and statistical manual of mental disorders (DSM-5) [687] are fairly complex. According to the first criterion, only an individual who is showing at least five the nine symptoms shown in Figure 16.5 may be considered to have major depressive disorder (MDD). MDD is one of the most prevalent mental disorders worldwide. It is estimated that at least 300 million people suffer from MDD [815]. The exact cause for MDD is unclear [816]. Neuroimaging studies provided some evidence that MDD is correlated to cerebral volume changes, and functional changes in the part of the brain associated with emotional processing and cognition [817].

Major Depressive Disorder
(at least 5 symptoms present)

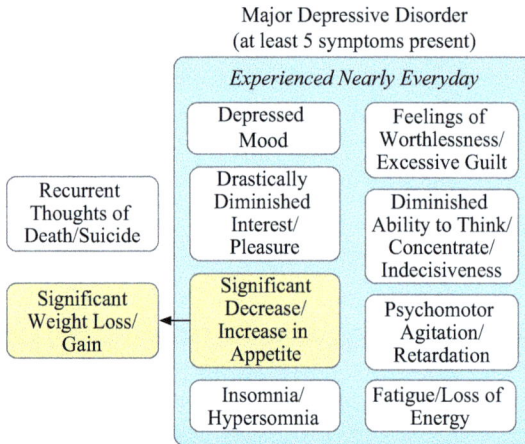

Figure 16.5 Main MDD symptoms

According to [818], patients with MDD rarely receive adequate treatment. Although a large fraction of patients with MDD are prescribed with psychiatric drugs, patients with MDD are likely to remain depressed. MDD could negatively impact the patients' physical health, career, and would lead to a significantly shorter lifespan [819,820]. In recent years, physical exercises have been increasingly recognized as an effective intervention to help treat patients with MDD and numerous studies have presented the outcomes of exercise-based intervention for patients with MDD [815,816,821,822] and the subject has been well-reviewed [818,823,824]. The hypothesized mechanics on how exercises could improve MDD symptoms [815,825] are rather similar to those we have introduced at the beginning of this chapter and shown in Figure 16.3.

Most exercise-based intervention programs incorporated aerobic exercises in some forms [823]. While all studies have reported a reduction in depressive symptoms at the end of the intervention, not all studies demonstrated that the exercise-based intervention had advantage over the method used in the control group.

16.3.1 Clinical assessments in MDD studies

MDD studies used a number of clinical scales for depression and functioning assessment, as well as several anthropometric and hemodynamic parameters.

- The Hamilton depression rating scale (HAMD-17) is a simple 21-item questionnaire assessing 17 variables about the individual, including depressed mood, suicide, work $ loss of interest, retardation, agitation, gastrointestinal symptoms, general somatic symptoms, hypochondriasis, insight, and loss of weight [353]. This scale was used in [816,826,827].
- The Beck depression inventory-II (BDI-II) [443] is a 21-item inventory assessing the severity of depression. However, it is not designed for diagnosis of MDD.

Early studies and some recent studies [828] used the original version of the inventory [829]. In general, recent studies [734,816,830] used BDI-II.

- The depression anxiety stress scales (DASS) contain 42 items and are used to measure the levels of depression, anxiety, and stress [831]. DASS-21 is a shorter version of DASS containing only 21 items [831]. DASS-21 was used in [832].
- The comprehensive psychopathological rating scale (CPRS) is a scale for rating the severity of psychiatric symptoms and behaviors [833]. CPRS consists of 65 items covering symptoms such as psychosis, mood, anxiety, and somatoform issues. This scale was used in early studies [828].
- The Global assessment of functioning (GAF) [834] is a 100-point scale. It is divided into ten 10-point intervals, with each interval having an anchor statement describing some level of symptoms and functioning. The higher the points, the lesser the severity of the disorder. The 91–100 range indicates superior functioning. The 81–90 range indicates good functioning. The 71–80 range indicates that no more than slight impairment and the symptoms are transient. The 61–70 range indicates some mild symptoms. The 51–60 range indicates moderate symptoms. The 41–50 range indicates serious symptoms. The 31–40 range indicates some major impairment. The 21–30 range indicates the presence of delusions or hallucinations. The 11–20 range indicates some degree of danger of hurt self or others. The 1–10 range indicates persistent danger of severely hurting self or others. The score 0 is given when there is inadequate information. This scale was used in [816].
- The PHQ-9 is subset of the patient health questionnaire (PHQ) for depression diagnosis and severity measurement [835]. It has only nine items and is significantly shorter than other depression scales. This scale was used in [836].
- The clinical global impression scale (CGI) was initially developed in studies for schizophrenic [837]. It consists of three subscales, severity of illness, global improvement, and efficacy index. The first two subscales each has a 7-point scale. This scale was used in [816].
- The ruminative response scale (RRS) [838] is a 22-item questionnaire used to assess ruminative though patterns. RRS has three subscales on depressive, brooding, and reflective ruminations. This scale was used in [839].
- Physiological measures. Several physiological parameters were measured to characterize the physical fitness of the participants. The body mass index (BMI) is often used as an anthropometric measure [836]. BMI is defined as the weight divided by squared height. Blood pressure and heart rate are common hemodynamic parameters. Cardiorespiratory fitness, i.e., maximum oxygen intake during peak exercise intensity, has also been used [836,839]. Some studies used blood samples to measure the total cholesterol, LDL, and HDL.

16.3.2   *Supporting studies*

In [839], the effects of an 8-week exercise-based intervention program on reducing depression in patients with MDD were evaluated. The experimental group engaged in three moderate-to-vigorous exercise sessions per week. The control group engaged in three sessions of light-intensity stretching sessions per week. The study measured the depression severity using BDI-II, ruminative thought patterns using RRS, cardiorespiratory fitness, and a test for cognitive control. The study showed that the experimental group exhibited stronger reduction in depressive symptoms, stronger reduction in rumination, and stronger cognitive control than the control group. However, there are no significant changes in cardiorespiratory fitness.

In [830], the efficacy of an exercise-based intervention on improving depressive symptoms in patients with MDD was evaluated in a randomized controlled clinical trial. Three groups were formed, one with aerobic exercise (with 14 participants), the second with stretching exercise (with 11 participants), and the third without any exercise (with 10 participants). The aerobic exercise group engaged in one supervised 30-min session of brisk walking or jogging per day for 10 consecutive days. Furthermore, the exercise intensity was measured by a heart rate monitoring device (with Polar S725 chest belt and wristwatch) and the heart rate was maintained within 65–75% of age-adjusted maximal heart rate in each session. The stretching exercise group also engaged in one 30-min session per day for 10 days, with each session involving only stretching activities. All three groups received antidepressants. The depression severity was measured using BDI-II. Both the aerobic group and the stretching group showed significant improvements in depressive symptoms, and the control group (without any exercise) showed no improvement post intervention. Furthermore, the aerobic exercise showed greater improvements than the stretching group.

In [826], the efficacy of two interventions on improving depression symptoms in patients with MDD was studied in a randomized trial. One intervention is based on aerobic exercise training (AET), and the other intervention is electroconvulsive therapy (ECT) [840]. In addition, the study measured the BDNF levels pre- and post-intervention, and examined the correlation between the increase in BDNF levels and depressive symptoms. The 60 participants were randomly assigned to one of the three groups: ECT group, AET group, and ECT+AET group. The ECT+AET group engaged in both ECT and AET. The AET group engaged in three supervised sessions per week for 4 weeks. Each session consists of cycling on a treadmill for about 40–45 min. The exercise intensity was monitored by oxygen intake (60–75% of the maximum value). The depression severity levels were measured using BDI and HAMD-17. The study found that the levels of depression severity were reduced for all three groups after the intervention. The ECT+AET exhibited the most significant reduction in depression severity. This is followed by the ECT group. However, the study found no correlation between the BDNF levels and depressive symptoms.

In [827], the effects of an 8-week aerobic exercise program on the depressive symptoms in patients with mild to severe depression were studied. The study showed significant reduction in depressive symptoms according to HAMD-17, but only moderate changes according to BDI-II.

In [832], the effects of an aerobic exercise program on helping patients with depressive symptoms were evaluated in a randomized controlled trial. Twenty-six participants with depressive symptoms were randomized into the experimental group and the control group. The experimental group engaged in three aerobic exercise sessions per week for 16 weeks. The control group did not engage in any exercise. Both groups received antidepressants treatment. Each aerobic exercise session lasted for 45–50 min and 30 min of which are aerobics measured by the heart rate. Participants could choose from a number of aerobic exercise options, including natural circuit workouts, jump ropes, brisk walking and dancing, fitness balls, and traditional games. The depressive symptoms were measured using BDI-II and DASS-21 [831]. The study observed significant reduction in depressive symptoms in the experimental group. The study also tested the hypothesis that aerobic exercise would improve the participants' self-esteem. However, the study found no difference in self-esteem between the experimental and control groups.

In [816], the effects of a 12-week home-exercise program on improving depressive symptoms in treatment-resistant patients with MDD were studied. Participants in the program engaged in five sessions per week (one of the sessions was supervised). Each session had 30–45 min of walks. The study showed that the exercise group had made improvement on all depression and functioning parameters assessed by several clinical scales, including HAMD-17, BDI-II, CGS, and GAF.

In [828], the effects of a 9-week randomized trial of using aerobic exercise-based intervention program on the reduction of depression levels in patients with depression were reported. The experimental group engaged in 1-h of supervised aerobic exercise training three times a week. The assessment was carried at the beginning of the trial, and after 3, 6, and 9 weeks of training. The original BDI [829] and a comprehensive psychopathological rating scale [833] were used to assess the depression severity. Furthermore, physical condition was assessed by measuring the maximum oxygen uptake during the ergometer bicycle testing [841].

16.3.3 Nonsupporting studies

In [836], the effects of a 12-week exercise-based intervention program on smoking abstinence (instead of reducing depression) in depressed female smokers were studied. The exercise duration and intensity was designed to increase gradually during the trial. The participants engaged in three exercise sessions per week. The session duration was 20 min during the first week, and it was extended to 30 min during the last week of the trial. During each session, a participant could select a preferred workout equipment in the gym. Each session has a period of vigorous exercise of 4 min during week one, and it was extended by 2–4 min per week. In addition to using PHQ-9 to measure the depression severity level, BMI and cardiorespiratory fitness were measured. The latter was done via measuring the oxygen consumed while the participant was doing a prescribed treadmill test to the maximal exertion. The study showed that although the intervention was effective in helping smoking abstinence, and the depression severity of the participants in both the experimental group and the control group (only received health education) was reduced at the end of the study. However, the improvement of the depression level in the experimental group was similar, which indicates that the exercise program has no advantage over the health education.

Similarly, another study [734] that we have discussed earlier in Section 16.1.2 showed that the depressive symptoms for the exercise-based experimental group were not significantly different from those of the control group at the end of the trial as measured by the BDI-II.

16.4 Exercise-based rehabilitation for patients with post-traumatic stress disorder

Post-traumatic stress disorder (PTSD) refers to a clinical condition of significant social and occupational impairment due to exposure to one or more traumatic events. According to the diagnostic and statistical manual of mental disorders (DSM-5) [687], PTSD many exhibit one or more of the following characteristic symptoms: (1) negative emotional and behavioral behaviors driven by the fear of the traumatic events (such as reliving the traumatic event or trying to avoid thinking about the traumatic event); (2) negative mood states and cognitions to the extent that one would lose the ability to experience pleasure (i.e., anhedonic); (3) antisocial, aggressive behavior towards other (i.e., externalizing symptoms); and (4) dissociative symptoms as a way to escape reality. It was estimated that between 3.5% and 10% of the population would experience PTSD in their lifetime [842,843]. Furthermore, patients who suffer from PTSD often have other issues, such as depression, obesity, diabetes, cardiovascular disease, and metabolic conditions [844,845].

As we have elaborated in previous sections, physical exercises are known to be an effective treatment for mental disorders. Studies of exercise-based intervention for PTSD have been received reviewed in [844], where nine observation studies and ten intervention studies were included.

In [846], a 12-week randomized controlled trial was conducted with 47 veterans who suffer from PTSD. The participants were randomly assigned to an integrative exercise group, or a waitlist control group. The integrative exercise intervention program combined aerobic and resistance exercise, and integrated with mindfulness-based practices. The program runs three 1-h session per week for 12 weeks. The experimental group exhibited a more significant reduction in PTSD symptom severity, a greater improvement in psychological quality of life, and some improvement in physical quality of life.

In [847], a 12-week randomized controlled trial was conducted to assess the efficacy of using physical exercises as an intervention to reduce the symptoms of PTSD. The control group received the usual care for patients with PTDS, including psychotherapy, pharmaceutical interventions, and group therapy. The experimental group received physical exercise training in addition to the usual care. The exercise intervention consists of three 30-min resistance training sessions per week and a walking program. At the end of the study, PTSD symptoms in the experimental group were significantly reduced compared with the control group. Furthermore, the experimental group exhibited reduced depressive symptoms, reduced waist circumference, and improved sleep quality.

In [848], a voluntary bicycle-exercise intervention program was offered to a group of military veterans who suffered from PTSD. At the end of the study, the PTSD symptoms and the sleep quality were measured. The study found that positive effect of the exercise program was significant only for participants who had poor sleep quality at the intake of the study where the greater reductions in hyperarousal symptoms were observed. The study did not provide the details of the intervention program such as the duration of the study.

In [849], a 2-week (6-session) randomized controlled trial was conducted with 33 patients with PTSD. The participants were randomly assigned to one of three groups. All three groups engaged in all the 6-session aerobic exercises (stationary biking). The three groups differed in the way attention was prompted, or the lack thereof, during the sessions. In the first group, participants received prompts directing their attention to the interoceptive effects of exercise. In the second group, participants watched a nature documentary to distract them from somatic arousal. The third group received no prompt to attention or distraction. The study found that the first group had the most dramatic reduction in PTSD symptoms, although PTSD symptom reduction was observed for all three groups. In a follow-up study [850], the same intervention was used to determine whether or not cardiorespiratory fitness can predict who will benefit most on the exercise-based intervention. The study concluded that those who have low pre-treatment levels of cardiorespiratory fitness enjoyed the most significant reduction in PTSD symptoms.

In [851], a 15-session aerobic exercise intervention program (in the form of moderate-intensity walking) was conducted to assess the effect of the exercise program for female adolescents with PTSD, depression, and anxiety. Immediate after the intervention, strong reduction in PTSD and trauma symptoms was detected. However, the severity of depression and anxiety was not changed by the intervention. The 1-month follow-up assessment had mixed results, indicating that the exercise intervention should be carried out continuously for it to be effective.

In [852], an 8-week aerobic exercise program was conducted with 15 adolescents who suffer from childhood PTSD. During the intervention program, the participants engaged in three 40-min aerobic exercise sessions per week for 8 weeks. The study confirmed that the exercise program had positive effects in reducing the levels of PTSD, anxiety, and depression of the participants. In [853], the efficacy of a 12-session aerobic exercise program on reducing PTSD symptoms and the level of depression and anxiety was reported. Significant reductions in PTSD symptoms, anxiety, and depression were observed following the exercise intervention.

Some studies examined the rehabilitation of those who were not diagnosed as having PTSD, but were exposed to some trauma events. Similar to the numerous studies on using aerobic exercise as a way to reduce PTSD symptoms, [845] confirmed that the exercise program significantly reduced the mood symptoms in the experimental group compared with the control group.

Conclusion

Physical exercises are a core component in all forms of rehabilitation. Technology is playing an increasingly important role in tracking and recognizing human movement, providing objective feedback in real-time to the patient, as well as providing guidance to the patient. Furthermore, technology has been used to provide a virtual or mixed reality environment for the patient to practice rehabilitation exercises that is more fun and engaging than traditional descriptive therapy programs. A general consensus is that technology can lower the cost of rehabilitation, increase the accessibility and convenience of rehabilitation, and enhances adherence to the prescribed rehabilitation program.

In this book, we presented the state-of-the-art research on motion tracking technology in rehabilitation. We showed that a wide-spectrum of sensing modalities have been used for motion tracking in rehabilitation. Except for basic range of motion tracking, typically sophisticated methods (i.e., rule based and machine-learning based) are required to objectively recognize the exercises and to automatically assess the quality of the exercises performed by the patients. The rule-based methods have the advantage of providing real-time and precise feedback to the patients. However, the rule-based methods require the expert input from clinicians on the specification of the rehabilitation exercises. Machine learning provides more flexibility than the rule-based approach because no detailed rules are needed. However, typically high quality training datasets are needed for supervised machine learning, which also requires deep involvement of the clinicians. It is also challenging for machine learning-based approach to provide precise feedback to the patients.

Exergames with a virtual or mixed reality interface could play a much bigger role in future rehabilitation because the integration of the gamification mechanics and the immersive environment have proven to be significantly more engaging and entertaining to the patients than traditional rehabilitation programs. Exergames could mitigate the adherence issue in the field of rehabilitation.

We have to acknowledge that the level of technology use in standard rehabilitation protocols is still very low. Typically, the only form of technology considered is assistive technology, which helps the patients to carry out activities of daily living. There is lack of mature technology-driven systems that can be readily used in standard rehabilitation programs. For a technology-driven solution to be clinically viable, providing objective assessment at the level of individual movement is insufficient. Such a solution must also be able to automatically provide accurate scores according to relevant clinical scales for each type of rehabilitation. While there are some studies towards this goal, there is a long way to go.

References

[1] Connolly J, Condell J, O'Flynn B, *et al.* IMU sensor-based electronic goniometric glove for clinical finger movement analysis. *IEEE Sensors Journal.* 2017;18(3):1273–1281.

[2] Chang HT, Chang JY. Sensor glove based on novel inertial sensor fusion control algorithm for 3-d real-time hand gestures measurements. *IEEE Transactions on Industrial Electronics.* 2019;67(1):658–666.

[3] Lin BS, Lee IJ, Hsiao PC, *et al.* An assessment system for post-stroke manual dexterity using principal component analysis and logistic regression. *IEEE Transactions on Neural Systems and Rehabilitation Engineering.* 2019;27(8):1626–1634.

[4] Lin BS, Lee IJ, Chen JL. Novel assembled sensorized glove platform for comprehensive hand function assessment by using inertial sensors and force sensing resistors. *IEEE Sensors Journal.* 2020;20(6):3379–3389.

[5] Borghetti M, Sardini E, Serpelloni M. Sensorized glove for measuring hand finger flexion for rehabilitation purposes. *IEEE Transactions on Instrumentation and Measurement.* 2013;62(12):3308–3314.

[6] Sbernini L, Quitadamo LR, Riillo F, *et al.* Sensory-glove-based open surgery skill evaluation. *IEEE Transactions on Human–Machine Systems.* 2018;48(2):213–218.

[7] Chuang WC, Hwang WJ, Tai TM, *et al.* Continuous finger gesture recognition based on flex sensors. *Sensors.* 2019;19(18):3986.

[8] Simone LK, Sundarrajan N, Luo X, *et al.* A low cost instrumented glove for extended monitoring and functional hand assessment. *Journal of Neuroscience Methods.* 2007;160(2):335–348.

[9] Oess NP, Wanek J, Curt A. Design and evaluation of a low-cost instrumented glove for hand function assessment. *Journal of Neuroengineering and Rehabilitation.* 2012;9(1):1–11.

[10] Wang L, Meydan T, Williams PI. A two-axis goniometric sensor for tracking finger motion. *Sensors.* 2017;17(4):770.

[11] Wang L, Meydan T, Williams P. Design and evaluation of a 3-D printed optical sensor for monitoring finger flexion. *IEEE Sensors Journal.* 2017;17(6):1937–1944.

[12] Kayaalp ME, Agres AN, Reichmann J, *et al.* Validation of a novel device for the knee monitoring of orthopaedic patients. *Sensors.* 2019;19(23):5193.

[13] Bland JM, Altman D. Statistical methods for assessing agreement between two methods of clinical measurement. *The Lancet.* 1986;327(8476): 307–310.

[14] Hanly S. Accelerometer Specifications: Deciphering an Accelerometer's Datasheet; 2018. https://blog.endaq.com/accelerometer-specifications-decoding-a-datasheet.

[15] Beauregard S, Haas H. Pedestrian dead reckoning: a basis for personal positioning. In: *Proceedings of the 3rd Workshop on Positioning, Navigation and Communication*; 2006. p. 27–35.

[16] Chang H, Xue L, Qin W, *et al.* An integrated MEMS gyroscope array with higher accuracy output. *Sensors*. 2008;8(4):2886–2899.

[17] Allan DW. Statistics of atomic frequency standards. *Proceedings of the IEEE*. 1966;54(2):221–230.

[18] El-Sheimy N, Hou H, Niu X. Analysis and modeling of inertial sensors using Allan variance. *IEEE Transactions on Instrumentation and Measurement*. 2007;57(1):140–149.

[19] Finlay CC, Maus S, Beggan C, *et al.* International geomagnetic reference field: the eleventh generation. *Geophysical Journal International*. 2010;183(3):1216–1230.

[20] Todaro MT, Sileo L, De Vittorio M. Magnetic field sensors based on microelectromechanical systems (MEMS) technology. In: K. Kuang (Eds.), *Magnetic Sensors—Principles and Applications*. Europe: InTech, 2012; p. 103–124.

[21] Herrera-May AL, Aguilera-Cortés LA, García-Ramírez PJ, *et al.* Resonant magnetic field sensors based on MEMS technology. *Sensors*. 2009;9(10):7785–7813.

[22] Abbasi-Kesbi R, Nikfarjam A, Memarzadeh-Tehran H. A patient-centric sensory system for in-home rehabilitation. *IEEE Sensors Journal*. 2016;17(2):524–533.

[23] Nguyen ND, Pham TH, Pathirana PN, *et al.* Sensing and characterization of the wrist using dart thrower's movement. *IEEE Sensors Journal*. 2018;18(10):4145–4153.

[24] Abdelhady M, van den Bogert AJ, Simon D. A high-fidelity wearable system for measuring lower-limb kinetics and kinematics. *IEEE Sensors Journal*. 2019;19(24):12482–12493.

[25] Ai QS, Chen L, Liu Q, *et al.* Rehabilitation assessment for lower limb disability based on multi-disciplinary approaches. *Australasian Physical & Engineering Sciences in Medicine*. 2014;37(2):355–365.

[26] Huang YP, Liu YY, Hsu WH, *et al.* Monitoring and assessment of rehabilitation progress on range of motion after total knee replacement by sensor-based system. *Sensors*. 2020;20(6):1703.

[27] Cortell-Tormo JM, García-Jaén M, Ruiz-Fernández D, *et al.* Lumbatex: a wearable monitoring system based on inertial sensors to measure and control the lumbar spine motion. *IEEE Transactions on Neural Systems and Rehabilitation Engineering*. 2019;27(8):1644–1653.

[28] He J, Chen S, Guo Z, *et al.* A comparative study of motion recognition methods for efficacy assessment of upper limb function. *International Journal of Adaptive Control and Signal Processing*. 2019;33(8):1248–1256.

[29] He Z, Liu T, Yi J. A wearable sensing and training system: towards gait rehabilitation for elderly patients with knee osteoarthritis. *IEEE Sensors Journal*. 2019;19(14):5936–5945.

[30] Lui J, Menon C. Would a thermal sensor improve arm motion classification accuracy of a single wrist-mounted inertial device? *Biomedical Engineering Online*. 2019;18(1):1–17.

[31] Bussmann J, Martens W, Tulen J, *et al.* Measuring daily behavior using ambulatory accelerometry: the activity monitor. *Behavior Research Methods, Instruments, & Computers*. 2001;33(3):349–356.

[32] Papi E, Osei-Kuffour D, Chen YMA, *et al.* Use of wearable technology for performance assessment: a validation study. *Medical Engineering & Physics*. 2015;37(7):698–704.

[33] González-Villanueva L, Cagnoni S, Ascari L. Design of a wearable sensing system for human motion monitoring in physical rehabilitation. *Sensors*. 2013;13(6):7735–7755.

[34] Hullfish TJ, Qu F, Stoeckl BD, *et al.* Measuring clinically relevant knee motion with a self-calibrated wearable sensor. *Journal of Biomechanics*. 2019;89:105–109.

[35] Motoi K, Taniguchi S, Baek M, *et al.* Development of a wearable gait monitoring system for evaluating efficacy of walking training in rehabilitation. *Sensors and Materials*. 2012;24(6):359–373.

[36] Wong WY, Wong MS. Trunk posture monitoring with inertial sensors. *European Spine Journal*. 2008;17(5):743–753.

[37] Liu X, Rajan S, Ramasarma N, *et al.* The use of a finger-worn accelerometer for monitoring of hand use in ambulatory settings. *IEEE Journal of Biomedical and Health Informatics*. 2019;23(2):599–606.

[38] Qiu S, Liu L, Wang Z, *et al.* Body sensor network-based gait quality assessment for clinical decision-support via multi-sensor fusion. *IEEE Access*. 2019;7:59884–59894.

[39] Qiu S, Wang H, Li J, *et al.* Towards wearable-inertial-sensor-based gait posture evaluation for subjects with unbalanced gaits. *Sensors*. 2020;20(4):1193.

[40] Ranganathan R, Wang R, Dong B, *et al.* Identifying compensatory movement patterns in the upper extremity using a wearable sensor system. *Physiological Measurement*. 2017;38(12):2222.

[41] Annegarn J, Spruit MA, Uszko-Lencer NH, *et al.* Objective physical activity assessment in patients with chronic organ failure: a validation study of a new single-unit activity monitor. *Archives of Physical Medicine and Rehabilitation*. 2011;92(11):1852–1857.

[42] Allseits E, Kim KJ, Bennett C, *et al.* A novel method for estimating knee angle using two leg-mounted gyroscopes for continuous monitoring with mobile health devices. *Sensors*. 2018;18(9):2759.

[43] Chiang CY, Chen KH, Liu KC, *et al.* Data collection and analysis using wearable sensors for monitoring knee range of motion after total knee arthroplasty. *Sensors*. 2017;17(2):418.

[44]   Horenstein RE, Lewis CL, Yan S, *et al.* Validation of magneto-inertial measuring units for measuring hip joint angles. *Journal of Biomechanics.* 2019;91:170–174.

[45]   Parrington L, Jehu DA, Fino PC, *et al.* Validation of an inertial sensor algorithm to quantify head and trunk movement in healthy young adults and individuals with mild traumatic brain injury. *Sensors.* 2018;18(12):4501.

[46]   Tulipani L, Boocock MG, Lomond KV, *et al.* Validation of an inertial sensor system for physical therapists to quantify movement coordination during functional tasks. *Journal of Applied Biomechanics.* 2018;34(1):23–30.

[47]   Argent R, Drummond S, Remus A, *et al.* Evaluating the use of machine learning in the assessment of joint angle using a single inertial sensor. *Journal of Rehabilitation and Assistive Technologies Engineering.* 2019;6:2055668319868544.

[48]   Baraka A, Shaban H, El-Nasr A, *et al.* Wearable accelerometer and semg-based upper limb bsn for tele-rehabilitation. *Applied Sciences.* 2019;9(14):2795.

[49]   Biswas D, Corda D, Baldus G, *et al.* Recognition of elementary arm movements using orientation of a tri-axial accelerometer located near the wrist. *Physiological Measurement.* 2014;35(9):1751.

[50]   Derungs A, Schuster-Amft C, Amft O. Longitudinal walking analysis in hemiparetic patients using wearable motion sensors: Is there convergence between body sides? *Frontiers in Bioengineering and Biotechnology.* 2018;6:57.

[51]   Seiter J, Derungs A, Schuster-Amft C, *et al.* Daily life activity routine discovery in hemiparetic rehabilitation patients using topic models. *Methods of Information in Medicine.* 2015;54(03):248–255.

[52]   Bai L, Pepper MG, Yan Y, *et al.* Low cost inertial sensors for the motion tracking and orientation estimation of human upper limbs in neurological rehabilitation. *IEEE Access.* 2020;8:54254–54268.

[53]   Bai L, Pepper MG, Yan Y, *et al.* Quantitative assessment of upper limb motion in neurorehabilitation utilizing inertial sensors. *IEEE Transactions on Neural Systems and Rehabilitation Engineering.* 2014;23(2):232–243.

[54]   Digo E, Pierro G, Pastorelli S, *et al.* Evaluation of spinal posture during gait with inertial measurement units. *Proceedings of the Institution of Mechanical Engineers, Part H: Journal of Engineering in Medicine.* 2020;234(10):1094–1105.

[55]   Gil-Agudo Á, de Los Reyes-Guzman A, Dimbwadyo-Terrer I, *et al.* A novel motion tracking system for evaluation of functional rehabilitation of the upper limbs. *Neural Regeneration Research.* 2013;8(19):1773.

[56]   Laudanski A, Brouwer B, Li Q. Activity classification in persons with stroke based on frequency features. *Medical Engineering & Physics.* 2015;37(2):180–186.

[57]   Müller P, Bégin MA, Schauer T, *et al.* Alignment-free, self-calibrating elbow angles measurement using inertial sensors. *IEEE Journal of Biomedical and Health Informatics.* 2016;21(2):312–319.

[58] Oubre B, Daneault JF, Jung HT, *et al.* Estimating upper-limb impairment level in stroke survivors using wearable inertial sensors and a minimally-burdensome motor task. *IEEE Transactions on Neural Systems and Rehabilitation Engineering.* 2020;28(3):601–611.

[59] Pérez R, Costa Ú, Torrent M, *et al.* Upper limb portable motion analysis system based on inertial technology for neurorehabilitation purposes. *Sensors.* 2010;10(12):10733–10751.

[60] Zhou H, Hu H, Tao Y. Inertial measurements of upper limb motion. *Medical and Biological Engineering and Computing.* 2006;44(6):479–487.

[61] Newman J, Zhou H, Hu H. Inertial sensors for motion detection of human upper limbs. *Sensor Review.* 2007;27(2):151–158.

[62] Bavan L, Surmacz K, Beard D, *et al.* Adherence monitoring of rehabilitation exercise with inertial sensors: a clinical validation study. *Gait & Posture.* 2019;70:211–217.

[63] Bleser G, Steffen D, Weber M, *et al.* A personalized exercise trainer for the elderly. *Journal of Ambient Intelligence and Smart Environments.* 2013;5(6):547–562.

[64] Bolink S, Naisas H, Senden R, *et al.* Validity of an inertial measurement unit to assess pelvic orientation angles during gait, sit–stand transfers and step-up transfers: comparison with an optoelectronic motion capture system. *Medical Engineering & Physics.* 2016;38(3):225–231.

[65] van Rooij WM, Senden R, Heyligers IC, *et al.* Physical functioning of low back pain patients: perceived physical functioning and functional capacity, but not physical activity is affected. *Disability and Rehabilitation.* 2015;37(24):2257–2263.

[66] Nguyen H, Lebel K, Bogard S, *et al.* Using inertial sensors to automatically detect and segment activities of daily living in people with Parkinson's disease. *IEEE Transactions on Neural Systems and Rehabilitation Engineering.* 2017;26(1):197–204.

[67] Patel S, Hughes R, Hester T, *et al.* A novel approach to monitor rehabilitation outcomes in stroke survivors using wearable technology. *Proceedings of the IEEE.* 2010;98(3):450–461.

[68] Schwenk M, Hauer K, Zieschang T, *et al.* Sensor-derived physical activity parameters can predict future falls in people with dementia. *Gerontology.* 2014;60(6):483–492.

[69] Seel T, Raisch J, Schauer T. IMU-based joint angle measurement for gait analysis. *Sensors.* 2014;14(4):6891–6909.

[70] Bittel AJ, Elazzazi A, Bittel DC. Accuracy and precision of an accelerometer-based smartphone app designed to monitor and record angular movement over time. *Telemedicine and e-Health.* 2016;22(4):302–309.

[71] Cheng Q, Juen J, Bellam S, *et al.* Predicting pulmonary function from phone sensors. *Telemedicine and e-Health.* 2017;23(11): 913–919.

[72] Kegelmeyer DA, Kostyk SK, Fritz NE, *et al.* Quantitative biomechanical assessment of trunk control in Huntington's disease reveals more

impairment in static than dynamic tasks. *Journal of the Neurological Sciences*. 2017;376:29–34.

[73] Steins D, Sheret I, Dawes H, *et al.* A smart device inertial-sensing method for gait analysis. *Journal of Biomechanics*. 2014;47(15):3780–3785.

[74] Kyritsis AI, Willems G, Deriaz M, *et al.* Gait pattern recognition using a smartwatch assisting postoperative physiotherapy. *International Journal of Semantic Computing*. 2019;13(02):245–257.

[75] Appelboom G, Taylor BE, Bruce E, *et al.* Mobile phone-connected wearable motion sensors to assess postoperative mobilization. *JMIR mHealth and uHealth*. 2015;3(3):e78.

[76] Braun S, Bjornson K, Dillon-Naftolin E, *et al.* Reliability of StepWatch activity monitor to measure locomotor activity in youth with lower limb salvage. *Pediatric Physical Therapy: The Official Publication of the Section on Pediatrics of the American Physical Therapy Association*. 2018;30(3):217.

[77] McCullagh R, Dillon C, O'Connell AM, *et al.* Step-count accuracy of 3 motion sensors for older and frail medical inpatients. *Archives of Physical Medicine and Rehabilitation*. 2017;98(2):295–302.

[78] Capio CM, Sit CH, Abernethy B. Physical activity measurement using MTI (actigraph) among children with cerebral palsy. *Archives of Physical Medicine and Rehabilitation*. 2010;91(8):1283–1290.

[79] Joseph C, Strömbäck B, Hagströmer M, *et al.* Accelerometry: a feasible method to monitor physical activity during sub-acute rehabilitation of persons with stroke. *Journal of Rehabilitation Medicine*. 2018;50(5):429–434.

[80] Furlanetto KC, Bisca GW, Oldemberg N, *et al.* Step counting and energy expenditure estimation in patients with chronic obstructive pulmonary disease and healthy elderly: accuracy of 2 motion sensors. *Archives of Physical Medicine and Rehabilitation*. 2010;91(2):261–267.

[81] McNamara RJ, Tsai LLY, Wootton SL, *et al.* Measurement of daily physical activity using the SenseWear Armband: compliance, comfort, adverse side effects and usability. *Chronic Respiratory Disease*. 2016;13(2):144–154.

[82] Miyamoto S, Minakata Y, Azuma Y, *et al.* Verification of a motion sensor for evaluating physical activity in COPD patients. *Canadian Respiratory Journal*. 2018;2018.

[83] Probst VS, Kovelis D, Hernandes NA, *et al.* Effects of 2 exercise training programs on physical activity in daily life in patients with COPD. *Respiratory Care*. 2011;56(11):1799–1807.

[84] Kanno I. Piezoelectric MEMS: ferroelectric thin films for MEMS applications. *Japanese Journal of Applied Physics*. 2018;57(4):040101.

[85] Chen W, Wang D, Xu B, *et al.* Multimode fiber tip Fabry-Perot cavity for highly sensitive pressure measurement. *Scientific Reports*. 2017;7(1):1–6.

[86] Urban F, Kadlec J, Vlach R, *et al.* Design of a pressure sensor based on optical fiber Bragg grating lateral deformation. *Sensors*. 2010;10(12):11212–11225.

[87] Suter M, Eichelberger P, Frangi J, *et al.* Measuring lumbar back motion during functional activities using a portable strain gauge sensor-based system:

a comparative evaluation and reliability study. *Journal of Biomechanics.* 2020;100:109593.

[88] Consmüller T, Rohlmann A, Weinland D, *et al.* Velocity of lordosis angle during spinal flexion and extension. *PLoS One.* 2012;7(11):e50135.

[89] Dideriksen J, Gizzi L, Petzke F, *et al.* Deterministic accessory spinal movement in functional tasks characterizes individuals with low back pain. *Clinical Neurophysiology.* 2014;125(8):1663–1668.

[90] Vaisy M, Gizzi L, Petzke F, *et al.* Measurement of lumbar spine functional movement in low back pain. *The Clinical Journal of Pain.* 2015;31(10): 876–885.

[91] Clark RA, Bryant AL, Pua Y, *et al.* Validity and reliability of the Nintendo Wii Balance Board for assessment of standing balance. *Gait & Posture.* 2010;31(3):307–310.

[92] Holmes JD, Jenkins ME, Johnson AM, *et al.* Validity of the Nintendo Wii® balance board for the assessment of standing balance in Parkinson's disease. *Clinical Rehabilitation.* 2013;27(4):361–366.

[93] Yang CX. Low-cost experimental system for center of mass and center of pressure measurement. *IEEE Access.* 2018;6:45021–45033.

[94] Okamoto M, Kasamatsu S. Storage medium storing a load detecting program and load detecting apparatus. Google Patents; 2016. US Patent 9,421,456.

[95] Bonde-Petersen F. A simple force platform. *European Journal of Applied Physiology and Occupational Physiology.* 1975;34(1):51–54.

[96] Raper DP, Witchalls J, Philips EJ, *et al.* Use of a tibial accelerometer to measure ground reaction force in running: a reliability and validity comparison with force plates. *Journal of Science and Medicine in Sport.* 2018;21(1):84–88.

[97] Mengarelli A, Verdini F, Cardarelli S, *et al.* Balance assessment during squatting exercise: a comparison between laboratory grade force plate and a commercial, low-cost device. *Journal of Biomechanics.* 2018;71: 264–270.

[98] Liu K, Yan J, Liu Y, *et al.* Noninvasive estimation of joint moments with inertial sensor system for analysis of STS rehabilitation training. *Journal of Healthcare Engineering.* 2018;2018:6570617.

[99] Nelson A, Koslakiewicz N, Almonroeder TG. Assessment of knee kinetic symmetry using force plate technology. *Journal of Sport Rehabilitation.* 2018;27(6).

[100] Gore S, Franklyn-Miller A, Richter C, *et al.* The effects of rehabilitation on the biomechanics of patients with athletic groin pain. *Journal of Biomechanics.* 2020;99:109474.

[101] Powers B, Alfano L, Miller N, *et al.* HEREDITARY NEUROPATHIES & ALS: P. 116 Determining the utility of the GAITRite walkway to assess foot drop in subjects with CMT1A. *Neuromuscular Disorders.* 2020;30:S80.

[102] Riis J, Byrgesen SM, Kragholm KH, *et al.* Validity of the GAITRite walkway compared to functional balance tests for fall risk assessment in geriatric outpatients. *Geriatrics.* 2020;5(4):77.

[103] Eid MM, Abdelbasset WK, Abdelaty FM, *et al.* Effect of physical therapy rehabilitation program combined with music on children with lower limb burns: a twelve-week randomized controlled study. *Burns.* 2020;47(5):1146–1152.

[104] Heeb R, Morgan K, Putnam M, *et al.* Factors influencing the community participation of people aging with long-term physical disabilities. *Archives of Physical Medicine and Rehabilitation.* 2020;101(11):e29.

[105] Vítečková S, Horáková H, Poláková K, *et al.* Agreement between the GAITRite® System and the Wearable Sensor BTS G-Walk® for measurement of gait parameters in healthy adults and Parkinson's disease patients. *PeerJ.* 2020;8:e8835.

[106] Ylinen J, Pennanen A, Weir A, *et al.* Effect of biomechanical footwear on upper and lower leg muscle activity in comparison with knee brace and normal walking. *Journal of Electromyography and Kinesiology.* 2021;57:102528.

[107] Grampp J, Willson J, Kernozek T. The plantar loading variations to uphill and downhill gradients during treadmill walking. *Foot & Ankle International.* 2000;21(3):227–231.

[108] Putti A, Arnold G, Cochrane L, *et al.* The Pedar® in-shoe system: repeatability and normal pressure values. *Gait & Posture.* 2007;25(3):401–405.

[109] Young C. The F-SCAN system of foot pressure analysis. *Clinics in Podiatric Medicine and Surgery.* 1993;10(3):455–461.

[110] Luo ZP, Berglund LJ, An KN, *et al.* Validation of F-scan pressure sensor system: a technical note. *Journal of Rehabilitation Research and Development.* 1998;35:186–186.

[111] Catalfamo P, Moser D, Ghoussayni S, *et al.* Detection of gait events using an F-scan in-shoe pressure measurement system. *Gait & Posture.* 2008;28(3):420–426.

[112] Kong PW, De Heer H. Wearing the F-scan mobile in-shoe pressure measurement system alters gait characteristics during running. *Gait & Posture.* 2009;29(1):143–145.

[113] Mohammed S, Same A, Oukhellou L, *et al.* Recognition of gait cycle phases using wearable sensors. *Robotics and Autonomous Systems.* 2016;75:50–59.

[114] Patrick K, Donovan L. Test–retest reliability of the Tekscan® F-Scan® 7 in-shoe plantar pressure system during treadmill walking in healthy recreationally active individuals. *Sports Biomechanics.* 2018;17(1):83–97.

[115] Speight S, Reidy S, Stephenson J, *et al.* Combining F-Scan and GAITRite® for use in clinical gait analysis: a validation study. ResearchSquare; 2020.

[116] Strzalkowski ND, Peters RM, Inglis JT, *et al.* Cutaneous afferent innervation of the human foot sole: what can we learn from single-unit recordings? *Journal of Neurophysiology.* 2018;120(3):1233–1246.

[117] Tahir AM, Chowdhury ME, Khandakar A, *et al.* A systematic approach to the design and characterization of a smart insole for detecting vertical ground reaction force (vGRF) in gait analysis. *Sensors.* 2020;20(4):957.

[118] Barnea A, Oprisan C, Olaru D. Force sensitive resistors calibration for use in gripping devices. *Mechanical Testing and Diagnosis.* 2012;2(3):18–27.

[119] De Rossi S, Lenzi T, Vitiello N, *et al.* Development of an in-shoe pressure-sensitive device for gait analysis. In: *2011 Annual International Conference of the IEEE Engineering in Medicine and Biology Society*. New York, NY: IEEE; 2011. p. 5637–5640.

[120] Parmar S, Khodasevych I, Troynikov O. Evaluation of flexible force sensors for pressure monitoring in treatment of chronic venous disorders. *Sensors*. 2017;17(8):1923.

[121] Vilarinho D, Theodosiou A, Leitão C, *et al.* POFBG-embedded cork insole for plantar pressure monitoring. *Sensors*. 2017;17(12):2924.

[122] Tao J, Dong M, Li L, *et al.* Real-time pressure mapping smart insole system based on a controllable vertical pore dielectric layer. *Microsystems & Nanoengineering*. 2020;6(1):1–10.

[123] Lin F, Wang A, Zhuang Y, *et al.* Smart insole: a wearable sensor device for unobtrusive gait monitoring in daily life. *IEEE Transactions on Industrial Informatics*. 2016;12(6):2281–2291.

[124] Howell AM, Kobayashi T, Hayes HA, *et al.* Kinetic gait analysis using a low-cost insole. *IEEE Transactions on Biomedical Engineering*. 2013;60(12):3284–3290.

[125] Lee SI, Park E, Huang A, *et al.* Objectively quantifying walking ability in degenerative spinal disorder patients using sensor equipped smart shoes. *Medical Engineering & Physics*. 2016;38(5):442–449.

[126] Kong K, Tomizuka M. A gait monitoring system based on air pressure sensors embedded in a shoe. *IEEE/ASME Transactions on Mechatronics*. 2009;14(3):358–370.

[127] Jung PG, Oh S, Lim G, *et al.* A mobile motion capture system based on inertial sensors and smart shoes. *Journal of Dynamic Systems, Measurement, and Control*. 2014;136(1).

[128] Zhang W, Tomizuka M, Byl N. A wireless human motion monitoring system for smart rehabilitation. *Journal of Dynamic Systems, Measurement, and Control*. 2016;138(11).

[129] Jung KC, Son JH, Chang SH. Self-powered smart shoes with tension-type ribbon harvesters and sensors. *Advanced Materials Technologies*. 2020;6:2000872.

[130] Zou Y, Libanori A, Xu J, *et al.* Triboelectric nanogenerator enabled smart shoes for wearable electricity generation. *Research*. 2020;2020:7158953.

[131] Yamada T, Hayamizu Y, Yamamoto Y, *et al.* A stretchable carbon nanotube strain sensor for human-motion detection. *Nature Nanotechnology*. 2011;6(5):296.

[132] Byko M. From electric corsets to self-cleaning pants: the materials science and engineering of textiles. *JOM*. 2005;57(7):14–18.

[133] Sardini E, Serpelloni M, Pasqui V. Wireless wearable T-shirt for posture monitoring during rehabilitation exercises. *IEEE Transactions on Instrumentation and Measurement*. 2014;64(2):439–448.

[134] Tognetti A, Lorussi F, Bartalesi R, *et al.* Wearable kinesthetic system for capturing and classifying upper limb gesture in post-stroke rehabilitation. *Journal of NeuroEngineering and Rehabilitation*. 2005;2(1):1–16.

[135]   Giorgino T, Tormene P, Lorussi F, *et al.* Sensor evaluation for wearable strain gauges in neurological rehabilitation. *IEEE Transactions on Neural Systems and Rehabilitation Engineering.* 2009;17(4):409–415.

[136]   Tormene P, Bartolo M, De Nunzio AM, *et al.* Estimation of human trunk movements by wearable strain sensors and improvement of sensor's placement on intelligent biomedical clothes. *Biomedical Engineering Online.* 2012;11(1):1–8.

[137]   Mattmann C, Amft O, Harms H, *et al.* Recognizing upper body postures using textile strain sensors. In: *2007 11th IEEE International Symposium on Wearable Computers.* New York, NY: IEEE; 2007. p. 29–36.

[138]   Bergmann JH, Anastasova-Ivanova S, Spulber I, *et al.* An attachable clothing sensor system for measuring knee joint angles. *IEEE Sensors Journal.* 2013;13(10):4090–4097.

[139]   Papi E, Spulber I, Kotti M, *et al.* Smart sensing system for combined activity classification and estimation of knee range of motion. *IEEE Sensors Journal.* 2015;15(10):5535–5544.

[140]   Esfahani MIM, Taghinezhad S, Mottaghitalab V, *et al.* Novel printed body worn sensor for measuring the human movement orientation. *Sensor Review.* 2016;35(3):321–331.

[141]   Esfahani MIM, Nussbaum MA. A "smart" undershirt for tracking upper body motions: task classification and angle estimation. *IEEE Sensors Journal.* 2018;18(18):7650–7658.

[142]   Poomsalood S, Muthumayandi K, Hambly K. Can stretch sensors measure knee range of motion in healthy adults? *Biomedical Human Kinetics.* 2019;11(1):1–8.

[143]   Hu S, Dai M, Dong T, *et al.* A textile sensor for long durations of human motion capture. *Sensors.* 2019;19(10):2369.

[144]   Lorussi F, Rocchia W, Scilingo EP, *et al.* Wearable, redundant fabric-based sensor arrays for reconstruction of body segment posture. *IEEE Sensors Journal.* 2004;4(6):807–818.

[145]   Giuffrida JP, Lerner A, Steiner R, *et al.* Upper-extremity stroke therapy task discrimination using motion sensors and electromyography. *IEEE Transactions on Neural Systems and Rehabilitation Engineering.* 2008;16(1): 82–90.

[146]   Reaz MBI, Hussain MS, Mohd-Yasin F. Techniques of EMG signal analysis: detection, processing, classification and applications. *Biological Procedures Online.* 2006;8(1):11–35.

[147]   Cram JR, Kasman GS, Holtz J. Introduction to surface electromyography. Aspen Publishers; 1998.

[148]   Finkelstein G. Mechanical neuroscience: Emil du Bois-Reymond's innovations in theory and practice. *Frontiers in Systems Neuroscience.* 2015; 9:133.

[149]   Fang C, He B, Wang Y, *et al.* EMG-centered multisensory based technologies for pattern recognition in rehabilitation: state of the art and challenges. *Biosensors.* 2020;10(8):85.

[150] Arozi M, Caesarendra W, Ariyanto M, *et al.* Pattern recognition of single-channel sEMG signal using PCA and ANN method to classify nine hand movements. *Symmetry.* 2020;12(4):541.

[151] Rissanen SM, Kankaanpää M, Meigal A, *et al.* Surface EMG and acceleration signals in Parkinson's disease: feature extraction and cluster analysis. *Medical & Biological Engineering & Computing.* 2008;46(9): 849–858.

[152] Malik OA, Senanayake SA, Zaheer D. An intelligent recovery progress evaluation system for ACL reconstructed subjects using integrated 3-D kinematics and EMG features. *IEEE Journal of Biomedical and Health Informatics.* 2015;19(2):453–463.

[153] ORIZIO C. Muscle sound: bases for the introduction of a mechanomyographic signal in muscle studies. *Critical Reviews in Biomedical Engineering.* 1993;21(3):201–243.

[154] Posatskiy A, Chau T. Design and evaluation of a novel microphone-based mechanomyography sensor with cylindrical and conical acoustic chambers. *Medical Engineering & Physics.* 2012;34(8):1184–1190.

[155] Woodward RB, Shefelbine SJ, Vaidyanathan R. Pervasive monitoring of motion and muscle activation: inertial and mechanomyography fusion. *IEEE/ASME Transactions on Mechatronics.* 2017;22(5):2022–2033.

[156] Wininger M, Kim NH, Craelius W. Pressure signature of forearm as predictor of grip force. *Journal of Rehabilitation Research & Development.* 2008;45(6).

[157] Ogris G, Kreil M, Lukowicz P. Using FSR based muscle activity monitoring to recognize manipulative arm gestures. In: *2007 11th IEEE International Symposium on Wearable Computers.* New York, NY: IEEE; 2007. p. 45–48.

[158] Xiao ZG, Menon C. Towards the development of a wearable feedback system for monitoring the activities of the upper-extremities. *Journal of Neuroengineering and Rehabilitation.* 2014;11(1):1–13.

[159] Amft O, Junker H, Lukowicz P, *et al.* Sensing muscle activities with body-worn sensors. In: *International Workshop on Wearable and Implantable Body Sensor Networks (BSN'06).* New York, NY: IEEE; 2006. p. 4.

[160] Kreil M, Ogris G, Lukowicz P. Muscle activity evaluation using force sensitive resistors. In: *2008 5th International Summer School and Symposium on Medical Devices and Biosensors.* New York, NY: IEEE; 2008. p. 107–110.

[161] Nissler C, Mouriki N, Castellini C, *et al.* OMG: introducing optical myography as a new human machine interface for hand amputees. In: *2015 IEEE International Conference on Rehabilitation Robotics (ICORR).* New York, NY: IEEE; 2015. p. 937–942.

[162] Olson E. AprilTag: a robust and flexible visual fiducial system. In: *2011 IEEE International Conference on Robotics and Automation.* New York, NY: IEEE; 2011. p. 3400–3407.

[163] Nissler C, Mouriki N, Castellini C. Optical myography: detecting finger movements by looking at the forearm. *Frontiers in Neurorobotics.* 2016;10:3.

[164] Hoerl AE, Kennard RW. Ridge regression: biased estimation for nonorthogonal problems. *Technometrics*. 1970;12(1):55–67.

[165] Nissler C, Badshah I, Castellini C, *et al.* Improving optical myography via convolutional neural networks. In: *Proceedings of the Myoelectric Controls and Upper Limb Prosthetics Symposium*. Institute of Biomedical Engineering, University of New Brunswick; 2017. p. 84.

[166] Wu YT, Fujiwara E, Suzuki CK. Evaluation of optical myography sensor as predictor of hand postures. *IEEE Sensors Journal*. 2019;19(13):5299–5306.

[167] Zalevsky Z, Shpunt A, Malzels A, *et al.* Method and system for object reconstruction; 2013. US Patent 8,400,494.

[168] Bamji CS, Mehta S, Thompson B, *et al.* IMpixel 65nm BSI 320MHz demodulated TOF Image sensor with 3μm global shutter pixels and analog binning. In: *2018 IEEE International Solid-State Circuits Conference (ISSCC)*. New York, NY: IEEE; 2018. p. 94–96.

[169] Lun R, Zhao W. A survey of applications and human motion recognition with Microsoft kinect. *International Journal of Pattern Recognition and Artificial Intelligence*. 2015;29(5):1555008.

[170] Schonauer C, Pintaric T, Kaufmann H, *et al.* Chronic pain rehabilitation with a serious game using multimodal input. In: *2011 International Conference on Virtual Rehabilitation (ICVR)*. New York, NY: IEEE; 2011. p. 1–8.

[171] Chang CY, Lange B, Zhang M, *et al.* Towards pervasive physical rehabilitation using Microsoft Kinect. In: *2012 6th International Conference on Pervasive Computing Technologies for Healthcare (PervasiveHealth)*. New York, NY: IEEE; 2012. p. 159–162.

[172] Clark RA, Pua YH, Fortin K, *et al.* Validity of the Microsoft Kinect for assessment of postural control. *Gait & Posture*. 2012;36(3):372–377.

[173] Clark RA, Pua YH, Bryant AL, *et al.* Validity of the Microsoft Kinect for providing lateral trunk lean feedback during gait retraining. *Gait & Posture*. 2013;38(4):1064–1066.

[174] Fern'ndez-Baena A, Susín A, Lligadas X. Biomechanical validation of upper-body and lower-body joint movements of kinect motion capture data for rehabilitation treatments. In: *Proceedings of the 4th International Conference on Intelligent Networking and Collaborative Systems*. New York, NY: IEEE; 2012. p. 656–661.

[175] Mobini A, Behzadipour S, Saadat Foumani M. Accuracy of Kinect's skeleton tracking for upper body rehabilitation applications. *Disability and Rehabilitation: Assistive Technology*. 2013;(0):1–9.

[176] Bonnechere B, Jansen B, Salvia P, *et al.* Validity and reliability of the Kinect within functional assessment activities: comparison with standard stereophotogrammetry. *Gait & Posture*. 2014;39(1):593–598.

[177] Galna B, Barry G, Jackson D, *et al.* Accuracy of the Microsoft Kinect sensor for measuring movement in people with Parkinson's disease. *Gait & Posture*. 2014;39(4):1062–1068.

[178] Obdrzalek S, Kurillo G, Ofli F, *et al.* Accuracy and robustness of Kinect pose estimation in the context of coaching of elderly population. In: *2012*

Annual International Conference of the Engineering in Medicine and Biology Society (EMBC). New York, NY: IEEE; 2012. p. 1188–1193.

[179] Nixon ME, Howard AM, Chen YP. Quantitative evaluation of the Microsoft Kinect TM for use in an upper extremity virtual rehabilitation environment. In: *2013 International Conference on Virtual Rehabilitation (ICVR)*. New York, NY: IEEE; 2013. p. 222–228.

[180] Zhao W, Espy DD, Reinthal MA, *et al.* A feasibility study of using a single kinect sensor for rehabilitation exercises monitoring: a rule based approach. In: *2014 IEEE Symposium on Computational Intelligence in Healthcare and e-health (CICARE)*. New York, NY: IEEE; 2014. p. 1–8.

[181] Zhao W, Espy DD, Reinthal MA. Assessment of sit-to-stand movements using a single kinect sensor: a preliminary study in healthy subjects. *International Journal of Healthcare Information Systems and Informatics (IJHISI)*. 2019;14(1):29–43.

[182] Mobini A, Behzadipour S, Saadat M. Test-retest reliability of Kinect's measurements for the evaluation of upper body recovery of stroke patients. *Biomedical Engineering Online*. 2015;14(1):1–13.

[183] Harte JM, Golby CK, Acosta J, *et al.* Chest wall motion analysis in healthy volunteers and adults with cystic fibrosis using a novel Kinect-based motion tracking system. *Medical & Biological Engineering & Computing*. 2016;54(11):1631–1640.

[184] Kurillo G, Chen A, Bajcsy R, *et al.* Evaluation of upper extremity reachable workspace using Kinect camera. *Technology and Health Care*. 2013;21(6): 641–656.

[185] Foreman MH, Engsberg JR. A virtual reality tool for measuring and shaping trunk compensation for persons with stroke: design and initial feasibility testing. *Journal of Rehabilitation and Assistive Technologies Engineering*. 2019;6:2055668318823673.

[186] Lee MH, Siewiorek DP, Smailagic A, *et al.* Interactive hybrid approach to combine machine and human intelligence for personalized rehabilitation assessment. In: *Proceedings of the ACM Conference on Health, Inference, and Learning*; 2020. p. 160–169.

[187] Mortazavi F, Nadian-Ghomsheh A. Continues online exercise monitoring and assessment system with visual guidance feedback for stroke rehabilitation. *Multimedia Tools and Applications*. 2019;78(22):32055–32085.

[188] Bijalwan V, Semwal VB, Singh G, *et al.* HDL-PSR: modelling spatio-temporal features using hybrid deep learning approach for post-stroke rehabilitation. *Neural Processing Letters*. 2022;p. 1–20.

[189] Su CJ. Personal rehabilitation exercise assistant with kinect and dynamic time warping. *International Journal of Information and Education Technology*. 2013;p. 448–454.

[190] Velloso E, Bulling A, Gellersen H. MotionMA: motion modelling and analysis by demonstration. In: *Proceedings of the SIGCHI Conference on Human Factors in Computing Systems*. London: ACM; 2013. p. 1309–1318.

[191] Antón D, Goñi A, Illarramendi A, *et al.* KiReS: a Kinect-based telerehabilitation system. In: *2013 IEEE 15th International Conference on e-health Networking, Applications and Services (Healthcom 2013).* New York, NY: IEEE; 2013. p. 444–448.

[192] Osgouei RH, Soulsby D, Bello F. Rehabilitation exergames: use of motion sensing and machine learning to quantify exercise performance in healthy volunteers. *JMIR Rehabilitation and Assistive Technologies.* 2020;7(2):e17289.

[193] Zhao W, Feng H, Lun R, *et al.* A Kinect-based rehabilitation exercise monitoring and guidance systems. In: *Proceedings of the 5th IEEE International Conference on Software Engineering and Service Science.* New York, NY: IEEE; 2014. p. 762–765.

[194] Zhao W, Reinthal MA, Espy DD, *et al.* Rule-based human motion tracking for rehabilitation exercises: realtime assessment, feedback, and guidance. *IEEE Access.* 2017;5:21382–21394.

[195] Zhao W, Lun R, Espy DD, *et al.* Realtime motion assessment for rehabilitation exercises: integration of kinematic modeling with fuzzy inference. *Journal of Artificial Intelligence and Soft Computing Research.* 2014;4(4):267–285.

[196] Zhang Z, Hong R, Lin A, *et al.* Automated and accurate assessment for postural abnormalities in patients with Parkinson's disease based on Kinect and machine learning. *Journal of NeuroEngineering and Rehabilitation.* 2021;18(1):1–10.

[197] Ustinova K, Perkins J, Leonard W, *et al.* Virtual reality game-based therapy for persons with TBI: a pilot study. In: *2013 International Conference on Virtual Rehabilitation (ICVR).* New York, NY: IEEE; 2013. p. 87–93.

[198] Cao L, Fan C, Wang H, *et al.* A novel combination model of convolutional neural network and long short-term memory network for upper limb evaluation using kinect-based system. *IEEE Access.* 2019;7:145227–145234.

[199] Capecci M, Ciabattoni L, Ferracuti F, *et al.* Collaborative design of a telerehabilitation system enabling virtual second opinion based on fuzzy logic. *IET Computer Vision.* 2018;12(4):502–512.

[200] Chiang AT, Chen Q, Wang Y, *et al.* Kinect-based in-home exercise system for lymphatic health and lymphedema intervention. *IEEE Journal of Translational Engineering in Health and Medicine.* 2018;6:1–13.

[201] Moreira R, Magalhães A, Oliveira HP. A kinect-based system to assess lymphedema impairments in breast cancer patients. In: *Iberian Conference on Pattern Recognition and Image Analysis.* New York, NY: Springer; 2015. p. 228–236.

[202] Moreira R, Magalhães A, Oliveira HP. A Kinect-based system for upper-body function assessment in breast cancer patients. *Journal of Imaging.* 2015;1(1):134–155.

[203] Dorado J, del Toro X, Santofimia MJ, *et al.* A computer-vision-based system for at-home rheumatoid arthritis rehabilitation. *International Journal of Distributed Sensor Networks.* 2019;15(9):1550147719875649.

[204] Chang YJ, Chen SF, Huang JD. A Kinect-based system for physical rehabilitation: a pilot study for young adults with motor disabilities. *Research in Developmental Disabilities*. 2011;32(6):2566–2570.

[205] Rahman MA, Hossain MS. m-Therapy: a multisensor framework for in-home therapy management: a social therapy of things perspective. *IEEE Internet of Things Journal*. 2017;5(4):2548–2556.

[206] Avola D, Cinque L, Foresti GL, *et al.* VRheab: a fully immersive motor rehabilitation system based on recurrent neural network. *Multimedia Tools and Applications*. 2018;77(19):24955–24982.

[207] Palacios-Navarro G, García-Magariño I, Ramos-Lorente P. A Kinect-based system for lower limb rehabilitation in Parkinson's disease patients: a pilot study. *Journal of Medical Systems*. 2015;39(9):1–10.

[208] Abdur Rahman M, Qamar AM, Ahmed MA, *et al.* Multimedia interactive therapy environment for children having physical disabilities. In: *Proceedings of the 3rd ACM Conference on International Conference on Multimedia Retrieval. ICMR '13*. New York, NY: ACM; 2013. p. 313–314. Available from: http://doi.acm.org/10.1145/2461466.2461522.

[209] Valdivia S, Blanco R, Uribe-Quevedo A, *et al.* Development and evaluation of two posture-tracking user interfaces for occupational health care. *Advances in Mechanical Engineering*. 2018;10(6):1687814018769489.

[210] Kim WS, Cho S, Baek D, *et al.* Upper extremity functional evaluation by Fugl-Meyer assessment scoring using depth-sensing camera in hemiplegic stroke patients. *PLoS One*. 2016;11(7):e0158640.

[211] Lange B, Chang CY, Suma E, *et al.* Development and evaluation of low cost game-based balance rehabilitation tool using the microsoft kinect sensor. In: *Proceedings of the Annual International Conference of the IEEE Engineering in Medicine and Biology Society (EMBC)*; 2011. p. 1831–1834.

[212] Lange B, Koenig S, McConnell E, *et al.* Interactive game-based rehabilitation using the Microsoft Kinect. In: *Proceedings of the IEEE Virtual Reality Short Papers and Posters*; 2012. p. 171–172.

[213] Da Gama A, Chaves T, Figueiredo L, *et al.* Guidance and movement correction based on therapeutics movements for motor rehabilitation support systems. In: *2012 14th Symposium on Virtual and Augmented Reality*. New York, NY: IEEE; 2012. p. 191–200.

[214] Wei W, McElroy C, Dey S. Towards on-demand virtual physical therapist: machine learning-based patient action understanding, assessment and task recommendation. *IEEE Transactions on Neural Systems and Rehabilitation Engineering*. 2019;27(9):1824–1835.

[215] Dao TT, Tannous H, Pouletaut P, *et al.* Interactive and connected rehabilitation systems for e-Health. *IRBM*. 2016;37(5–6):289–296.

[216] Capecci M, Ceravolo MG, Ferracuti F, *et al.* The KIMORE dataset: KInematic assessment of MOvement and clinical scores for remote monitoring of physical REhabilitation. *IEEE Transactions on Neural Systems and Rehabilitation Engineering*. 2019;27(7):1436–1448.

[217]   Brooke J. SUS—A quick and dirty usability scale. *Usability Evaluation in Industry*. 1996;189(194):4–7.

[218]   Brunnstrom S. Motor testing procedures in hemiplegia: based on sequential recovery stages. *Physical Therapy*. 1966;46(4):357–375.

[219]   Ryan RM, Deci EL. Self-determination theory and the facilitation of intrinsic motivation, social development, and well-being. *American Psychologist*. 2000;55(1):68.

[220]   Sauro J, Dumas JS. Comparison of three one-question, post-task usability questionnaires. In: *Proceedings of the SIGCHI Conference on Human Factors in Computing Systems*; 2009. p. 1599–1608.

[221]   Zhao W, Lun R, Espy DD, *et al.* Rule based realtime motion assessment for rehabilitation exercises. In: *Proceedings of the IEEE Symposium on Computational Intelligence in Healthcare and e-health*. New York, NY: IEEE; 2014. p. 133–140.

[222]   Pedro LM, de Paula Caurin GA. Kinect evaluation for human body movement analysis. In: *2012 4th IEEE RAS EMBS International Conference on Biomedical Robotics and Biomechatronics (BioRob)*; 2012. p. 1856–1861.

[223]   Huang MC, Xu W, Su Y, *et al.* SmartGlove for upper extremities rehabilitative gaming assessment. In: *Proceedings of the 5th International Conference on PErvasive Technologies Related to Assistive Environments. PETRA '12*. New York, NY: ACM; 2012. p. 20:1–20:4.

[224]   Bo APL, Hayashibe M, Poignet P. Joint angle estimation in rehabilitation with inertial sensors and its integration with Kinect. In: *2011 Annual International Conference of the Engineering in Medicine and Biology Society, EMBC*. New York, NY: IEEE; 2011. p. 3479–3483.

[225]   Tamei T, Orito Y, Funaya H, *et al.* Kinect-based posturography for in-home rehabilitation of balance disorders. *APSIPA Transactions on Signal and Information Processing*. 2015;4:e17.

[226]   Zhao W, Yang S, Luo X. Towards rehabilitation at home after total knee replacement. *Tsinghua Science and Technology*. 2021;26(6): 791–799.

[227]   Chua J, Ong LY, Leow MC. Telehealth using PoseNet-based system for in-home rehabilitation. *Future Internet*. 2021;13(7):173.

[228]   Rick SR, Bhaskaran S, Sun Y, *et al.* NeuroPose: geriatric rehabilitation in the home using a webcam and pose estimation. In: *Proceedings of the 24th International Conference on Intelligent User Interfaces: Companion*; 2019. p. 105–106.

[229]   Biebl JT, Rykala M, Strobel M, *et al.* App-based feedback for rehabilitation exercise correction in patients with knee or hip osteoarthritis: prospective cohort study. *Journal of Medical Internet Research*. 2021;23(7):e26658.

[230]   Fasko M, Zhao W, Yang S, *et al.* Towards human activity recognition and objective performance assessment in human patient simulation: a case study. In: *2020 IEEE International Conference on Systems, Man, and Cybernetics (SMC)*. New York, NY: IEEE; 2020. p. 2702–2707.

[231] Kendall A, Grimes M, Cipolla R. Posenet: a convolutional network for real-time 6-dof camera relocalization. In: *Proceedings of the IEEE International Conference on computer Vision*; 2015. p. 2938–2946.

[232] Schettino LF, Adamovich SV, Hening W, *et al.* Hand preshaping in Parkinson's disease: effects of visual feedback and medication state. *Experimental Brain Research*. 2006;168(1):186–202.

[233] Térémetz M, Colle F, Hamdoun S, *et al.* A novel method for the quantification of key components of manual dexterity after stroke. *Journal of Neuroengineering and Rehabilitation*. 2015;12(1):1–16.

[234] Nishiyama M, Watanabe K. Wearable sensing glove with embedded heterocore fiber-optic nerves for unconstrained hand motion capture. *IEEE Transactions on Instrumentation and Measurement*. 2009;58(12):3995–4000.

[235] Wise S, Gardner W, Sabelman E, *et al.* Evaluation of a fiber optic glove for semi-automated goniometric measurements. *Journal of Rehabilitation Research and Development*. 1990;27(4).

[236] Lin BS, Lee IJ, Chen JL. Novel assembled sensorized glove platform for comprehensive hand function assessment by using inertial sensors and force sensing resistors. *IEEE Sensors Journal*. 2019;20(6):3379–3389.

[237] Kortier HG, Sluiter VI, Roetenberg D, *et al.* Assessment of hand kinematics using inertial and magnetic sensors. *Journal of Neuroengineering and Rehabilitation*. 2014;11(1):1–15.

[238] Dipietro L, Sabatini AM, Dario P. A survey of glove-based systems and their applications. *IEEE Transactions on Systems, Man, and Cybernetics, Part C (Applications and Reviews)*. 2008;38(4):461–482.

[239] Dipietro L, Sabatini AM, Dario P. Evaluation of an instrumented glove for hand-movement acquisition. *Journal of Rehabilitation Research and Development*. 2003;40(2):179–190.

[240] Jack D, Boian R, Merians AS, *et al.* Virtual reality-enhanced stroke rehabilitation. *IEEE Transactions on Neural Systems and Rehabilitation Engineering*. 2001;9(3):308–318.

[241] Adamovich SV, Merians AS, Boian R, *et al.* A virtual reality-based exercise system for hand rehabilitation post-stroke. *Presence: Teleoperators & Virtual Environments*. 2005;14(2):161–174.

[242] Gracia-Ibáñez V, Vergara M, Buffi JH, *et al.* Across-subject calibration of an instrumented glove to measure hand movement for clinical purposes. *Computer Methods in Biomechanics and Biomedical Engineering*. 2017;20(6):587–597.

[243] Hernandez-Rebollar JL, Kyriakopoulos N, Lindeman RW. The Accele-Glove: a whole-hand input device for virtual reality. In: *ACM SIGGRAPH 2002 Conference Abstracts and Applications*; 2002. p. 259–259.

[244] Saggio G, Riillo F, Sbernini L, *et al.* Resistive flex sensors: a survey. *Smart Materials and Structures*. 2015;25(1):013001.

[245] Kim JS, Kim BK, Jang M, *et al.* Wearable hand module and real-time tracking algorithms for measuring finger joint angles of different hand sizes with high accuracy using FBG strain sensor. *Sensors*. 2020;20(7):1921.

[246] Jha CK, Gajapure K, Chakraborty AL. Design and evaluation of an FBG sensor-based glove to simultaneously monitor flexure of ten finger joints. *IEEE Sensors Journal*. 2021;21(6):7620–7620.

[247] Da Silva AF, Gonçalves AF, Mendes PM, *et al.* FBG sensing glove for monitoring hand posture. *IEEE Sensors Journal*. 2011;11(10):2442–2448.

[248] Fujiwara E, dos Santos MFM, Suzuki CK. Flexible optical fiber bending transducer for application in glove-based sensors. *IEEE Sensors Journal*. 2014;14(10):3631–3636.

[249] Lim KY, Goh FYK, Dong W, *et al.* A wearable, self-calibrating, wireless sensor network for body motion processing. In: *2008 IEEE International Conference on Robotics and Automation*. New York, NY: IEEE; 2008. p. 1017–1022.

[250] Li K, Chen IM, Yeo SH, *et al.* Development of finger-motion capturing device based on optical linear encoder. *Journal of Rehabilitation Research & Development*. 2011;48(1).

[251] Kadouche R, Mokhtari M, Maier M. Modeling of the residual capability for people with severe motor disabilities: analysis of hand posture. In: *International Conference on User Modeling*. New York, NY: Springer; 2005. p. 231–235.

[252] Ramsden E. *Hall-Effect Sensors: Theory and Application*. New York, NY: Elsevier; 2011.

[253] Hamilton NP, Weimar W, Luttgens K. *Kinesiology: Scientific Basis of Human Motion*. Brown & Benchmark; 2011.

[254] Euston M, Coote P, Mahony R, *et al.* A complementary filter for attitude estimation of a fixed-wing UAV. In: *2008 IEEE/RSJ International Conference on Intelligent Robots and Systems*. New York, NY: IEEE; 2008. p. 340–345.

[255] Meinhold RJ, Singpurwalla ND. Understanding the Kalman filter. *The American Statistician*. 1983;37(2):123–127.

[256] Portney LG, Watkins MP. *Foundations of Clinical Research: Applications to Practice*. Upper Saddle River, NJ: Pearson/Prentice Hall; 2009.

[257] Shrout PE, Fleiss JL. Intraclass correlations: uses in assessing rater reliability. *Psychological Bulletin*. 1979;86(2):420.

[258] Koo TK, Li MY. A guideline of selecting and reporting intraclass correlation coefficients for reliability research. *Journal of Chiropractic Medicine*. 2016;15(2):155–163.

[259] Zaki R, Bulgiba A, Ismail R, *et al.* Statistical methods used to test for agreement of medical instruments measuring continuous variables in method comparison studies: a systematic review. *PLoS One*. 2012;7(5):e37908.

[260] Bartlett J, Frost C. Reliability, repeatability and reproducibility: analysis of measurement errors in continuous variables. *Ultrasound in Obstetrics and Gynecology: The Official Journal of the International Society of Ultrasound in Obstetrics and Gynecology*. 2008;31(4):466–475.

[261] Cooper G, Sheret I, McMillian L, *et al.* Inertial sensor-based knee flexion/extension angle estimation. *Journal of Biomechanics*. 2009;42(16):2678–2685.

[262] Schulze M, Calliess T, Gietzelt M, *et al.* Development and clinical validation of an unobtrusive ambulatory knee function monitoring system with inertial 9DoF sensors. In: *2012 Annual International Conference of the IEEE Engineering in Medicine and Biology Society*. New York, NY: IEEE; 2012. p. 1968–1971.

[263] Nüesch C, Roos E, Pagenstert G, *et al.* Measuring joint kinematics of treadmill walking and running: comparison between an inertial sensor based system and a camera-based system. *Journal of Biomechanics*. 2017;57:32–38.

[264] Jakob C, Kugler P, Hebenstreit F, *et al.* Estimation of the knee flexion-extension angle during dynamic sport motions using body-worn inertial sensors. In: *Proceedings of the 8th International Conference on Body Area Networks*; 2013. p. 289–295.

[265] Britannica. Student's t-test. In: *Encyclopedia Britannica*. Britannica; 2020.

[266] Moser BK, Stevens GR. Homogeneity of variance in the two-sample means test. *The American Statistician*. 1992;46(1):19–21.

[267] Lee J, Koh D, Ong C. Statistical evaluation of agreement between two methods for measuring a quantitative variable. *Computers in Biology and Medicine*. 1989;19(1):61–70.

[268] Guttman L. A basis for analyzing test-retest reliability. *Psychometrika*. 1945;10(4):255–282.

[269] Weir JP. Quantifying test-retest reliability using the intraclass correlation coefficient and the SEM. *The Journal of Strength & Conditioning Research*. 2005;19(1):231–240.

[270] McGraw KO, Wong SP. Forming inferences about some intraclass correlation coefficients. *Psychological Methods*. 1996;1(1):30.

[271] Altman DG, Bland JM. Measurement in medicine: the analysis of method comparison studies. *Journal of the Royal Statistical Society: Series D (The Statistician)*. 1983;32(3):307–317.

[272] Liao Y, Vakanski A, Xian M, *et al.* A review of computational approaches for evaluation of rehabilitation exercises. *Computers in Biology and Medicine*. 2020;119:103687.

[273] Boukhennoufa I, Zhai X, Utti V, *et al.* Wearable sensors and machine learning in post-stroke rehabilitation assessment: a systematic review. *Biomedical Signal Processing and Control*. 2022;71:103197.

[274] Balestra N, Sharma G, Riek LM, *et al.* Automatic identification of upper extremity rehabilitation exercise type and dose using body-worn sensors and machine learning: a pilot study. *Digital Biomarkers*. 2021;5(2):158–166.

[275] Meng Z, Zhang M, Guo C, *et al.* Recent progress in sensing and computing techniques for human activity recognition and motion analysis. *Electronics*. 2020;9(9):1357.

[276] Lee SI, Adans-Dester CP, Grimaldi M, *et al.* Enabling stroke rehabilitation in home and community settings: a wearable sensor-based approach for upper-limb motor training. *IEEE Journal of Translational Engineering in Health and Medicine*. 2018;6:1–11.

[277]   Yang G, Deng J, Pang G, *et al.* An IoT-enabled stroke rehabilitation system based on smart wearable armband and machine learning. *IEEE Journal of Translational Engineering in Health and Medicine.* 2018;6:1–10.

[278]   Lin JFS, Kulić D. Online segmentation of human motion for automated rehabilitation exercise analysis. *IEEE Transactions on Neural Systems and Rehabilitation Engineering.* 2013;22(1):168–180.

[279]   Fod A, Matarić MJ, Jenkins OC. Automated derivation of primitives for movement classification. *Autonomous Robots.* 2002;12(1):39–54.

[280]   Lucas A, Hermiz J, Labuzetta J, *et al.* Use of accelerometry for long term monitoring of stroke patients. *IEEE Journal of Translational Engineering in Health and Medicine.* 2019;7:1–10.

[281]   Shensa MJ. The discrete wavelet transform: wedding the a trous and Mallat algorithms. *IEEE Transactions on Signal Processing.* 1992;40(10): 2464–2482.

[282]   Jolliffe IT. Principal component analysis: a beginner's guide—I. Introduction and application. *Weather.* 1990;45(10):375–382.

[283]   Roweis ST, Saul LK. Nonlinear dimensionality reduction by locally linear embedding. *Science.* 2000;290(5500):2323–2326.

[284]   Balasubramanian M, Schwartz EL. The isomap algorithm and topological stability. *Science.* 2002;295(5552):7–7.

[285]   Coifman RR, Lafon S, Lee AB, *et al.* Geometric diffusions as a tool for harmonic analysis and structure definition of data: diffusion maps. *Proceedings of the National Academy of Sciences.* 2005;102(21):7426–7431.

[286]   Belkin M, Niyogi P. Laplacian eigenmaps for dimensionality reduction and data representation. *Neural Computation.* 2003;15(6):1373–1396.

[287]   Donoho DL, Grimes C. Hessian eigenmaps: locally linear embedding techniques for high-dimensional data. *Proceedings of the National Academy of Sciences.* 2003;100(10):5591–5596.

[288]   Weiss Y. Segmentation using eigenvectors: a unifying view. In: *Proceedings of the Seventh IEEE International Conference on Computer Vision*, vol. 2. New York, NY: IEEE; 1999. p. 975–982.

[289]   Weinberger KQ, Saul LK. Unsupervised learning of image manifolds by semidefinite programming. *International Journal of Computer Vision.* 2006;70(1):77–90.

[290]   Hall MA. Correlation-based Feature Selection for Machine Learning [PhD Thesis]. The University of Waikato; 1999.

[291]   Peng H, Long F, Ding C. Feature selection based on mutual information criteria of max-dependency, max-relevance, and min-redundancy. *IEEE Transactions on Pattern Analysis and Machine Intelligence.* 2005;27(8):1226–1238.

[292]   Tibshirani R. Regression shrinkage and selection via the lasso. *Journal of the Royal Statistical Society: Series B (Methodological).* 1996;58(1):267–288.

[293]   Houmanfar R, Karg M, Kulić D. Movement analysis of rehabilitation exercises: distance metrics for measuring patient progress. *IEEE Systems Journal.* 2014;10(3):1014–1025.

[294] Vapnik V. *The Nature of Statistical Learning Theory*. New York, NY: Springer Science & Business Media; 1999.

[295] Cortes C, Vapnik V. Support-vector networks. *Machine Learning*. 1995;20(3):273–297.

[296] Duan KB, Keerthi SS. Which is the best multiclass SVM method? An empirical study. In: *International Workshop on Multiple Classifier Systems*. New York, NY: Springer; 2005. p. 278–285.

[297] Crammer K, Singer Y. On the algorithmic implementation of multiclass kernel-based vector machines. *Journal of Machine Learning Research*. 2001;2:265–292.

[298] Quinlan JR. Induction of decision trees. *Machine Learning*. 1986;1(1): 81–106.

[299] Anderson JA. *An Introduction to Neural Networks*. New York, NY: MIT Press; 1995.

[300] Hochreiter S, Schmidhuber J. Long short-term memory. *Neural Computation*. 1997;9(8): 1735–1780.

[301] Chung J, Gulcehre C, Cho K, *et al.* Empirical evaluation of gated recurrent neural networks on sequence modeling; 2014. arXiv preprint arXiv:14123555.

[302] Moya Rueda F, Grzeszick R, Fink GA, *et al.* Convolutional neural networks for human activity recognition using body-worn sensors. In: *Informatics*. vol. 5. Multidisciplinary Digital Publishing Institute; 2018. p. 26.

[303] Ignatov A. Real-time human activity recognition from accelerometer data using Convolutional Neural Networks. *Applied Soft Computing*. 2018;62:915–922.

[304] Avilés-Cruz C, Ferreyra-Ramírez A, Zúñiga-López A, *et al.* Coarse-fine convolutional deep-learning strategy for human activity recognition. *Sensors*. 2019;19(7):1556.

[305] Huang J, Lin S, Wang N, *et al.* TSE-CNN: a two-stage end-to-end CNN for human activity recognition. *IEEE Journal of Biomedical and Health Informatics*. 2019;24(1):292–299.

[306] Tang Y, Teng Q, Zhang L, *et al.* Layer-wise training convolutional neural networks with smaller filters for human activity recognition using wearable sensors. *IEEE Sensors Journal*. 2020;21(1):581–592.

[307] Arifoglu D, Bouchachia A. Activity recognition and abnormal behaviour detection with recurrent neural networks. *Procedia Computer Science*. 2017;110:86–93.

[308] Zebin T, Sperrin M, Peek N, *et al.* Human activity recognition from inertial sensor time-series using batch normalized deep LSTM recurrent networks. In: *2018 40th Annual International Conference of the IEEE Engineering in Medicine and Biology Society (EMBC)*. New York, NY: IEEE; 2018. p. 1–4.

[309] Chung S, Lim J, Noh KJ, *et al.* Sensor data acquisition and multimodal sensor fusion for human activity recognition using deep learning. *Sensors*. 2019;19(7):1716.

[310]   Zhao Y, Yang R, Chevalier G, *et al.* Deep residual bidir-LSTM for human activity recognition using wearable sensors. *Mathematical Problems in Engineering.* 2018;2018:7316954.

[311]   Xia K, Huang J, Wang H. LSTM-CNN architecture for human activity recognition. *IEEE Access.* 2020;8:56855–56866.

[312]   Capela NA, Lemaire ED, Baddour N. Feature selection for wearable smartphone-based human activity recognition with able bodied, elderly, and stroke patients. *PLoS One.* 2015;10(4):e0124414.

[313]   Chen PW, Baune NA, Zwir I, *et al.* Measuring activities of daily living in stroke patients with motion machine learning algorithms: a pilot study. *International Journal of Environmental Research and Public Health.* 2021;18(4):1634.

[314]   Chae SH, Kim Y, Lee KS, *et al.* Development and clinical evaluation of a web-based upper limb home rehabilitation system using a smartwatch and machine learning model for chronic stroke survivors: prospective comparative study. *JMIR mHealth and uHealth.* 2020;8(7):e17216.

[315]   Moncada-Torres A, Leuenberger K, Gonzenbach R, *et al.* Activity classification based on inertial and barometric pressure sensors at different anatomical locations. *Physiological Measurement.* 2014;35(7):1245.

[316]   Kaku A, Parnandi A, Venkatesan A, *et al.* Towards data-driven stroke rehabilitation via wearable sensors and deep learning. In: *Machine Learning for Healthcare Conference. PMLR*; 2020. p. 143–171.

[317]   Ar FM, Jaasko L, Leyman I, *et al.* The post-stroke hemiplegic patient. 1. A method for evaluation of physical performance. *Scandinavian Journal of Rehabilitation Medicine.* 1975;7(1):13–31.

[318]   Adans-Dester C, Hankov N, O'Brien A, *et al.* Enabling precision rehabilitation interventions using wearable sensors and machine learning to track motor recovery. *NPJ Digital Medicine.* 2020;3(1):1–10.

[319]   Wolf SL, Catlin PA, Ellis M, *et al.* Assessing Wolf motor function test as outcome measure for research in patients after stroke. *Stroke.* 2001;32(7):1635–1639.

[320]   Yu L, Xiong D, Guo L, *et al.* A remote quantitative Fugl-Meyer assessment framework for stroke patients based on wearable sensor networks. *Computer Methods and Programs in Biomedicine.* 2016;128:100–110.

[321]   Heldner MR, Jung S, Zubler C, *et al.* Outcome of patients with occlusions of the internal carotid artery or the main stem of the middle cerebral artery with NIHSS score of less than 5: comparison between thrombolysed and non-thrombolysed patients. *Journal of Neurology, Neurosurgery & Psychiatry.* 2015;86(7):755–760.

[322]   Bestall J, Paul E, Garrod R, *et al.* Usefulness of the Medical Research Council (MRC) dyspnoea scale as a measure of disability in patients with chronic obstructive pulmonary disease. *Thorax.* 1999;54(7): 581–586.

[323]   Park E, Lee K, Han T, *et al.* Automatic grading of stroke symptoms for rapid assessment using optimized machine learning and 4-limb

kinematics: clinical validation study. *Journal of Medical Internet Research*. 2020;22(9):e20641.

[324] Denehy L, de Morton NA, Skinner EH, *et al.* A physical function test for use in the intensive care unit: validity, responsiveness, and predictive utility of the physical function ICU test (scored). *Physical Therapy*. 2013;93(12):1636–1645.

[325] Mannini A, Trojaniello D, Cereatti A, *et al.* A machine learning framework for gait classification using inertial sensors: application to elderly, post-stroke and Huntington's disease patients. *Sensors*. 2016;16(1):134.

[326] Hsu WC, Sugiarto T, Lin YJ, *et al.* Multiple-wearable-sensor-based gait classification and analysis in patients with neurological disorders. *Sensors*. 2018;18(10):3397.

[327] Wang FC, Chen SF, Lin CH, *et al.* Detection and classification of stroke gaits by deep neural networks employing inertial measurement units. *Sensors*. 2021;21(5):1864.

[328] Bochniewicz EM, Emmer G, McLeod A, *et al.* Measuring functional arm movement after stroke using a single wrist-worn sensor and machine learning. *Journal of Stroke and Cerebrovascular Diseases*. 2017;26(12):2880–2887.

[329] Akdoğan E, Taçgın E, Adli MA. Knee rehabilitation using an intelligent robotic system. *Journal of Intelligent Manufacturing*. 2009;20(2):195–202.

[330] Wang Q, Turaga P, Coleman G, *et al.* SomaTech: an exploratory interface for altering movement habits. In: *CHI'14 Extended Abstracts on Human Factors in Computing Systems*. London: ACM; 2014. p. 1765–1770.

[331] Salarian A, Russmann H, Vingerhoets FJ, *et al.* Ambulatory monitoring of physical activities in patients with Parkinson's disease. *IEEE Transactions on Biomedical Engineering*. 2007;54(12):2296–2299.

[332] Massé F, Gonzenbach RR, Arami A, *et al.* Improving activity recognition using a wearable barometric pressure sensor in mobility-impaired stroke patients. *Journal of Neuroengineering and Rehabilitation*. 2015;12(1):1–15.

[333] Bedregal BC, Costa AC, Dimuro GP. Fuzzy rule-based hand gesture recognition. In: *Ifip International Conference on Artificial Intelligence in Theory and Practice*. New York, NY: Springer; 2006. p. 285–294.

[334] Hachaj T, Ogiela MR. Rule-based approach to recognizing human body poses and gestures in real time. *Multimedia Systems*. 2014;20(1):81–99.

[335] Bond S, Laddu DR, Ozemek C, *et al.* Exergaming and virtual reality for health: implications for cardiac rehabilitation. *Current Problems in Cardiology*. 2021;46(3):100472.

[336] Bainbridge E, Bevans S, Keeley B, *et al.* The effects of the Nintendo Wii Fit on community-dwelling older adults with perceived balance deficits: a pilot study. *Physical & Occupational Therapy in Geriatrics*. 2011;29(2):126–135.

[337] Berg K, Wood-Dauphine S, Williams J, *et al.* Measuring balance in the elderly: preliminary development of an instrument. *Physiotherapy Canada*. 1989;41(6):304–311.

[338] Esculier JF, Vaudrin J, Beriault P, *et al.* Home-based balance training programme using Wii Fit with balance board for Parkinsons's disease: a pilot study. *Journal of Rehabilitation Medicine.* 2012;44(2):144–150.

[339] Tinetti ME. Performance-oriented assessment of mobility problems in elderly patients. *Journal of the American Geriatrics Society.* 1986;34(2): 119–126.

[340] Broadbent S, Crowley-McHattan Z, Zhou S, *et al.* The effect of the Nintendo Wii Fit on exercise capacity and gait in an elderly woman with CREST syndrome. *International Journal of Therapy and Rehabilitation.* 2014;21(11):539–546.

[341] Podsiadlo D, Richardson S. The timed "Up & Go": a test of basic functional mobility for frail elderly persons. *Journal of the American Geriatrics Society.* 1991;39(2):142–148.

[342] Mhatre PV, Vilares I, Stibb SM, *et al.* Wii Fit balance board playing improves balance and gait in Parkinson disease. *Pm&r.* 2013;5(9):769–777.

[343] Shumway-Cook A, Woollacott MH. Theory and practical applications. In: *Motor Control.* Baltimore, MD: Williams & Wilkins; 1995.

[344] Powell LE, Myers AM. The activities-specific balance confidence (ABC) scale. *The Journals of Gerontology Series A: Biological Sciences and Medical Sciences.* 1995;50(1):M28–M34.

[345] Sheikh JI, Yesavage JA. Geriatric Depression Scale (GDS): recent evidence and development of a shorter version. *Clinical Gerontologist: The Journal of Aging and Mental Health.* 1986;5(1–2):165–173.

[346] Agmon M, Perry CK, Phelan E, *et al.* A pilot study of Wii Fit exergames to improve balance in older adults. *Journal of Geriatric Physical Therapy.* 2011;34(4):161–167.

[347] Hakim RM, Salvo CJ, Balent A, *et al.* Case report: a balance training program using the Nintendo Wii Fit to reduce fall risk in an older adult with bilateral peripheral neuropathy. *Physiotherapy Theory and Practice.* 2015;31(2):130–139.

[348] Herz NB, Mehta SH, Sethi KD, *et al.* Nintendo Wii rehabilitation ("Wii-hab") provides benefits in Parkinson's disease. *Parkinsonism & Related Disorders.* 2013;19(11):1039–1042.

[349] Nouri F, Lincoln N. An extended activities of daily living scale for stroke patients. *Clinical Rehabilitation.* 1987;1(4):301–305.

[350] MDSTF on Rating Scales for Parkinson's Disease. The unified Parkinson's disease rating scale (UPDRS): status and recommendations. *Movement Disorders.* 2003;18(7):738–750.

[351] Earhart GM, Cavanaugh JT, Ellis T, *et al.* The 9-hole PEG test of upper extremity function: average values, test-retest reliability, and factors contributing to performance in people with Parkinson disease. *Journal of Neurologic Physical Therapy.* 2011;35(4):157–163.

[352] Tiffin J, Asher EJ. The Purdue Pegboard: norms and studies of reliability and validity. *Journal of Applied Psychology.* 1948;32(3):234.

[353] Hamilton M. A rating scale for depression. *Journal of Neurology, Neurosurgery, and Psychiatry.* 1960;23(1):56.

[354] Peto V, Jenkinson C, Fitzpatrick R, *et al.* The development and validation of a short measure of functioning and well being for individuals with Parkinson's disease. *Quality of Life Research.* 1995;4(3):241–248.

[355] Joo Y, Xu D, Thia E, *et al.* A feasibility study using interactive commercial off-the-shelf computer gaming in upper limb rehabilitation in patients after stroke. *Journal of Rehabilitation Medicine.* 2010;42(5):437–441.

[356] Fugel-Meyer A, Jaasko L, Leyman I, *et al.* The post-stroke hemiplegic patient1, a method for evaluation of physical performance. *Scandinavian Journal of Rehabilitation Medicine.* 1975;7:13–31.

[357] Charalambous CP. Interrater reliability of a modified Ashworth scale of muscle spasticity. In: *Classic Papers in Orthopaedics. New York, NY:* Springer; 2014. p. 415–417.

[358] Clark R, Kraemer T. Clinical use of Nintendo Wii™ bowling simulation to decrease fall risk in an elderly resident of a nursing home: a case report. *Journal of Geriatric Physical Therapy.* 2009;32(4):174–180.

[359] Klompstra L, Jaarsma T, Strömberg A. Exergaming to increase the exercise capacity and daily physical activity in heart failure patients: a pilot study. *BMC Geriatrics.* 2014;14(1):1–9.

[360] Nicholson VP, McKean M, Lowe J, *et al.* Six weeks of unsupervised Nintendo Wii Fit gaming is effective at improving balance in independent older adults. *Journal of Aging and Physical Activity.* 2015;23(1):153–158.

[361] Delbaere K, T Smith S, Lord SR. Development and initial validation of the iconographical falls efficacy scale. *Journals of Gerontology Series A: Biomedical Sciences and Medical Sciences.* 2011;66(6):674–680.

[362] Park EC, Kim SG, Lee CW. The effects of virtual reality game exercise on balance and gait of the elderly. *Journal of Physical Therapy Science.* 2015;27(4):1157–1159.

[363] Cho GH, Hwangbo G, Shin HS. The effects of virtual reality-based balance training on balance of the elderly. *Journal of Physical Therapy Science.* 2014;26(4):615–617.

[364] Khasnis A, Gokula R. Romberg's test. *Journal of Postgraduate Medicine.* 2003;49(2):169–172.

[365] Jorgensen MG, Laessoe U, Hendriksen C, *et al.* Efficacy of Nintendo Wii training on mechanical leg muscle function and postural balance in community-dwelling older adults: a randomized controlled trial. *Journals of Gerontology Series A: Biomedical Sciences and Medical Sciences.* 2013;68(7):845–852.

[366] Kempen GI, Yardley L, Van Haastregt JC, *et al.* The Short FES-I: a shortened version of the falls efficacy scale-international to assess fear of falling. *Age and Ageing.* 2008;37(1):45–50.

[367] Jones CJ, Rikli RE, Beam WC. A 30-s chair-stand test as a measure of lower body strength in community-residing older adults. *Research Quarterly for Exercise and Sport.* 1999;70(2):113–119.

[368] Rendon AA, Lohman EB, Thorpe D, *et al.* The effect of virtual reality gaming on dynamic balance in older adults. *Age and Ageing.* 2012;41(4):549–552.

[369] Rose DJ, Jones CJ, Lucchese N. Predicting the probability of falls in community-residing older adults using the 8-foot up-and-go: a new measure of functional mobility. *Journal of Aging and Physical Activity.* 2002;10(4):466–475.

[370] Maillot P, Perrot A, Hartley A. Effects of interactive physical-activity video-game training on physical and cognitive function in older adults. *Psychology and Aging.* 2012;27(3):589.

[371] Bieryla KA, Dold NM. Feasibility of Wii Fit training to improve clinical measures of balance in older adults. *Clinical Interventions in Aging.* 2013;8:775.

[372] Bateni H. Changes in balance in older adults based on use of physical therapy vs the Wii Fit gaming system: a preliminary study. *Physiotherapy.* 2012;98(3):211–216.

[373] Hsu JK, Thibodeau R, Wong SJ, *et al.* A "Wii" bit of fun: the effects of adding Nintendo Wii® Bowling to a standard exercise regimen for residents of long-term care with upper extremity dysfunction. *Physiotherapy Theory and Practice.* 2011;27(3):185–193.

[374] Pompeu JE, dos Santos Mendes FA, da Silva KG, *et al.* Effect of Nintendo Wii™-based motor and cognitive training on activities of daily living in patients with Parkinson's disease: a randomised clinical trial. *Physiotherapy.* 2012;98(3):196–204.

[375] Imam B, Miller WC, Finlayson H, *et al.* A randomized controlled trial to evaluate the feasibility of the Wii Fit for improving walking in older adults with lower limb amputation. *Clinical Rehabilitation.* 2017;31(1):82–92.

[376] Saposnik G, Cohen LG, Mamdani M, *et al.* Efficacy and safety of non-immersive virtual reality exercising in stroke rehabilitation (EVREST): a randomised, multicentre, single-blind, controlled trial. *The Lancet Neurology.* 2016;15(10):1019–1027.

[377] Ramnath U, Rauch L, Lambert EV, *et al.* Efficacy of interactive video gaming in older adults with memory complaints: a cluster-randomized exercise intervention. *PLoS One.* 2021;16(5):e0252016.

[378] Türkbey TA, Kutlay S, Gök H. Clinical feasibility of Xbox KinectTM training for stroke rehabilitation: a single-blind randomized controlled pilot study. *Journal of Rehabilitation Medicine.* 2017;49(1):22–29.

[379] Desrosiers J, Bravo G, Hébert R, *et al.* Validation of the Box and Block Test as a measure of dexterity of elderly people: reliability, validity, and norms studies. *Archives of Physical Medicine and Rehabilitation.* 1994;75(7):751–755.

[380] Wolf SL, Lecraw DE, Barton LA, *et al.* Forced use of hemiplegic upper extremities to reverse the effect of learned nonuse among chronic stroke and head-injured patients. *Experimental Neurology.* 1989;104(2):125–132.

[381] Granger CV, Hamilton BB, Keith RA, *et al.* Advances in functional assessment for medical rehabilitation. *Topics in Geriatric Rehabilitation.* 1986;1(3):59–74.

[382] Yavuzer G, Senel A, Atay M, *et al.* "Playstation eyetoy games" improve upper extremity-related motor functioning in subacute stroke: a randomized

controlled clinical trial. *European Journal of Physical and Rehabilitation Medicine*. 2008;44(3):237–44.

[383] Keith RA. The functional independence measure: a new tool for rehabilitation. *Advances in Clinical Rehabilitation*. 1987;2:6–18.

[384] Goršič M, Cikajlo I, Novak D. Competitive and cooperative arm rehabilitation games played by a patient and unimpaired person: effects on motivation and exercise intensity. *Journal of Neuroengineering and Rehabilitation*. 2017;14(1):1–18.

[385] Wittmann F, Held JP, Lambercy O, *et al.* Self-directed arm therapy at home after stroke with a sensor-based virtual reality training system. *Journal of Neuroengineering and Rehabilitation*. 2016;13(1):1–10.

[386] Wittmann F, Lambercy O, Gonzenbach RR, *et al.* Assessment-driven arm therapy at home using an IMU-based virtual reality system. In: *2015 IEEE International Conference on Rehabilitation Robotics (ICORR)*. New York, NY: IEEE; 2015. p. 707–712.

[387] Bonnechère B, Jansen B, Haack I, *et al.* Automated functional upper limb evaluation of patients with Friedreich ataxia using serious games rehabilitation exercises. *Journal of Neuroengineering and Rehabilitation*. 2018;15(1):1–9.

[388] Trombetta M, Henrique PPB, Brum MR, *et al.* Motion Rehab AVE 3D: a VR-based exergame for post-stroke rehabilitation. *Computer Methods and Programs in Biomedicine*. 2017;151:15–20.

[389] Dukes PS, Hayes A, Hodges LF, *et al.* Punching ducks for post-stroke neurorehabilitation: system design and initial exploratory feasibility study. In: *2013 IEEE Symposium on 3D User Interfaces (3DUI)*. New York, NY: IEEE; 2013. p. 47–54.

[390] Merriman NA, Whyatt C, Setti A, *et al.* Successful balance training is associated with improved multisensory function in fall-prone older adults. *Computers in Human Behavior*. 2015;45:192–203.

[391] Young W, Ferguson S, Brault S, *et al.* Assessing and training standing balance in older adults: a novel approach using the "Nintendo Wii" Balance Board. *Gait & Posture*. 2011;33(2):303–305.

[392] Tinetti ME, Richman D, Powell L. Falls efficacy as a measure of fear of falling. *Journal of Gerontology*. 1990;45(6):P239–P243.

[393] Shams L, Kamitani Y, Shimojo S. What you see is what you hear. *Nature*. 2000;408(6814):788–788.

[394] Ianculescu M, Andrei B, Alexandru A. A smart assistance solution for remotely monitoring the orthopaedic rehabilitation process using wearable technology: Re. flex system. *Studies in Informatics and Control*. 2019;28(3):317–326.

[395] O'Sullivan SB, Schmitz TJ, Fulk G. *Physical Rehabilitation*. 7th ed. FA Davis; 2019.

[396] Adams JA. A closed-loop theory of motor learning. *Journal of Motor Behavior*. 1971;3(2):111–150.

[397] Schmidt RA. A schema theory of discrete motor skill learning. *Psychological Review*. 1975;82(4):225.

[398] Seron P, Oliveros MJ, Gutierrez-Arias R, *et al.* Effectiveness of telere-habilitation in physical therapy: a rapid overview. *Physical Therapy.* 2021;101(6):pzab053.

[399] Moral-Munoz JA, Zhang W, Cobo MJ, *et al.* Smartphone-based systems for physical rehabilitation applications: a systematic review. *Assistive Technology.* 2021;33(4):223–236.

[400] Gil MJV, Gonzalez-Medina G, Lucena-Anton D, *et al.* Augmented reality in physical therapy: systematic review and meta-analysis. *JMIR Serious Games.* 2021;9(4):e30985.

[401] Milgram P, Kishino F. A taxonomy of mixed reality visual displays. *IEICE Transactions on Information and Systems.* 1994;77(12):1321–1329.

[402] Zhao W, Matcham W, Mclennan C, *et al.* Minimizing errors in the nursing profession with technology-enhanced education and training. In: *Proceedings of the 2019 IEEE SmartWorld, Ubiquitous Intelligence & Computing, Advanced & Trusted Computing, Scalable Computing & Communications, Cloud & Big Data Computing, Internet of People and Smart City Innovation (SmartWorld/SCALCOM/UIC/ATC/CBDCom/IOP/SCI).* New York, NY: IEEE; 2019. p. 196–201.

[403] Colomer C, Llorens R, Noé E, *et al.* Effect of a mixed reality-based intervention on arm, hand, and finger function on chronic stroke. *Journal of Neuroengineering and Rehabilitation.* 2016;13(1):1–11.

[404] Jeon S, Kim J. Effects of augmented-reality-based exercise on muscle parameters, physical performance, and exercise self-efficacy for older adults. *International Journal of Environmental Research and Public Health.* 2020;17(9):3260.

[405] Janssen S, de Ruyter van Steveninck J, Salim HS, *et al.* The effects of augmented reality visual cues on turning in place in Parkinson's disease patients with freezing of gait. *Frontiers in Neurology.* 2020;11:185.

[406] Nutt JG, Bloem BR, Giladi N, *et al.* Freezing of gait: moving forward on a mysterious clinical phenomenon. *The Lancet Neurology.* 2011;10(8):734–744.

[407] Ginis P, Nackaerts E, Nieuwboer A, *et al.* Cueing for people with Parkinson's disease with freezing of gait: a narrative review of the state-of-the-art and novel perspectives. *Annals of Physical and Rehabilitation Medicine.* 2018;61(6):407–413.

[408] Ku J, Kim YJ, Cho S, *et al.* Three-dimensional augmented reality system for balance and mobility rehabilitation in the elderly: a randomized controlled trial. *Cyberpsychology, Behavior, and Social Networking.* 2019;22(2):132–141.

[409] Ku J, Kim YJ, Cho S, *et al.* Utility of a three-dimensional interactive augmented reality program for balance and mobility rehabilitation in the elderly: a feasibility study. *Annals of Rehabilitation Medicine.* 2015;39(3): 462–472.

[410] Espay AJ, Baram Y, Dwivedi AK, *et al.* At-home training with closed-loop augmented-reality cueing device for improving gait in patients with

Parkinson disease. *Journal of Rehabilitation Research and Development.* 2010;47(6):573–581.

[411]   Ortiz-Catalan M, Gumundsdottir RA, Kristoffersen MB, *et al.* Phantom motor execution facilitated by machine learning and augmented reality as treatment for phantom limb pain: a single group, clinical trial in patients with chronic intractable phantom limb pain. *The Lancet.* 2016;388(10062):2885–2894.

[412]   Kim IC, Lee BH. Effects of augmented reality with functional electric stimulation on muscle strength, balance and gait of stroke patients. *Journal of Physical Therapy Science.* 2012;24(8):755–762.

[413]   Jung GU, Moon TH, Park GW, *et al.* Use of augmented reality-based training with EMG-triggered functional electric stimulation in stroke rehabilitation. *Journal of Physical Therapy Science.* 2013;25(2): 147–151.

[414]   Lee CH, Kim Y, Lee BH. Augmented reality-based postural control training improves gait function in patients with stroke: randomized controlled trial. *Hong Kong Physiotherapy Journal.* 2014;32(2):51–57.

[415]   Mousavi Hondori H, Khademi M. A review on technical and clinical impact of microsoft kinect on physical therapy and rehabilitation. *Journal of Medical Engineering.* 2014;2014:846514.

[416]   Lun R, Zhao W. Kinect applications in healthcare. In: *Advanced Methodologies and Technologies in Medicine and Healthcare.* Hershey, PA: IGI Global; 2019. p. 391–402.

[417]   Wiederhold B, Riva G. Balance recovery through virtual stepping exercises using Kinect skeleton tracking: a follow-up study with chronic stroke patients. *Annual Review of Cybertherapy and Telemedicine 2012: Advanced Technologies in the Behavioral, Social and Neurosciences.* 2012;181:108–112.

[418]   Exell T, Freeman C, Meadmore K, *et al.* Goal orientated stroke rehabilitation utilising electrical stimulation, iterative learning and microsoft kinect. In: *2013 IEEE 13th International Conference on Rehabilitation Robotics (ICORR).* New York, NY: IEEE; 2013. p. 1–6.

[419]   Lee G. Effects of training using video games on the muscle strength, muscle tone, and activities of daily living of chronic stroke patients. *Journal of Physical Therapy Science.* 2013;25(5):595–597.

[420]   Sin H, Lee G. Additional virtual reality training using Xbox Kinect in stroke survivors with hemiplegia. *American Journal of Physical Medicine & Rehabilitation.* 2013;92(10):871–880.

[421]   Bao X, Mao Y, Lin Q, *et al.* Mechanism of Kinect-based virtual reality training for motor functional recovery of upper limbs after subacute stroke. *Neural Regeneration Research.* 2013;8(31):2904.

[422]   Galna B, Jackson D, Schofield G, *et al.* Retraining function in people with Parkinson's disease using the Microsoft kinect: game design and pilot testing. *Journal of Neuroengineering and Rehabilitation.* 2014;11(1): 1–12.

[423]   Pompeu JE, Arduini L, Botelho A, *et al.* Feasibility, safety and outcomes of playing Kinect Adventures!™ for people with Parkinson's disease: a pilot study. *Physiotherapy*. 2014;100(2):162–168.

[424]   Luna-Oliva L, Ortiz-Gutiérrez RM, Cano-de la Cuerda R, *et al.* Kinect Xbox 360 as a therapeutic modality for children with cerebral palsy in a school environment: a preliminary study. *NeuroRehabilitation*. 2013;33(4):513–521.

[425]   Chang YJ, Han WY, Tsai YC. A Kinect-based upper limb rehabilitation system to assist people with cerebral palsy. *Research in Developmental Disabilities*. 2013;34(11):3654–3659.

[426]   Holmes H, Wood J, Jenkins S, *et al.* Xbox Kinect™ represents high intensity exercise for adults with cystic fibrosis. *Journal of Cystic Fibrosis*. 2013;12(6):604–608.

[427]   Ortiz-Gutiérrez R, Cano-de-la Cuerda R, Galán-del Río F, *et al.* A telerehabilitation program improves postural control in multiple sclerosis patients: a Spanish preliminary study. *International Journal of Environmental Research and Public Health*. 2013;10(11):5697–5710.

[428]   Ulaşlı AM, Türkmen U, Toktaş H, *et al.* The complementary role of the Kinect virtual reality game training in a patient with metachromatic leukodystrophy. *PM&R*. 2014;6(6):564–567.

[429]   Parry I, Carbullido C, Kawada J, *et al.* Keeping up with video game technology: objective analysis of Xbox Kinect™ and PlayStation 3 Move™ for use in burn rehabilitation. *Burns*. 2014;40(5):852–859.

[430]   Ilg W, Schatton C, Schicks J, *et al.* Video game–based coordinative training improves ataxia in children with degenerative ataxia. *Neurology*. 2012;79(20):2056–2060.

[431]   Stone EE, Skubic M. Capturing habitual, in-home gait parameter trends using an inexpensive depth camera. In: *2012 Annual International Conference of the IEEE Engineering in Medicine and Biology Society*. New York, NY: IEEE; 2012. p. 5106–5109.

[432]   Pu F, Sun S, Wang L, *et al.* Investigation of key factors affecting the balance function of older adults. *Aging Clinical and Experimental Research*. 2015;27(2):139–147.

[433]   Afzal MR, Oh MK, Choi HY, *et al.* A novel balance training system using multimodal biofeedback. *Biomedical Engineering Online*. 2016;15(1):1–11.

[434]   Chiu YL, Tsai YJ, Lin CH, *et al.* Evaluation of a smartphone-based assessment system in subjects with chronic ankle instability. *Computer Methods and Programs in Biomedicine*. 2017;139:191–195.

[435]   Chan MH, Keung DT, Lui SY, *et al.* A validation study of a smartphone application for functional mobility assessment of the elderly. *Hong Kong Physiotherapy Journal*. 2016;35:1–4.

[436]   Csuka M, McCarty DJ. Simple method for measurement of lower extremity muscle strength. *The American Journal of Medicine*. 1985;78(1):77–81.

[437] Drozdzowska B, Wiktor K, Pluskiewicz W. Functional status and prevalence of falls and fractures in population-based sample of postmenopausal women from the RAC-OST-POL Study. *International Journal of Clinical Practice.* 2013;67(7):673–681.

[438] Karakaya MG, Bilgin SC, Ekici G, *et al.* Functional mobility, depressive symptoms, level of independence, and quality of life of the elderly living at home and in the nursing home. *Journal of the American Medical Directors Association.* 2009;10(9):662–666.

[439] Kammerlander C, Riedmüller P, Gosch M, *et al.* Functional outcome and mortality in geriatric distal femoral fractures. *Injury.* 2012;43(7): 1096–1101.

[440] Choi YH, Ku J, Lim H, *et al.* Mobile game-based virtual reality rehabilitation program for upper limb dysfunction after ischemic stroke. *Restorative Neurology and Neuroscience.* 2016;34(3):455–463.

[441] Shah S, Vanclay F, Cooper B. Improving the sensitivity of the Barthel Index for stroke rehabilitation. *Journal of Clinical Epidemiology.* 1989;42(8): 703–709.

[442] Group TE. EuroQol – a new facility for the measurement of health-related quality of life. *Health Policy.* 1990;16(3):199–208.

[443] Beck AT, Steer RA, Brown G. Beck depression inventory–II. *Psychological Assessment.* San Antonio, TX: Psychological Corporation; 1996.

[444] Ginis P, Nieuwboer A, Dorfman M, *et al.* Feasibility and effects of home-based smartphone-delivered automated feedback training for gait in people with Parkinson's disease: a pilot randomized controlled trial. *Parkinsonism & Related Disorders.* 2016;22:28–34.

[445] Casamassima F, Ferrari A, Milosevic B, *et al.* A wearable system for gait training in subjects with Parkinson's disease. *Sensors.* 2014;14(4): 6229–6246.

[446] Mazilu S, Blanke U, Dorfman M, *et al.* A wearable assistant for gait training for Parkinson's disease with freezing of gait in out-of-the-lab environments. *ACM Transactions on Interactive Intelligent Systems (TiiS).* 2015;5(1):1–31.

[447] Matera G, Boonyasirikool C, Saggini R, *et al.* The new smartphone application for wrist rehabilitation. *The Journal of Hand Surgery (Asian-Pacific Volume).* 2016;21(01):2–7.

[448] Mourcou Q, Fleury A, Diot B, *et al.* iProprio: a Smartphone-based system to measure and improve proprioceptive function. In: *2016 38th Annual International Conference of the IEEE Engineering in Medicine and Biology Society (EMBC).* New York, NY: IEEE; 2016. p. 2622–2625.

[449] Ongvisatepaiboon K, Vanijja V, Chignell M, *et al.* Smartphone-based audio-biofeedback system for shoulder joint tele-rehabilitation. *Journal of Medical Imaging and Health Informatics.* 2016;6(4):1127–1134.

[450] Shin DC. Smartphone-based visual feedback trunk control training for gait ability in stroke patients: a single-blind randomized controlled trial. *Technology and Health Care.* 2020;28(1):45–55.

[451] Creek J. *The Core Concepts of Occupational Therapy: A Dynamic Framework for Practice*. London: Jessica Kingsley Publishers; 2010.

[452] Association AOT, *et al.* Standards of practice for occupational therapy. *The American Journal of Occupational Therapy.* 2021;75(Supplement_3):7513410030.

[453] Association AOT, *et al.* Occupational therapy practice framework: Domain et process. Am J Occup Ther. 2020;74(Supplement 2):7412410010p1–7412410010p87.

[454] Zhao W, Pillai JA, Leverenz JB, *et al.* Technology-facilitated detection of mild cognitive impairment: a review. In: *2018 IEEE International Conference on Electro/Information Technology (EIT)*. New York, NY: IEEE; 2018. p. 0284–0289.

[455] Fimland MS, Vasseljen O, Gismervik S, *et al.* Occupational rehabilitation programs for musculoskeletal pain and common mental health disorders: study protocol of a randomized controlled trial. *BMC Public Health.* 2014;14(1):1–9.

[456] Hayes SC, Strosahl KD, Wilson KG. *Acceptance and Commitment Therapy an Experiential Approach to Behavior Change*. New York, NY: The Guilford Press; 1999.

[457] Masselink CE. Considering technology in the occupational therapy practice framework. *The Open Journal of Occupational Therapy.* 2018;6(3):6.

[458] Hung KN G, Fong KN. Effects of telerehabilitation in occupational therapy practice: a systematic review. *Hong Kong Journal of Occupational Therapy.* 2019;32(1):3–21.

[459] Anson D. Assistive technology. In: Pendleton HM, Schultz-Krohn W (Eds), *Pedretti's Occupational Therapy-e-book: Practice Skills for Physical Dysfunction*. New York, NY: Elsevier Health Sciences; 2017. p. 416–433.

[460] Lawson S, Tang Z, Feng J. Supporting stroke motor recovery through a mobile application: a pilot study. *The American Journal of Occupational Therapy.* 2017;71(3):7103350010p1–7103350010p5.

[461] Bergquist T, Gehl C, Lepore S, *et al.* Internet-based cognitive rehabilitation in individuals with acquired brain injury: a pilot feasibility study. *Brain Injury.* 2008;22(11):891–897.

[462] Ng EM, Polatajko HJ, Marziali E, *et al.* Telerehabilitation for addressing executive dysfunction after traumatic brain injury. *Brain Injury.* 2013;27(5):548–564.

[463] Hegel MT, Lyons KD, Hull JG, *et al.* Feasibility study of a randomized controlled trial of a telephone-delivered problem-solving–occupational therapy intervention to reduce participation restrictions in rural breast cancer survivors undergoing chemotherapy. *Psycho-Oncology.* 2011;20(10): 1092–1101.

[464] Breeden LE. Occupational therapy home safety intervention via telehealth. *International Journal of Telerehabilitation.* 2016;8(1):29.

[465] Reifenberg G, Gabrosek G, Tanner K, *et al.* Feasibility of pediatric game-based neurorehabilitation using telehealth technologies: a case

report. *The American Journal of Occupational Therapy.* 2017;71(3): 7103190040p1–7103190040p8.

[466] Ferre CL, Brandão M, Surana B, *et al.* Caregiver-directed home-based intensive bimanual training in young children with unilateral spastic cerebral palsy: a randomized trial. *Developmental Medicine & Child Neurology.* 2017;59(5):497–504.

[467] Golomb MR, McDonald BC, Warden SJ, *et al.* In-home virtual reality videogame telerehabilitation in adolescents with hemiplegic cerebral palsy. *Archives of Physical Medicine and Rehabilitation.* 2010;91(1):1–8.

[468] Criss MJ. School-based telerehabilitation in occupational therapy: using telerehabilitation technologies to promote improvements in student performance. *International Journal of Telerehabilitation.* 2013; 5(1):39.

[469] Yuen HK, Pope C. Oral home telecare for adults with tetraplegia: a feasibility study. *Special Care in Dentistry.* 2009;29(5):204–209.

[470] Linder SM, Rosenfeldt AB, Bay RC, *et al.* Improving quality of life and depression after stroke through telerehabilitation. *The American Journal of Occupational Therapy.* 2015;69(2):6902290020p1–6902290020p10.

[471] Boehm N, Muehlberg H, Stube JE. Managing poststroke fatigue using telehealth: a case report. *The American Journal of Occupational Therapy.* 2015;69(6):6906350020p1–6906350020p7.

[472] Hermann VH, Herzog M, Jordan R, *et al.* Telerehabilitation and electrical stimulation: an occupation-based, client-centered stroke intervention. *The American Journal of Occupational Therapy.* 2010;64(1):73–81.

[473] Trombini M, Ferraro F, Iaconi G, *et al.* A study protocol for occupational rehabilitation in multiple sclerosis. *Sensors.* 2021;21(24):8436.

[474] Lun R, Gordon C, Zhao W. The design and implementation of a Kinect-based framework for selective human activity tracking. In: *Proceedings of the IEEE International Conference on Systems, Man, and Cybernetics.* New York, NY: IEEE; 2016. p. 002890–002895.

[475] Zhao W, Lun R, Gordon C, *et al.* A human-centered activity tracking system: toward a healthier workplace. *IEEE Transactions on Human–Machine Systems.* 2017;47(3):343–355.

[476] Yang S, Hans A, Zhao W, *et al.* Indoor localization and human activity tracking with multiple kinect sensors. In: *Smart Assisted Living.* New York, NY: Springer; 2020. p. 23–42.

[477] Zhao W, Perish J. Monitoring activities of daily living with a mobile app and Bluetooth beacons. In: *Proceedings of the IEEE Symposium Series on Computational Intelligence (SSCI).* New York, NY: IEEE; 2021. p. 1–8.

[478] Ali M, Lyden P, Brady M. Aphasia and dysarthria in acute stroke: recovery and functional outcome. *International Journal of Stroke.* 2015;10(3): 400–406.

[479] Poole ML, Vogel AP. Linking motor speech function and dementia. In: *Genetics, Neurology, Behavior, and Diet in Dementia.* New York, NY: Elsevier; 2020. p. 665–676.

[480]   Ramig L, Halpern A, Spielman J, *et al.* Speech treatment in Parkinson's disease: randomized controlled trial (RCT). *Movement Disorders.* 2018;33(11):1777–1791.

[481]   Urban P, Wicht S, Vukurevic G, *et al.* Dysarthria in acute ischemic stroke: lesion topography, clinicoradiologic correlation, and etiology. *Neurology.* 2001;56(8):1021–1027.

[482]   Morgan AT, Murray E, Liegeois FJ. Interventions for childhood apraxia of speech. *Cochrane Database of Systematic Reviews.* 2018;(5).

[483]   Kummer AW, Lee L. Evaluation and treatment of resonance disorders. *Language, Speech, and Hearing Services in Schools.* 1996;27(3):271–281.

[484]   Cler GJ, Mittelman T, Braden MN, *et al.* Video game rehabilitation of velopharyngeal dysfunction: a case series. *Journal of Speech, Language, and Hearing Research.* 2017;60(6S):1800–1809.

[485]   Tompkins CA. Rehabilitation for cognitive-communication disorders in right hemisphere brain damage. *Archives of Physical Medicine and Rehabilitation.* 2012;93(1):S61–S69.

[486]   Turkstra LS, Politis AM, Forsyth R. Cognitive–communication disorders in children with traumatic brain injury. *Developmental Medicine & Child Neurology.* 2015;57(3):217–222.

[487]   Hewetson R, Cornwell P, Shum D. Social participation following right hemisphere stroke: influence of a cognitive-communication disorder. *Aphasiology.* 2018;32(2):164–182.

[488]   Lanzi AM, Saylor AK, Cohen ML. Survey results of speech-language pathologists working with cognitive-communication disorders: improving practices for mild cognitive impairment and early-stage dementia from Alzheimer's disease. *American Journal of Speech-Language Pathology.* 2022;p. 1–19.

[489]   Bayles KA, Tomoeda CK. *Cognitive-Communication disorders of Dementia.* San Diego, CA: Plural Publishing; 2007.

[490]   Leonard LB. Is expressive language disorder an accurate diagnostic category? *American Journal of Speech-Language Pathology.* 2009;18(2): 115–123.

[491]   Lewis BA, Freebairn LA, Taylor HG. Follow-up of children with early expressive phonology disorders. *Journal of Learning Disabilities.* 2000;33(5):433–444.

[492]   Mohan M, Bajaj G, Deshpande A, *et al.* Child, parent, and play – an insight into these dimensions among children with and without receptive expressive language disorder using video-based analysis. *Psychology Research and Behavior Management.* 2021;14:971.

[493]   Anderson C. *Essentials of Linguistics.* Hamilton: McMaster University; 2018.

[494]   Fox CM, Ramig LO, Ciucci MR, *et al.* The science and practice of LSVT/LOUD: neural plasticity-principled approach to treating individuals with Parkinson disease and other neurological disorders. In: *Seminars in Speech and Language.* vol. 27. New York, NY: Thieme Medical Publishers, Inc.; 2006. p. 283–299.

[495] De Swart BJ, Willemse S, Maassen B, *et al*. Improvement of voicing in patients with Parkinson's disease by speech therapy. *Neurology*. 2003;60(3):498–500.

[496] Maier A, Hönig F, Bocklet T, *et al*. Automatic detection of articulation disorders in children with cleft lip and palate. *The Journal of the Acoustical Society of America*. 2009;126(5):2589–2602.

[497] Chen YPP, Johnson C, Lalbakhsh P, *et al*. Systematic review of virtual speech therapists for speech disorders. *Computer Speech & Language*. 2016;37:98–128.

[498] Furlong L, Erickson S, Morris ME. Computer-based speech therapy for childhood speech sound disorders. *Journal of Communication Disorders*. 2017;68:50–69.

[499] Ahmad A, Ahlin K, Mozelius P. Technology-enhanced speech and language relearning for stroke patients-defining requirements for a software application development. In: *11th Scandinavian Conference on Information Systems (SCIS2020)*, Sundsvall, Sweden; 2020.

[500] Egaji OA, Asghar I, Griffiths M, *et al*. Digital speech therapy for the aphasia patients: challenges, opportunities and solutions. In: *Proceedings of the 9th International Conference on Information Communication and Management*; 2019. p. 85–88.

[501] Abad A, Pompili A, Costa A, *et al*. Automatic word naming recognition for an on-line aphasia treatment system. *Computer Speech & Language*. 2013;27(6):1235–1248.

[502] Vuuren Sv, Cherney LR. A virtual therapist for speech and language therapy. In: *International Conference on Intelligent Virtual Agents*. New York, NY: Springer; 2014. p. 438–448.

[503] Massaro DW, Light J. Using visible speech to train perception and production of speech for individuals with hearing loss. *Journal of Speech, Language, and Hearing Research*. 2004;47:304–320.

[504] Schipor OA, Pentiuc SG, Schipor MD. Improving computer based speech therapy using a fuzzy expert system. *Computing and Informatics*. 2010;29(2):303–318.

[505] Stacey PC, Raine CH, O'Donoghue GM, *et al*. Effectiveness of computer-based auditory training for adult users of cochlear implants. *International Journal of Audiology*. 2010;49(5):347–356.

[506] Duval J, Rubin Z, Segura EM, *et al*. SpokeIt: building a mobile speech therapy experience. In: *Proceedings of the 20th International Conference on Human-Computer Interaction with Mobile Devices and Services*; 2018. p. 1–12.

[507] Thompson CK, Choy JJ, Holland A, *et al*. Sentactics®: computer-automated treatment of underlying forms. *Aphasiology*. 2010;24(10):1242–1266.

[508] Shtern M, Haworth MB, Yunusova Y, *et al*. A game system for speech rehabilitation. In: *International Conference on Motion in Games*. New York, NY: Springer; 2012. p. 43–54.

[509] Yunusova Y, Kearney E, Kulkarni M, *et al*. Game-based augmented visual feedback for enlarging speech movements in Parkinson's disease.

Journal of Speech, Language, and Hearing Research. 2017;60(6S): 1818–1825.

[510] Beijer L, Rietveld A, Ruiter M, *et al.* Preparing an E-learning-based Speech Therapy (EST) efficacy study: identifying suitable outcome measures to detect within-subject changes of speech intelligibility in dysarthric speakers. *Clinical Linguistics & Phonetics*. 2014;28(12):927–950.

[511] Byun TM, Campbell H, Carey H, *et al.* Enhancing intervention for residual rhotic errors via app-delivered biofeedback: a case study. *Journal of Speech, Language, and Hearing Research*. 2017;60(6S):1810–1817.

[512] Ganzeboom M, Bakker M, Beijer L, *et al.* Speech training for neurological patients using a serious game. *British Journal of Educational Technology*. 2018;49(4):761–774.

[513] Hair A, Ballard KJ, Markoulli C, *et al.* A longitudinal evaluation of tablet-based child speech therapy with Apraxia World. *ACM Transactions on Accessible Computing (TACCESS)*. 2021;14(1):1–26.

[514] Shute VJ. Focus on formative feedback. *Review of Educational Research*. 2008;78(1):153–189.

[515] Maas E, Robin DA, Hula SNA, *et al.* Principles of motor learning in treatment of motor speech disorders. *American Journal of Speech-Language Pathology*; 2008;17(3):277–298.

[516] Schmidt RA, Lee TD, Winstein C, *et al. Motor Control and Learning: A Behavioral Emphasis*. Champaign, IL: Human Kinetics; 2018.

[517] McCarthy B, Casey D, Devane D, *et al.* Pulmonary rehabilitation for chronic obstructive pulmonary disease. *Cochrane Database of Systematic Reviews*. 2015;(2).

[518] Spruit MA, Singh SJ, Garvey C, *et al.* An official American Thoracic Society/European Respiratory Society statement: key concepts and advances in pulmonary rehabilitation. *American Journal of Respiratory and Critical Care Medicine*. 2013;188(8):e13–e64.

[519] Garcia-Aymerich J, Lange P, Benet M, *et al.* Regular physical activity reduces hospital admission and mortality in chronic obstructive pulmonary disease: a population based cohort study. *Thorax*. 2006;61(9):772–778.

[520] Spruit MA, Wouters EF. Organizational aspects of pulmonary rehabilitation in chronic respiratory diseases. *Respirology*. 2019;24(9):838–843.

[521] Troosters T, Blondeel A, Janssens W, *et al.* The past, present and future of pulmonary rehabilitation. *Respirology*. 2019;24(9):830–837.

[522] Borg GA. Psychophysical bases of perceived exertion. *Medicine & Science in Sports & Exercise*. 1982;14(5):377–381.

[523] Williams N. The Borg rating of perceived exertion (RPE) scale. *Occupational Medicine*. 2017;67(5):404–405.

[524] Borg G. *Borg's Perceived Exertion and Pain Scales*. Champaign, IL: Human Kinetics; 1998.

[525] Horowitz MB, Littenberg B, Mahler DA. Dyspnea ratings for prescribing exercise intensity in patients with COPD. *Chest*. 1996;109(5): 1169–1175.

[526] Betschart M, Rezek S, Unger I, *et al.* Feasibility of an outpatient training program after COVID-19. *International Journal of Environmental Research and Public Health*. 2021;18(8):3978.

[527] Hayden MC, Limbach M, Schuler M, *et al.* Effectiveness of a three-week inpatient pulmonary rehabilitation program for patients after COVID-19: a prospective observational study. *International Journal of Environmental Research and Public Health*. 2021;18(17):9001.

[528] Borg G. Ratings of perceived exertion and heart rates during short-term cycle exercise and their use in a new cycling strength test. *International Journal of Sports Medicine*. 1982;3(03):153–158.

[529] Barrett B, Locken K, Maberry R, *et al.* The Wisconsin Upper Respiratory Symptom Survey (WURSS): a new research instrument for assessing the common cold. *The Journal of Family Practice*. 2002;51(3):265–265.

[530] Mohamed AA, Alawna M. The effect of aerobic exercise on immune biomarkers and symptoms severity and progression in patients with COVID-19: a randomized control trial. *Journal of Bodywork and Movement Therapies*. 2021;28:425–432.

[531] Gift AG, Narsavage G. Validity of the numeric rating scale as a measure of dyspnea. *American Journal of Critical Care*. 1998;7(3):200–204.

[532] Hsu KY, Lin JR, Lin MS, *et al.* The modified Medical Research Council dyspnoea scale is a good indicator of health-related quality of life in patients with chronic obstructive pulmonary disease. *Singapore Medical Journal*. 2013;54(6):321–327.

[533] Calvo-Paniagua J, Díaz-Arribas MJ, Valera-Calero JA, *et al.* A tele-health primary care rehabilitation program improves self-perceived exertion in COVID-19 survivors experiencing Post-COVID fatigue and dyspnea: a quasi-experimental study. *PLoS One*. 2022;17(8):e0271802.

[534] Kidd D, Stewart G, Baldry J, *et al.* The functional independence measure: a comparative validity and reliability study. *Disability and Rehabilitation*. 1995;17(1):10–14.

[535] Büsching G, Zhang Z, Schmid JP, *et al.* Effectiveness of pulmonary rehabilitation in severe and critically Ill COVID-19 patients: a controlled study. *International Journal of Environmental Research and Public Health*. 2021;18(17):8956.

[536] Spielmanns M, Pekacka-Egli AM, Schoendorf S, *et al.* Effects of a comprehensive pulmonary rehabilitation in severe post-COVID-19 patients. *International Journal of Environmental Research and Public Health*. 2021;18(5):2695.

[537] Linn BS, Linn MW, Gurel L. Cumulative illness rating scale. *Journal of the American Geriatrics Society*. 1968;16(5):622–626.

[538] Miller MD, Paradis CF, Houck PR, *et al.* Rating chronic medical illness burden in geropsychiatric practice and research: application of the Cumulative Illness Rating Scale. *Psychiatry Research*. 1992;41(3):237–248.

[539] Jones P, Quirk F, Baveystock C. The St George's respiratory questionnaire. *Respiratory Medicine*. 1991;85:25–31.

[540]   Mazzoleni S, Montagnani G, Vagheggini G, *et al.* Interactive videogame as rehabilitation tool of patients with chronic respiratory diseases: preliminary results of a feasibility study. *Respiratory Medicine.* 2014;108(10):1516–1524.

[541]   Bennett K. Measuring health state preference and utilities: rating scale, time tradeoff, and standard gamble techniques. *Quality of Life and Pharmacoeconomics in Clinical Trials.* 1996. p. 253–265.

[542]   Puhan MA, Behnke M, Devereaux P, *et al.* Measurement of agreement on health-related quality of life changes in response to respiratory rehabilitation by patients and physicians – a prospective study. *Respiratory Medicine.* 2004;98(12):1195–1202.

[543]   Schünemann HJ, Griffith L, Jaeschke R, *et al.* Evaluation of the minimal important difference for the feeling thermometer and the St. George's Respiratory Questionnaire in patients with chronic airflow obstruction. *Journal of Clinical Epidemiology.* 2003;56(12):1170–1176.

[544]   Enright PL. The six-minute walk test. *Respiratory Care.* 2003;48(8): 783–785.

[545]   Dun Y, Liu C, Ripley-Gonzalez JW, *et al.* Six-month outcomes and effect of pulmonary rehabilitation among patients hospitalized with COVID-19: a retrospective cohort study. *Annals of Medicine.* 2021;53(1): 2099–2109.

[546]   Guralnik JM, Simonsick EM, Ferrucci L, *et al.* A short physical performance battery assessing lower extremity function: association with self-reported disability and prediction of mortality and nursing home admission. *Journal of Gerontology.* 1994;49(2):M85–M94.

[547]   Udina C, Ars J, Morandi A, *et al.* Rehabilitation in adult post-COVID-19 patients in post-acute care with therapeutic exercise. *The Journal of Frailty & Aging.* 2021;10(3):297–300.

[548]   Holden MK, Gill KM, Magliozzi MR, *et al.* Clinical gait assessment in the neurologically impaired: reliability and meaningfulness. *Physical Therapy.* 1984;64(1):35–40.

[549]   Geddes EL, O'Brien K, Reid WD, *et al.* Inspiratory muscle training in adults with chronic obstructive pulmonary disease: an update of a systematic review. *Respiratory Medicine.* 2008;102(12):1715–1729.

[550]   Agustí A. Systemic effects of chronic obstructive pulmonary disease: what we know and what we don't know (but should). *Proceedings of the American Thoracic Society.* 2007;4(7):522–525.

[551]   Guarascio AJ, Ray SM, Finch CK, *et al.* The clinical and economic burden of chronic obstructive pulmonary disease in the USA. *ClinicoEconomics and Outcomes Research: CEOR.* 2013;5:235.

[552]   Spencer LM, McKeough ZJ. Maintaining the benefits following pulmonary rehabilitation: achievable or not? *Respirology.* 2019;24(9):909–915.

[553]   Tsutsui M, Gerayeli F, Sin DD. Pulmonary rehabilitation in a post-COVID-19 world: telerehabilitation as a new standard in patients with COPD. *International Journal of Chronic Obstructive Pulmonary Disease.* 2021;16:379.

[554] McNamara RJ, Dale M, McKeough ZJ. Innovative strategies to improve the reach and engagement in pulmonary rehabilitation. *Journal of Thoracic Disease*. 2019;11(Suppl 17):S2192.

[555] Wouters EF, Posthuma R, Koopman M, *et al.* An update on pulmonary rehabilitation techniques for patients with chronic obstructive pulmonary disease. *Expert Review of Respiratory Medicine*. 2020;14(2):149–161.

[556] Colombo V, Aliverti A, Sacco M. Virtual reality for COPD rehabilitation: a technological perspective. *Pulmonology*. 2022;28(2):119–133.

[557] Fekete M, Fazekas-Pongor V, Balazs P, *et al.* Role of new digital technologies and telemedicine in pulmonary rehabilitation. *Wiener klinische Wochenschrift*. 2021;133(21):1201–1207.

[558] Rodriguez-Blanco C, Gonzalez-Gerez JJ, Bernal-Utrera C, *et al.* Short-term effects of a conditioning telerehabilitation program in confined patients affected by COVID-19 in the acute phase. A pilot randomized controlled trial. *Medicina*. 2021;57(7):684.

[559] Vasilopoulou M, Papaioannou AI, Kaltsakas G, *et al.* Home-based maintenance tele-rehabilitation reduces the risk for acute exacerbations of COPD, hospitalisations and emergency department visits. *European Respiratory Journal*. 2017;49(5):1602129.

[560] Zanaboni P, Lien LA, Hjalmarsen A, *et al.* Long-term telerehabilitation of COPD patients in their homes: interim results from a pilot study in Northern Norway. *Journal of Telemedicine and Telecare*. 2013;19(7):425–429.

[561] Bernal-Utrera C, Anarte-Lazo E, De-La-Barrera-Aranda E, *et al.* Perspectives and attitudes of patients with COVID-19 toward a telerehabilitation programme: a qualitative study. *International Journal of Environmental Research and Public Health*. 2021;18(15):7845.

[562] Hameed F, Palatulan E, Jaywant A, *et al.* Outcomes of a COVID-19 recovery program for patients hospitalized with SARS-CoV-2 infection in New York City: a prospective cohort study. *PM&R*. 2021;13(6):609–617.

[563] Wootton SL, McKeough Z, Ng CL, *et al.* Effect on health-related quality of life of ongoing feedback during a 12-month maintenance walking programme in patients with COPD: a randomized controlled trial. *Respirology*. 2018;23(1):60–67.

[564] Steele BG, Belza B, Cain KC, *et al.* A randomized clinical trial of an activity and exercise adherence intervention in chronic pulmonary disease. *Archives of Physical Medicine and Rehabilitation*. 2008;89(3):404–412.

[565] Nguyen HQ, Gill DP, Wolpin S, *et al.* Pilot study of a cell phone-based exercise persistence intervention post-rehabilitation for COPD. *International Journal of Chronic Obstructive Pulmonary Disease*. 2009;4:301.

[566] Parent AA, Gosselin-Boucher V, Houle-Peloquin M, *et al.* Pilot project: physiologic responses to a high-intensity active video game with COPD patients – tools for home rehabilitation. *The Clinical Respiratory Journal*. 2018;12(5):1927–1936.

[567] Rutkowski S, Rutkowska A, Jastrzebski D, *et al.* Effect of virtual reality-based rehabilitation on physical fitness in patients with chronic

obstructive pulmonary disease. *Journal of Human Kinetics*. 2019;69(1): 149–157.

[568]   Rutkowski S, Rutkowska A, Kiper P, *et al.* Virtual reality rehabilitation in patients with chronic obstructive pulmonary disease: a randomized controlled trial. *International Journal of Chronic Obstructive Pulmonary Disease*. 2020;15:117.

[569]   LeGear T, LeGear M, Preradovic D, *et al.* Does a N intendo W ii exercise program provide similar exercise demands as a traditional pulmonary rehabilitation program in adults with COPD? *The Clinical Respiratory Journal*. 2016;10(3):303–310.

[570]   Makhabah D, Suradi S, Doewes M. The role of interactive game-based system in pulmonary rehabilitation of patients with COPD. *European Respiratory Journal*. 2015;46(suppl 59).

[571]   Sutanto Y, Makhabah D, Aphridasari J, *et al.* Videogame assisted exercise training in patients with chronic obstructive pulmonary disease: a preliminary study. *Pulmonology*. 2019;25(5):275–282.

[572]   Wardini R, Dajczman E, Yang N, *et al.* Using a virtual game system to innovate pulmonary rehabilitation: safety, adherence and enjoyment in severe chronic obstructive pulmonary disease. *Canadian Respiratory Journal*. 2013;20(5):357–361.

[573]   Dikken JB, Beijnum BJFv, Hofs DH, *et al.* An integrated virtual group training system for COPD patients at home. In: *International Joint Conference on Biomedical Engineering Systems and Technologies*. New York, NY: Springer; 2014. p. 346–359.

[574]   Colombo V, Mondellini M, Gandolfo A, *et al.* Usability and acceptability of a virtual reality-based system for endurance training in elderly with chronic respiratory diseases. In: *International Conference on Virtual Reality and Augmented Reality*. New York, NY: Springer; 2019. p. 87–96.

[575]   Jung T, Moorhouse N, Shi X, *et al.* A virtual reality-supported intervention for pulmonary rehabilitation of patients with chronic obstructive pulmonary disease: mixed methods study. *Journal of Medical Internet Research*. 2020;22(7):e14178.

[576]   Qin Y, Vincent CJ, Bianchi-Berthouze N, *et al.* AirFlow: designing immersive breathing training games for COPD. In: *CHI'14 Extended Abstracts on Human Factors in Computing Systems*; 2014. p. 2419–2424.

[577]   Stafford M, Lin F, Xu W. Flappy breath: a smartphone-based breath exergame. In: *2016 IEEE First International Conference on Connected Health: Applications, Systems and Engineering Technologies (CHASE)*. New York, NY: IEEE; 2016. p. 332–333.

[578]   Stamenova V, Liang K, Yang R, *et al.* Technology-enabled self-management of chronic obstructive pulmonary disease with or without asynchronous remote monitoring: randomized controlled trial. *Journal of Medical Internet Research*. 2020;22(7):e18598.

[579]   Petkov J, Harvey P, Battersby M. The internal consistency and construct validity of the partners in health scale: validation of a patient rated

chronic condition self-management measure. *Quality of Life Research*. 2010;19(7):1079–1085.

[580] Bartlett YK, Webb TL, Hawley MS. Using persuasive technology to increase physical activity in people with chronic obstructive pulmonary disease by encouraging regular walking: a mixed-methods study exploring opinions and preferences. *Journal of Medical Internet Research*. 2017;19(4):e6616.

[581] Oinas-Kukkonen H, Harjumaa M. Persuasive systems design: Key issues, process model, and system features. *Communications of the Association for Information Systems*. 2009;24(1):28.

[582] Voncken-Brewster V, Tange H, de Vries H, *et al.* A randomized controlled trial evaluating the effectiveness of a web-based, computer-tailored self-management intervention for people with or at risk for COPD. *International Journal of Chronic Obstructive Pulmonary Disease*. 2015;10:1061.

[583] Burkow TM, Vognild LK, Johnsen E, *et al.* Comprehensive pulmonary rehabilitation in home-based online groups: a mixed method pilot study in COPD. *BMC Research Notes*. 2015;8(1):1–11.

[584] Gloeckl R, Leitl D, Jarosch I, *et al.* Benefits of pulmonary rehabilitation in COVID-19: a prospective observational cohort study. *ERJ Open Research*. 2021;7(2):00108-2021.

[585] Besnier F, Bérubé B, Malo J, *et al.* Cardiopulmonary rehabilitation in long-COVID-19 patients with persistent breathlessness and fatigue: the COVID-Rehab Study. *International Journal of Environmental Research and Public Health*. 2022;19(7):4133.

[586] Stavrou VT, Tourlakopoulos KN, Vavougios GD, *et al.* Eight weeks unsupervised pulmonary rehabilitation in previously hospitalized of SARS-CoV-2 infection. *Journal of Personalized Medicine*. 2021;11(8):806.

[587] Guyatt GH, Berman LB, Townsend M, *et al.* A measure of quality of life for clinical trials in chronic lung disease. *Thorax*. 1987;42(10):773–778.

[588] Al Chikhanie Y, Veale D, Schoeffler M, *et al.* Effectiveness of pulmonary rehabilitation in COVID-19 respiratory failure patients post-ICU. *Respiratory Physiology & Neurobiology*. 2021;287:103639.

[589] Carterette EC, Friedman MP (Eds). *Handbook of Perception and Cognition*. 2nd ed. San Diego, CA: Academic Press; 1999.

[590] Baars BJ, Gage NM. *Cognition, Brain, and Consciousness: Introduction to Cognitive Neuroscience*. London: Academic Press; 2010.

[591] Audiffren M, André N. The exercise-cognition relationship: a virtuous circle. *Journal of Sport and Health Science*. 2019;8(4):339–347.

[592] Kramer AF, Erickson KI, Colcombe SJ. Exercise, cognition, and the aging brain. *Journal of Applied Physiology*. 2006;101(4):1237–1242.

[593] Miyake A, Friedman NP, Emerson MJ, *et al.* The unity and diversity of executive functions and their contributions to complex "frontal lobe" tasks: a latent variable analysis. *Cognitive Psychology*. 2000;41(1): 49–100.

[594] Miller E, Wallis J. Executive function and higher-order cognition: definition and neural substrates. *Encyclopedia of Neuroscience*. 2009;4:99–104.

[595]   Kramer AF, Hahn S, Cohen NJ, *et al.* Ageing, fitness and neurocognitive function. *Nature.* 1999;400(6743):418–419.

[596]   Colcombe SJ, Kramer AF, Erickson KI, *et al.* Cardiovascular fitness, cortical plasticity, and aging. *Proceedings of the National Academy of Sciences.* 2004;101(9):3316–3321.

[597]   Smiley-Oyen AL, Lowry KA, Francois SJ, *et al.* Exercise, fitness, and neurocognitive function in older adults: the "selective improvement" and "cardiovascular fitness" hypotheses. *Annals of Behavioral Medicine.* 2008;36(3):280–291.

[598]   Voelcker-Rehage C, Godde B, Staudinger UM. Cardiovascular and coordination training differentially improve cognitive performance and neural processing in older adults. *Frontiers in Human Neuroscience.* 2011;5:26.

[599]   Stroth S, Hille K, Spitzer M, *et al.* Aerobic endurance exercise benefits memory and affect in young adults. *Neuropsychological Rehabilitation.* 2009;19(2):223–243.

[600]   Hawkins HL, Kramer AF, Capaldi D. Aging, exercise, and attention. *Psychology and Aging.* 1992;7(4):643.

[601]   Hötting K, Röder B. Beneficial effects of physical exercise on neuroplasticity and cognition. *Neuroscience & Biobehavioral Reviews.* 2013;37(9):2243–2257.

[602]   Eisenberg N. Temperamental effortful control (self-regulation). In: *Encyclopedia on Early Childhood Development.* Centre of Excellence for Early Childhood Development; 2005. p. 1–5.

[603]   Rothbart MK, Bates JE. Temperament. In: Damon W EN (Ed.), *Handbook of Child Psychology: Social, Emotional, and Personality development.* vol. 3, 5th ed. New York, NY: John Wiley & Sons, Inc.; 1998.

[604]   Rakesh G, Szabo ST, Alexopoulos GS, *et al.* Strategies for dementia prevention: latest evidence and implications. *Therapeutic Advances in Chronic Disease.* 2017;8(8-9):121–136.

[605]   Håkansson K, Ledreux A, Daffner K, *et al.* BDNF responses in healthy older persons to 35 minutes of physical exercise, cognitive training, and mindfulness: associations with working memory function. *Journal of Alzheimer's Disease.* 2017;55(2):645–657.

[606]   Fahimi A, Baktir MA, Moghadam S, *et al.* Physical exercise induces structural alterations in the hippocampal astrocytes: exploring the role of BDNF-TrkB signaling. *Brain Structure and Function.* 2017;222(4):1797–1808.

[607]   Isaacs KR, Anderson BJ, Alcantara AA, *et al.* Exercise and the brain: angiogenesis in the adult rat cerebellum after vigorous physical activity and motor skill learning. *Journal of Cerebral Blood Flow & Metabolism.* 1992;12(1):110–119.

[608]   Stimpson NJ, Davison G, Javadi AH. Joggin' the noggin: towards a physiological understanding of exercise-induced cognitive benefits. *Neuroscience & Biobehavioral Reviews.* 2018;88:177–186.

[609]   Manuela Crispim Nascimento C, Rodrigues Pereira J, Pires de Andrade L, *et al.* Physical exercise in MCI elderly promotes reduction of

pro-inflammatory cytokines and improvements on cognition and BDNF peripheral levels. *Current Alzheimer Research*. 2014;11(8):799–805.

[610] Rogers RL, Meyer JS, Mortel KF. After reaching retirement age physical activity sustains cerebral perfusion and cognition. *Journal of the American Geriatrics Society*. 1990;38(2):123–128.

[611] Blumenthal JA, Emery CF, Madden DJ, *et al.* Effects of exercise training on cardiorespiratory function in men and women 60 years of age. *The American Journal of Cardiology*. 1991;67(7):633–639.

[612] Van Praag H, Christie BR, Sejnowski TJ, *et al.* Running enhances neuro-genesis, learning, and long-term potentiation in mice. *Proceedings of the National Academy of Sciences*. 1999;96(23):13427–13431.

[613] Yaffe K, Barnes D, Nevitt M, *et al.* A prospective study of physical activity and cognitive decline in elderly women: women who walk. *Archives of Internal Medicine*. 2001;161(14):1703–1708.

[614] Etgen T, Sander D, Huntgeburth U, *et al.* Physical activity and incident cognitive impairment in elderly persons: the INVADE study. *Archives of Internal Medicine*. 2010;170(2):186–193.

[615] Singh A, Uijtdewilligen L, Twisk JW, *et al.* Physical activity and per-formance at school: a systematic review of the literature including a methodological quality assessment. *Archives of Pediatrics & Adolescent Medicine*. 2012;166(1):49–55.

[616] Mora-Gonzalez J, Esteban-Cornejo I, Cadenas-Sanchez C, *et al.* Physical fitness, physical activity, and the executive function in children with overweight and obesity. *The Journal of Pediatrics*. 2019;208:50–56.

[617] Logan NE, Raine LB, Drollette ES, *et al.* The differential relationship of an afterschool physical activity intervention on brain function and cognition in children with obesity and their normal weight peers. *Pediatric Obesity*. 2021;16(2):e12708.

[618] Visier-Alfonso ME, Álvarez-Bueno C, Sánchez-López M, *et al.* Fitness and executive function as mediators between physical activity and aca-demic achievement: mediators between physical activity and academic achievement. *Journal of Sports Sciences*. 2021;39(14):1576–1584.

[619] Coe DP, Pivarnik JM, Womack CJ, *et al.* Effect of physical education and activity levels on academic achievement in children. *Medicine and Science in Sports and Exercise*. 2006;38(8):1515.

[620] Lipošek S, Planinšec J, Leskošek B, *et al.* Physical activity of university students and its relation to physical fitness and academic success. *Annales Kinesiologiae*. 2018;9(2):89–104.

[621] Donnelly JE, Greene JL, Gibson CA, *et al.* Physical Activity Across the Curriculum (PAAC): a randomized controlled trial to promote physical activity and diminish overweight and obesity in elementary school children. *Preventive Medicine*. 2009;49(4):336–341.

[622] Stevens TA, To Y, Stevenson SJ, *et al.* The importance of physical activity and physical education in the prediction of academic achievement. *Journal of Sport Behavior*. 2008;31(4):9418004.

[623]   Fredericks CR, Kokot SJ, Krog S. Using a developmental movement programme to enhance academic skills in grade 1 learners. *South African Journal for Research in Sport, Physical Education and Recreation*. 2006;28(1):29–42.

[624]   Sallis JF, McKenzie TL, Kolody B, *et al.* Effects of health-related physical education on academic achievement: Project SPARK. *Research Quarterly for Exercise and Sport*. 1999;70(2):127–134.

[625]   Ahamed Y, MacDonald H, Reed K, *et al.* School-based physical activity does not compromise children's academic performance. *Medicine and Science in Sports and Exercise*. 2007;39(2):371–376.

[626]   Ruiz JR, Castro-Piñero J, España-Romero V, *et al.* Field-based fitness assessment in young people: the ALPHA health-related fitness test battery for children and adolescents. *British Journal of Sports Medicine*. 2011;45(6):518–524.

[627]   Delis DC, Kaplan E, Kramer JH. *Delis-Kaplan Executive Function System*; 2001.

[628]   Wilson BA, Alderman N, Burgess PW, *et al. BADS: Behavioural Assessment of the Dysexecutive Syndrome*. London, England: Pearson; 1996.

[629]   Krohne H, Egloff B, Kohlmann C, *et al.* Investigations with a German version of the positive and negative affect schedule (PANAS). *Diagnostica*. 1996;42(2):139–156.

[630]   Hurd MD, Martorell P, Delavande A, *et al.* Monetary costs of dementia in the United States. *New England Journal of Medicine*. 2013;368(14):1326–1334.

[631]   World Health Organization. *Dementia Fact Sheet*. Geneva, Switzerland: World Health Organization; March 2023. Available from: https://www.who.int/news-room/fact-sheets/detail/dementia.

[632]   Cummings JL, Mega M, Gray K, *et al.* The Neuropsychiatric Inventory comprehensive assessment of psychopathology in dementia. *Neurology*. 1994;44(12):2308–2308.

[633]   Edwards ER, Spira AP, Barnes DE, *et al.* Neuropsychiatric symptoms in mild cognitive impairment: differences by subtype and progression to dementia. *International Journal of Geriatric Psychiatry*. 2009;24(7):716–722.

[634]   Rosenberg PB, Mielke MM, Appleby BS, *et al.* The association of neuropsychiatric symptoms in MCI with incident dementia and Alzheimer disease. *The American Journal of Geriatric Psychiatry*. 2013;21(7):685–695.

[635]   Pillai JA, Bonner-Jackson A. Review of information and communication technology devices for monitoring functional and cognitive decline in Alzheimer's disease clinical trials. *Journal of Healthcare Engineering*. 2015;6(1):71–84.

[636]   Petersen RC, Smith GE, Waring SC, *et al.* Mild cognitive impairment: clinical characterization and outcome. *Archives of Neurology*. 1999;56(3):303–308.

[637]   Jekel K, Damian M, Wattmo C, *et al.* Mild cognitive impairment and deficits in instrumental activities of daily living: a systematic review. *Alzheimer's Research & Therapy*. 2015;7(1):17.

[638] Kottorp A, Nygård L. Development of a short-form assessment for detection of subtle activity limitations: can use of everyday technology distinguish between MCI and Alzheimer's disease? *Expert Review of Neurotherapeutics*. 2011;11(5):647–655.

[639] Hedman A, Nygård L, Almkvist O, *et al.* Amount and type of everyday technology use over time in older adults with cognitive impairment. *Scandinavian Journal of Occupational Therapy*. 2015;22(3):196–206.

[640] Seelye AM, Schmitter-Edgecombe M, Cook DJ, *et al.* Naturalistic assessment of everyday activities and prompting technologies in mild cognitive impairment. *Journal of the International Neuropsychological Society*. 2013;19(4):442–452.

[641] Jekel K, Damian M, Storf H, *et al.* Development of a proxy-free objective assessment tool of instrumental activities of daily living in mild cognitive impairment using smart home technologies. *Journal of Alzheimer's Disease*. 2016;52(2):509–517.

[642] Sacco G, Joumier V, Darmon N, *et al.* Detection of activities of daily living impairment in Alzheimer's disease and mild cognitive impairment using information and communication technology. *Clinical Interventions in Aging*. 2012;7:539.

[643] Vermeersch S, Gorus E, Cornelis E, *et al.* An explorative study of the relationship between functional and cognitive decline in older persons with mild cognitive impairment and Alzheimer's disease. *British Journal of Occupational Therapy*. 2015;78(3):166–174.

[644] Rosenberg L, Kottorp A, Winblad B, *et al.* Perceived difficulty in everyday technology use among older adults with or without cognitive deficits. *Scandinavian Journal of Occupational Therapy*. 2009;16(4):216–226.

[645] Hedman A, Nygård L, Almkvist O, *et al.* Patterns of functioning in older adults with mild cognitive impairment: a two-year study focusing on everyday technology use. *Aging & Mental Health*. 2013;17(6):679–688.

[646] Ryd C, Nygård L, Malinowsky C, *et al.* Associations between performance of activities of daily living and everyday technology use among older adults with mild stage Alzheimer's disease or mild cognitive impairment. *Scandinavian Journal of Occupational Therapy*. 2015;22(1):33–42.

[647] Melrose RJ, Brommelhoff JA, Narvaez T, *et al.* The use of Information and Communication technology when completing instrumental activities of daily living. *Computers in Human Behavior*. 2016;63:471–479.

[648] Realdon O, Rossetto F, Nalin M, *et al.* Technology-enhanced multi-domain at home continuum of care program with respect to usual care for people with cognitive impairment: the Ability-TelerehABILITation study protocol for a randomized controlled trial. *BMC Psychiatry*. 2016; 16(1):425.

[649] Zhao W, Luo X, Qiu T. Smart healthcare. *Applied Sciences*. 2017;7(11): 1176.

[650] Zhao W, Luo X, Liu H, *et al.* Scientific programming towards a smart world. *Scientific Programming*. 2017;2017:3706232.

[651] Zhao W, Ma J, Wang KIK, *et al.* Report of the 2017 IEEE Cyber Science and Technology Congress. *Applied Sciences*. 2017;7(12):1299. Available from: http://www.mdpi.com/2076-3417/7/12/1299.

[652] Luo X, Deng J, Wang W, *et al.* A quantized kernel learning algorithm using a minimum kernel risk-sensitive loss criterion and bilateral gradient technique. *Entropy*. 2017;19(7):365.

[653] Luo X, Xu Y, Wang W, *et al.* Towards enhancing stacked extreme learning machine with sparse autoencoder by correntropy. *Journal of the Franklin Institute*. 2017. DOI: 10.1016/j.jfranklin.2017.08.014.

[654] König A, Crispim-Junior CF, Covella AGU, *et al.* Ecological assessment of autonomy in instrumental activities of daily living in dementia patients by the means of an automatic video monitoring system. *Frontiers in Aging Neuroscience*. 2015;7:98.

[655] König A, Crispim Junior CF, Derreumaux A, *et al.* Validation of an automatic video monitoring system for the detection of instrumental activities of daily living in dementia patients. *Journal of Alzheimer's Disease*. 2015;44(2):675–685.

[656] Bathrinarayanan V, Fosty B, Konig A, *et al.* Evaluation of a monitoring system for event recognition of older people. In: *2013 10th IEEE International Conference on Advanced Video and Signal Based Surveillance*. New York, NY: IEEE; 2013. p. 165–170.

[657] Cao Y, Tao L, Xu G. An event-driven context model in elderly health monitoring. In: *2009 Symposia and Workshops on Ubiquitous, Autonomic and Trusted Computing*. New York, NY: IEEE; 2009. p. 120–124.

[658] Romdhane R, Crispim CF, Bremond F, *et al.* Activity recognition and uncertain knowledge in video scenes. In: *2013 10th IEEE International Conference on Advanced Video and Signal Based Surveillance (AVSS)*. New York, NY: IEEE; 2013. p. 377–382.

[659] Riboni D, Bettini C, Civitarese G, *et al.* Smartfaber: recognizing fine-grained abnormal behaviors for early detection of mild cognitive impairment. *Artificial Intelligence in Medicine*. 2016;67:57–74.

[660] Stavropoulos TG, Meditskos G, Kontopoulos E, *et al.* Multi-sensing monitoring and knowledge-driven analysis for dementia assessment. *International Journal of E-Health and Medical Communications (IJEHMC)*. 2015;6(4):77–92.

[661] Andreadis S, Stavropoulos TG, Meditskos G, *et al.* Dem@ home: ambient intelligence for clinical support of people living with dementia. In: *International Semantic Web Conference*. New York, NY: Springer; 2016. p. 357–368.

[662] Lazarou I, Karakostas A, Stavropoulos TG, *et al.* A novel and intelligent home monitoring system for care support of elders with cognitive impairment. *Journal of Alzheimer's Disease*. 2016;54(4):1561–1591.

[663] Zhao W. A concise tutorial on human motion tracking and recognition with Microsoft Kinect. *Science China Information Sciences*. 2016;59(9): 93101.

[664] Jakkula V and Cook DJ. Anomaly detection using temporal data mining in a smart home environment. *Methods of Information in Medicine*. 2008;47(1):70–75.

[665] Nute D. Defeasible logic. In: *International Conference on Applications of Prolog*. New York, NY: Springer; 2001. p. 151–169.

[666] Sonia, Ghose A. Unobtrusive and pervasive monitoring of geriatric subjects for early screening of mild cognitive impairment. In: *Proceedings of the IEEE PerCom*; 2018.

[667] Skubic M, Guevara RD, Rantz M. Testing classifiers for embedded health assessment. In: Donnelly M, Paggetti C, Nugent C, *et al.* (Eds), *Impact Analysis of Solutions for Chronic Disease Prevention and Management*. Berlin, Heidelberg: Springer Berlin Heidelberg; 2012. p. 198–205.

[668] Duchêne F, Garbay C, Rialle V. Learning recurrent behaviors from heterogeneous multivariate time-series. *Artificial Intelligence in Medicine*. 2007;39(1):25–47.

[669] Rashidi P, Cook DJ. COM: a method for mining and monitoring human activity patterns in home-based health monitoring systems. *ACM Transactions on Intelligent Systems and Technology*. 2013;4(4):64.

[670] Virone G, Noury N, Demongeot J. A system for automatic measurement of circadian activity deviations in telemedicine. *IEEE Transactions on Biomedical Engineering*. 2002;49(12):1463–1469.

[671] Juarez JM, Ochotorena JM, Campos M, *et al.* Spatiotemporal data visualisation for homecare monitoring of elderly people. *Artificial Intelligence in Medicine*. 2015;65(2):97–111.

[672] Pavel M, Hayes TL, Adami A, *et al.* Unobtrusive assessment of mobility. In: *28th Annual International Conference of the Engineering in Medicine and Biology Society, 2006. EMBS'06*. New York, NY: IEEE; 2006. p. 6277–6280.

[673] Pavel M, Adami A, Morris M, *et al.* Mobility assessment using event-related responses. In: *1st Transdisciplinary Conference on Distributed Diagnosis and Home Healthcare, 2006. D2H2*. New York, NY: IEEE; 2006. p. 71–74.

[674] Akl A, Snoek J, Mihailidis A. Unobtrusive detection of mild cognitive impairment in older adults through home monitoring. *IEEE Journal of Biomedical and Health Informatics*. 2017;21(2):339–348.

[675] Van Erven T, Harremos P. Rényi divergence and Kullback-Leibler divergence. *IEEE Transactions on Information Theory*. 2014;60(7):3797–3820.

[676] Akl A, Chikhaoui B, Mattek N, *et al.* Clustering home activity distributions for automatic detection of mild cognitive impairment in older adults. *Journal of Ambient Intelligence and Smart Environments*. 2016;8(4):437–451.

[677] Satt A, Sorin A, Toledo-Ronen O, *et al.* Evaluation of speech-based protocol for detection of early-stage dementia. In: *INTERSPEECH*; 2013. p. 1692–1696.

[678] Kaye J, Mattek N, Dodge HH, *et al.* Unobtrusive measurement of daily computer use to detect mild cognitive impairment. *Alzheimer's & Dementia: The Journal of the Alzheimer's Association*. 2014;10(1):10–17.

[679]   Klimova B, Valis M, Kuca K. Potential of mobile technologies and applications in the detection of mild cognitive impairment among older generation groups. *Social Work in Health Care*. 2017;56(7):588–599.

[680]   Groot C, Hooghiemstra AM, Raijmakers PG, *et al.* The effect of physical activity on cognitive function in patients with dementia: a meta-analysis of randomized control trials. *Ageing Research Reviews*. 2016;25:13–23.

[681]   Green CS, Bavelier D. The cognitive neuroscience of video games. *Digital Media: Transformations in Human Communication*. 2006;1(1):211–223.

[682]   González-Ortega D, Díaz-Pernas FJ, Martínez-Zarzuela M, *et al.* A Kinect-based system for cognitive rehabilitation exercises monitoring. *Computer Methods and Programs in Biomedicine*. 2014;113(2):620–631.

[683]   Guilford JP. A system of the psychomotor abilities. *The American Journal of Psychology*. 1958;71(1):164–174.

[684]   Shatil E. Does combined cognitive training and physical activity training enhance cognitive abilities more than either alone? A four-condition randomized controlled trial among healthy older adults. *Frontiers in Aging Neuroscience*. 2013;5:8.

[685]   Hughes TF, Flatt JD, Fu B, *et al.* Interactive video gaming compared with health education in older adults with mild cognitive impairment: a feasibility study. *International Journal of Geriatric Psychiatry*. 2014;29(9):890–898.

[686]   *The US Mental Health Market: $225.1 Billion in Spending in 2019: An OPEN MINDS Market Intelligence Report*; 2020. https://openminds.com/intelligence-report/the-u-s-mental-health-market-225-1-billion-in-spending-in-2019-an-open-minds-market-intelligence-report/. Available from: https://openminds.com/intelligence-report/the-u-s-mental-health-market-225-1-billion-in-spending-in-2019-an-open-minds-market-intelligence-rep ort/.

[687]   American Psychiatric Association, DSM-5 Task Force. *Diagnostic and Statistical Manual of Mental Disorders: DSM-5*, 5th ed. Washington, DC: American Psychiatric Publishing, Inc.; 2013.

[688]   Abrams R. *Electroconvulsive Therapy*. Oxford: Oxford University Press; 2002.

[689]   Espinoza RT, Kellner CH. Electroconvulsive therapy. *New England Journal of Medicine*. 2022;386(7):667–672.

[690]   Baer RA. Mindfulness training as a clinical intervention: a conceptual and empirical review. *Clinical Psychology: Science and Practice*. 2003;10(2):125.

[691]   Siegling A, Petrides K. Measures of trait mindfulness: convergent validity, shared dimensionality, and linkages to the five-factor model. *Frontiers in Psychology*. 2014;5:1164.

[692]   Langer EJ. *Mindfulness*. Lebanon: Da Capo Lifelong Books; 2014.

[693]   Bodner T, Langer E. Individual differences in mindfulness: The mindfulness/mindlessness scale. In: *In Poster Presented at the 13th Annual American Psychological Society Convention*, Toronto, Ontario, Canada; 2001.

[694] Mothes H, Klaperski S, Seelig H, *et al.* Regular aerobic exercise increases dispositional mindfulness in men: a randomized controlled trial. *Mental Health and Physical Activity.* 2014;7(2):111–119.

[695] Byrne A, Byrne D. The effect of exercise on depression, anxiety and other mood states: a review. *Journal of Psychosomatic Research.* 1993;37(6):565–574.

[696] Mikkelsen K, Stojanovska L, Polenakovic M, *et al.* Exercise and mental health. *Maturitas.* 2017;106:48–56.

[697] McQuaid JR, Marx BP, Rosen MI, *et al.* Mental health assessment in rehabilitation research. *Journal of Rehabilitation Research and Development.* 2012;49(1):121–138.

[698] Picard RW. *Affective Computing.* London: MIT Press; 2000.

[699] Park CK, Lee S, Park HJ, *et al.* Autonomic function, voice, and mood states. *Clinical Autonomic Research.* 2011;21(2):103–110.

[700] Task Force of the European Society of Cardiology and the North American Society of Pacing and Electrophysiology. Heart rate variability: standards of measurement, physiological interpretation, and clinical use. *Circulation.* 1996;93(5):1043–1065.

[701] de Santos Sierra A, Ávila CS, Casanova JG, *et al.* A stress-detection system based on physiological signals and fuzzy logic. *IEEE Transactions on Industrial Electronics.* 2011;58(10):4857–4865.

[702] Freeman D, Bradley J, Antley A, *et al.* Virtual reality in the treatment of persecutory delusions: randomised controlled experimental study testing how to reduce delusional conviction. *The British Journal of Psychiatry.* 2016;209(1):62–67.

[703] Shek AC, Biondi A, Ballard D, *et al.* Technology-based interventions for mental health support after stroke: a systematic review of their acceptability and feasibility. *Neuropsychological Rehabilitation.* 2021;31(3):432–452.

[704] Turner AP, Kivlahan DR, Haselkorn JK. Exercise and quality of life among people with multiple sclerosis: looking beyond physical functioning to mental health and participation in life. *Archives of Physical Medicine and Rehabilitation.* 2009;90(3):420–428.

[705] Kritikos TK, Smith K, Holmbeck GN. Mental health guidelines for the care of people with spina bifida. *Journal of Pediatric Rehabilitation Medicine.* 2020;13(4):525–534.

[706] Brady SM, Fenton SA, Metsios GS, *et al.* Different types of physical activity are positively associated with indicators of mental health and psychological wellbeing in rheumatoid arthritis during COVID-19. *Rheumatology International.* 2021;41(2):335–344.

[707] Paluska SA, Schwenk TL. Physical activity and mental health. *Sports Medicine.* 2000;29(3):167–180.

[708] Daley AJ. Exercise therapy and mental health in clinical populations: is exercise therapy a worthwhile intervention? *Advances in Psychiatric Treatment.* 2002;8(4):262–270.

[709] Peluso MAM, De Andrade LHSG. Physical activity and mental health: the association between exercise and mood. *Clinics.* 2005;60(1):61–70.

[710]   Tendzegolskis Z, Viru A, Orlova E. Exercise-induced changes of endorphin contents in hypothalamus, hypophysis, adrenals and blood plasma. *International Journal of Sports Medicine*. 1991;12(05):495–497.

[711]   Sharifi M, Hamedinia M, Hosseini-Kakhak S. The effect of an exhaustive aerobic, anaerobic and resistance exercise on serotonin, beta-endorphin and BDnf in students. *Physical Education of Students*. 2018;(5): 272–277.

[712]   Balchin R, Linde J, Blackhurst D, *et al*. Sweating away depression? The impact of intensive exercise on depression. *Journal of Affective disorders*. 2016;200:218–221.

[713]   Raglin JS, Morgan WP. Influence of exercise and quiet rest on state anxiety and blood pressure. *Medicine & Science in Sports & Exercise*. 1987;19(5):456–463.

[714]   Raglin JS. Exercise and mental health. *Sports Medicine*. 1990;9(6): 323–329.

[715]   deVries HA. Tranquilizer effect of exercise: a critical review. *The Physician and Sportsmedicine*. 1981;9(11):46–55.

[716]   Youngstedt SD, Dishman RK, Cureton KJ, *et al*. Does body temperature mediate anxiolytic effects of acute exercise? *Journal of Applied Physiology*. 1993;74(2):825–831.

[717]   Craft LL, Perna FM. The benefits of exercise for the clinically depressed. *Primary Care Companion to the Journal of Clinical Psychiatry*. 2004; 6(3):104.

[718]   Brown GC, Murphy MP, Duchen MR, *et al*. Roles of mitochondria in human disease. *Essays in Biochemistry*. 2010;47:115–137.

[719]   Wei YH, Ma YS, Lee HC, *et al*. Mitochondrial theory of aging matures roles of mtDNA mutation and oxidative stress in human aging. *Chinese Medical Journal (Tai pei)*. 2001;64:259–270.

[720]   Bansal Y, Kuhad A. Mitochondrial dysfunction in depression. *Current Neuropharmacology*. 2016;14(6):610–618.

[721]   Klein IL, van de Loo KF, Smeitink JA, *et al*. Cognitive functioning and mental health in mitochondrial disease: a systematic scoping review. *Neuroscience & Biobehavioral Reviews*. 2021;125:57–77.

[722]   Broskey NT, Greggio C, Boss A, *et al*. Skeletal muscle mitochondria in the elderly: effects of physical fitness and exercise training. *The Journal of Clinical Endocrinology & Metabolism*. 2014;99(5):1852–1861.

[723]   Lipton JO, Sahin M. The neurology of mTOR. *Neuron*. 2014;84(2):275–291.

[724]   Hoeffer CA, Klann E. mTOR signaling: at the crossroads of plasticity, memory and disease. *Trends in Neurosciences*. 2010;33(2):67–75.

[725]   Lloyd BA, Hake HS, Ishiwata T, *et al*. Exercise increases mTOR signaling in brain regions involved in cognition and emotional behavior. *Behavioural Brain Research*. 2017;323:56–67.

[726]   Wipfli B, Landers D, Nagoshi C, *et al*. An examination of serotonin and psychological variables in the relationship between exercise and mental health. *Scandinavian Journal of Medicine & Science in Sports*. 2011;21(3):474–481.

[727]   Spencer RL, Deak T. A users guide to HPA axis research. *Physiology & Behavior*. 2017;178:43–65.

[728]   Akil H, Haskett RF, Young EA, *et al.* Multiple HPA profiles in endogenous depression: effect of age and sex on cortisol and beta-endorphin. *Biological Psychiatry*. 1993;33(2):73–85.

[729]   Lesch KP, Laux G, Schulte HM, *et al.* Corticotropin and cortisol response to human CRH as a probe for HPA system integrity in major depressive disorder. *Psychiatry Research*. 1988;24(1):25–34.

[730]   Dienstbier RA. Behavioral correlates of sympathoadrenal reactivity: the toughness model. *Medicine & Science in Sports & Exercise*. 1991;23(7):846–852.

[731]   Brosse AL, Sheets ES, Lett HS, *et al.* Exercise and the treatment of clinical depression in adults. *Sports Medicine*. 2002;32(12):741–760.

[732]   Dantzer R, O'connor JC, Freund GG, *et al.* From inflammation to sickness and depression: when the immune system subjugates the brain. *Nature Reviews Neuroscience*. 2008;9(1):46–56.

[733]   Gleeson M, Bishop NC, Stensel DJ, *et al.* The anti-inflammatory effects of exercise: mechanisms and implications for the prevention and treatment of disease. *Nature Reviews Immunology*. 2011;11(9):607–615.

[734]   Euteneuer F, Dannehl K, Del Rey A, *et al.* Immunological effects of behavioral activation with exercise in major depression: an exploratory randomized controlled trial. *Translational Psychiatry*. 2017;7(5):e1132–e1132.

[735]   Miller KJ, Mesagno C, McLaren S, *et al.* Exercise, mood, self-efficacy, and social support as predictors of depressive symptoms in older adults: direct and interaction effects. *Frontiers in Psychology*. 2019;10:2145.

[736]   Festinger L, Maccoby N. On resistance to persuasive communications. *The Journal of Abnormal and Social Psychology*. 1964;68(4):359.

[737]   Craft LL. Exercise and clinical depression: examining two psychological mechanisms. *Psychology of Sport and Exercise*. 2005;6(2):151–171.

[738]   Ryan MP. The antidepressant effects of physical activity: mediating self-esteem and self-efficacy mechanisms. *Psychology & Health*. 2008;23(3):279–307.

[739]   Lindsay-Smith G, O'Sullivan G, Eime R, *et al.* A mixed methods case study exploring the impact of membership of a multi-activity, multicentre community group on social wellbeing of older adults. *BMC Geriatrics*. 2018;18(1):1–14.

[740]   Glass TA, De Leon CFM, Bassuk SS, *et al.* Social engagement and depressive symptoms in late life: longitudinal findings. *Journal of Aging and Health*. 2006;18(4):604–628.

[741]   Wahlberg T. Cognitive theories and symptomology of autism. In: Wahlberg T, Obiakor F, Burkhardt S, *et al.* (Eds), *Autistic Spectrum Disorders: Educational and Clinical Interventions (Advances in Special Education)*, vol. 14. JAI, An Imprint of Elsevier Science; 2001. p. 301–302.

[742]   Happé FG. Communicative competence and theory of mind in autism: a test of relevance theory. *Cognition*. 1993;48(2):101–119.

[743]   Dawson G, Toth K, Abbott R, *et al.* Early social attention impairments in autism: social orienting, joint attention, and attention to distress. *Developmental Psychology.* 2004;40(2):271.

[744]   Ozonoff S, Pennington BF, Rogers SJ. Executive function deficits in high-functioning autistic individuals: relationship to theory of mind. *Journal of child Psychology and Psychiatry.* 1991;32(7):1081–1105.

[745]   Happé FG. Central coherence and theory of mind in autism: reading homographs in context. *British Journal of Developmental Psychology.* 1997;15(1):1–12.

[746]   Toscano CV, Barros L, Lima AB, *et al.* Neuroinflammation in autism spectrum disorders: exercise as a "pharmacological" tool. *Neuroscience & Biobehavioral Reviews.* 2021;129:63–74.

[747]   Liu X, Wu Q, Zhao W, *et al.* Technology-facilitated diagnosis and treatment of individuals with autism spectrum disorder: an engineering perspective. *Applied Sciences.* 2017;7(10):1051.

[748]   Liu X, Zhao W. Buddy: a virtual life coaching system for children and adolescents with high functioning autism. In: *2017 IEEE 15th Intl Conf on Dependable, Autonomic and Secure Computing, 15th Intl Conf on Pervasive Intelligence and Computing, 3rd Intl Conf on Big Data Intelligence and Computing and Cyber Science and Technology Congress (DASC/PiCom/DataCom/CyberSciTech).* New York, NY: IEEE; 2017. p. 293–298.

[749]   Zhao W, Liu X, Qiu T, *et al.* Virtual avatar-based life coaching for children with autism spectrum disorder. Computer. 2020;53(2):26–34.

[750]   Jarrold W, Mundy P, Gwaltney M, *et al.* Social attention in a virtual public speaking task in higher functioning children with autism. *Autism Research.* 2013;6(5):393–410.

[751]   Lorenzo G, Lledó A, Pomares J, *et al.* Design and application of an immersive virtual reality system to enhance emotional skills for children with autism spectrum disorders. *Computers & Education.* 2016;98:192–205.

[752]   Cheng Y, Huang CL, Yang CS. Using a 3D immersive virtual environment system to enhance social understanding and social skills for children with autism spectrum disorders. *Focus on Autism and Other Developmental Disabilities.* 2015;30(4):222–236.

[753]   Maskey M, Lowry J, Rodgers J, *et al.* Reducing specific phobia/fear in young people with autism spectrum disorders (ASDs) through a virtual reality environment intervention. *PLoS One.* 2014;9(7):e100374.

[754]   Lahiri U, Bekele E, Dohrmann E, *et al.* Design of a virtual reality based adaptive response technology for children with autism. *IEEE Transactions on Neural Systems and Rehabilitation Engineering.* 2013;21(1):55–64.

[755]   Carter EJ, Williams DL, Hodgins JK, *et al.* Are children with autism more responsive to animated characters? A study of interactions with humans and human-controlled avatars. *Journal of Autism and Developmental Disorders.* 2014;44(10):2475–2485.

[756]   Amaral CP, Simões MA, Castelo-Branco MS. Neural signals evoked by stimuli of increasing social scene complexity are detectable at the single-trial level and right lateralized. *PLoS One.* 2015;10(3):e0121970.

[757]   Kuriakose S, Lahiri U. Understanding the psycho-physiological implications of interaction with a virtual reality-based system in adolescents with autism: a feasibility study. *IEEE Transactions on Neural Systems and Rehabilitation Engineering.* 2015;23(4):665–675.

[758]   Lahiri U, Bekele E, Dohrmann E, *et al.* A physiologically informed virtual reality based social communication system for individuals with autism. *Journal of Autism and Developmental Disorders.* 2015;45(4):919–931.

[759]   Yang YD, Allen T, Abdullahi SM, *et al.* Brain responses to biological motion predict treatment outcome in young adults with autism receiving virtual reality social cognition training: preliminary findings. *Behaviour Research and Therapy.* 2017;93:55–66.

[760]   Wallace S, Parsons S, Bailey A. Self-reported sense of presence and responses to social stimuli by adolescents with ASD in a collaborative virtual reality environment. *Journal of Intellectual & Developmental Disability.* 2017;42(2):131–141.

[761]   Forbes PA, Pan X, Hamilton AFdC. Reduced mimicry to virtual reality avatars in Autism Spectrum Disorder. *Journal of Autism and developmental Disorders.* 2016;46(12):3788–3797.

[762]   Didehbani N, Allen T, Kandalaft M, *et al.* Virtual reality social cognition training for children with high functioning autism. *Computers in Human Behavior.* 2016;62:703–711.

[763]   Ke F, Im T. Virtual-reality-based social interaction training for children with high-functioning autism. *The Journal of Educational Research.* 2013;106(6):441–461.

[764]   Wang M, Reid D. Using the virtual reality-cognitive rehabilitation approach to improve contextual processing in children with autism. *The Scientific World Journal.* 2013;2013.

[765]   Miller HL, Bugnariu NL. Level of immersion in virtual environments impacts the ability to assess and teach social skills in autism spectrum disorder. *Cyberpsychology, Behavior, and Social Networking.* 2016;19(4):246–256.

[766]   Gotham K, Risi S, Pickles A, *et al.* The Autism Diagnostic Observation Schedule: revised algorithms for improved diagnostic validity. *Journal of Autism and Developmental Disorders.* 2007;37(4):613–627.

[767]   Bekele E, Zheng Z, Swanson A, *et al.* Understanding how adolescents with autism respond to facial expressions in virtual reality environments. *IEEE Transactions on Visualization and Computer Graphics.* 2013;19(4):711–720.

[768]   Jouen AL, Narzisi A, Xavier J, *et al.* GOLIAH (Gaming Open Library for Intervention in Autism at Home): a 6-month single blind matched controlled exploratory study. *Child and Adolescent Psychiatry and Mental Health.* 2017;11(1):17.

[769] Rutter M, Bailey A, Berument S, *et al. Autism Diagnostic Interview— Revised.* Los Angeles, CA: Western Psychological Services; 2003.

[770] Ehlers S, Gillberg C, Wing L. A screening questionnaire for Asperger syndrome and other high-functioning autism spectrum disorders in school age children. *Journal of Autism and Developmental Disorders.* 1999;29(2):129–141.

[771] Rutter M, Le Couteur A, Lord C, *et al. Autism Diagnostic Interview— Revised.* Los Angeles, CA: Western Psychological Services; 2003.

[772] Constantino JN, Gruber CP. *Social Responsiveness scale: SRS-2.* Torrance, CA: Western Psychological Services; 2012.

[773] Bass MM, Duchowny CA, Llabre MM. The effect of therapeutic horseback riding on social functioning in children with autism. *Journal of Autism and Developmental Disorders.* 2009;39(9):1261–1267.

[774] Gilliam JE. *GARS: Gilliam Autism Rating Scale,* 2nd ed. Austin, TX: Pro-Ed Inc.; 2006.

[775] Ward SC, Whalon K, Rusnak K, *et al.* The association between therapeutic horseback riding and the social communication and sensory reactions of children with autism. *Journal of Autism and Developmental Disorders.* 2013;43(9):2190–2198.

[776] Bahrami F, Movahedi A, Marandi SM, *et al.* Kata techniques training consistently decreases stereotypy in children with autism spectrum disorder. *Research in Developmental Disabilities.* 2012;33(4):1183–1193.

[777] Movahedi A, Bahrami F, Marandi SM, *et al.* Improvement in social dysfunction of children with autism spectrum disorder following long term Kata techniques training. *Research in Autism Spectrum Disorders.* 2013;7(9):1054–1061.

[778] Schopler E, Reichler RJ, Renner BR. *The Childhood Autism Rating Scale (CARS).* Los Angeles, CA: WPS; 2010.

[779] Kern JK, Fletcher CL, Grannemann BD, *et al.* Prospective trial of equine-assisted activities in autism spectrum disorder. *Alternative Therapies in Health and Medicine.* 2011;17(3):14.

[780] Caputo G, Ippolito G, Mazzotta M, *et al.* Effectiveness of a multisystem aquatic therapy for children with autism spectrum disorders. *Journal of Autism and Developmental Disorders.* 2018;48(6):1945–1956.

[781] Mundy P, Novotny S, Swain-Lerro L, *et al.* Joint-attention and the social phenotype of school-aged children with ASD. *Journal of Autism and Developmental Disorders.* 2017;47(5):1423–1435.

[782] Jyoti V, Lahiri U. Human-computer interaction based joint attention cues: implications on functional and physiological measures for children with autism spectrum disorder. *Computers in Human Behavior.* 2020;104:106163.

[783] Sparrow SS, Cicchetti DV. *The Vineland Adaptive Behavior Scales.* Boston, MA: Allyn & Bacon; 1989.

[784] Carter AS, Volkmar FR, Sparrow SS, *et al.* The Vineland Adaptive Behavior Scales: supplementary norms for individuals with autism. *Journal of Autism and Developmental disorders.* 1998;28(4):287–302.

[785] Gabriels RL, Agnew JA, Holt KD, *et al.* Pilot study measuring the effects of therapeutic horseback riding on school-age children and adolescents with autism spectrum disorders. *Research in Autism Spectrum Disorders.* 2012;6(2):578–588.

[786] Battaglia G, Agrò G, Cataldo P, *et al.* Influence of a specific aquatic program on social and gross motor skills in adolescents with autism spectrum disorders: three case reports. *Journal of Functional Morphology and Kinesiology.* 2019;4(2):27.

[787] Dunn W. *The Sensory Profile: Examiner's Manual.* San Antonio, TX: Psychological Corporation; 1999.

[788] Aman MG, Singh NN, Stewart AW, *et al.* The aberrant behavior checklist: a behavior rating scale for the assessment of treatment effects. *American Journal of Mental Deficiency.* 1985;89(5):485–491.

[789] Bruininks RH, Bruininks BD. *BOT2: Bruininks-Oseretsky Test of Motor Proficiency.* Pearson, Assessments; 2005.

[790] Conners CK. *Conners 3.* Toronto: Multi-Health Systems; 2008.

[791] March JS, Parker JD, Sullivan K, *et al.* The Multidimensional Anxiety Scale for Children (MASC): factor structure, reliability, and validity. *Journal of the American Academy of Child & Adolescent Psychiatry.* 1997;36(4):554–565.

[792] Wechsler D. *Wechsler Abbreviated Scale of Intelligence.* New York, NY: The Psychological Corporation: Harcourt Brace & Company; 1999.

[793] Dunn LM, Dunn LM. *Autism Diagnostic Interview—Revised.* Circle Pines, MN: American Guidance Service; 1997.

[794] Wang K, Julier S, Cho Y. Attention-based applications in extended reality to support autism: a systematic review. *IEEE Access.* 2022;10: 15574–15593.

[795] Knudsen EI. Fundamental components of attention. *Annual Review of Neuroscience.* 2007;30(1):57–78.

[796] Meindl JN, Cannella-Malone HI. Initiating and responding to joint attention bids in children with autism: a review of the literature. *Research in Developmental Disabilities.* 2011;32(5):1441–1454.

[797] Chen F, Wang L, Peng G, *et al.* Development and evaluation of a 3-D virtual pronunciation tutor for children with autism spectrum disorders. *PLoS One.* 2019;14(1):e0210858.

[798] Bono V, Narzisi A, Jouen AL, *et al.* GOLIAH: a gaming platform for home-based intervention in autism—principles and design. *Frontiers in Psychiatry.* 2016;7:70.

[799] Heyes C. Causes and consequences of imitation. *Trends in Cognitive Sciences.* 2001;5(6):253–261.

[800] Zhang L, Wade J, Bian D, *et al.* Cognitive load measurement in a virtual reality-based driving system for autism intervention. *IEEE Transactions on Affective Computing.* 2017;8(2):176–189.

[801] Serret S, Hun S, Iakimova G, *et al.* Facing the challenge of teaching emotions to individuals with low- and high-functioning autism using a new Serious game: a pilot study. *Molecular Autism.* 2014;5(1):37.

[802] Bremer E, Crozier M, Lloyd M. A systematic review of the behavioural outcomes following exercise interventions for children and youth with autism spectrum disorder. *Autism*. 2016;20(8):899–915.

[803] Teh EJ, Vijayakumar R, Tan TXJ, *et al.* Effects of physical exercise interventions on stereotyped motor behaviours in children with ASD: a meta-analysis. *Journal of Autism and Developmental Disorders*. 2022;52(7):2934–2957.

[804] Liang X, Li R, Wong SH, *et al.* The effects of exercise interventions on executive functions in children and adolescents with autism spectrum disorder: a systematic review and meta-analysis. *Sports Medicine*. 2021;52:1–14.

[805] Habib K, Montreuil T, Bertone A. Social learning through structured exercise for students with autism spectrum disorders. *Review Journal of Autism and Developmental Disorders*. 2018;5(3):285–293.

[806] Tse AC. Brief report: impact of a physical exercise intervention on emotion regulation and behavioral functioning in children with autism spectrum disorder. *Journal of Autism and Developmental Disorders*. 2020;50(11):4191–4198.

[807] Levinson LJ, Reid G. The effects of exercise intensity on the stereotypic behaviors of individuals with autism. *Adapted Physical Activity Quarterly*. 1993;10(3):255–268.

[808] Nicholson H, Kehle TJ, Bray MA, *et al.* The effects of antecedent physical activity on the academic engagement of children with autism spectrum disorder. *Psychology in the Schools*. 2011;48(2):198–213.

[809] Oriel KN, George CL, Peckus R, *et al.* The effects of aerobic exercise on academic engagement in young children with autism spectrum disorder. *Pediatric Physical Therapy*. 2011;23(2):187–193.

[810] Prupas A, Reid G. Effects of exercise frequency on stereotypic behaviors of children with developmental disabilities. *Education and Training in Mental Retardation and Developmental Disabilities*. 2001;36: 196–206.

[811] Rosenblatt LE, Gorantla S, Torres JA, *et al.* Relaxation response–based yoga improves functioning in young children with autism: a pilot study. *The Journal of Alternative and Complementary Medicine*. 2011;17(11): 1029–1035.

[812] Murphy KL, Hennebach KR. A systematic review of swimming programs for individuals with autism spectrum disorders. *Journal of Disability Studies*. 2020;6(1):26–32.

[813] Pan CY. Effects of water exercise swimming program on aquatic skills and social behaviors in children with autism spectrum disorders. *Autism*. 2010;14(1):9–28.

[814] Merrell KW. *School Social Behavior Scales*, 2nd ed. Eugene, OR: Assessment Intervention Resources; 2002.

[815] Ignácio ZM, da Silva RS, Plissari ME, *et al.* Physical exercise and neuroinflammation in major depressive disorder. *Molecular Neurobiology*. 2019;56(12):8323–8335.

[816] Mota-Pereira J, Silverio J, Carvalho S, *et al.* Moderate exercise improves depression parameters in treatment-resistant patients with major depressive disorder. *Journal of Psychiatric Research.* 2011;45(8):1005–1011.

[817] Zhang K, Zhu Y, Zhu Y, *et al.* Molecular, functional, and structural imaging of major depressive disorder. *Neuroscience Bulletin.* 2016;32(3):273–285.

[818] Belvederi Murri M, Ekkekakis P, Magagnoli M, *et al.* Physical exercise in major depression: reducing the mortality gap while improving clinical outcomes. *Frontiers in Psychiatry.* 2019;9:762.

[819] Cuijpers P, Vogelzangs N, Twisk J, *et al.* Comprehensive meta-analysis of excess mortality in depression in the general community versus patients with specific illnesses. *American Journal of Psychiatry.* 2014;171(4):453–462.

[820] Machado MO, Veronese N, Sanches M, *et al.* The association of depression and all-cause and cause-specific mortality: an umbrella review of systematic reviews and meta-analyses. *BMC Medicine.* 2018;16(1):1–13.

[821] Gourgouvelis J, Yielder P, Clarke ST, *et al.* Exercise leads to better clinical outcomes in those receiving medication plus cognitive behavioral therapy for major depressive disorder. *Frontiers in Psychiatry.* 2018;9:37.

[822] Saeed SA, Cunningham K, Bloch RM. Depression and anxiety disorders: benefits of exercise, yoga, and meditation. *American Family Physician.* 2019;99(10):620–627.

[823] Krogh J, Hjorthøj C, Speyer H, *et al.* Exercise for patients with major depression: a systematic review with meta-analysis and trial sequential analysis. *BMJ Open.* 2017;7(9):e014820.

[824] Morres ID, Hatzigeorgiadis A, Stathi A, *et al.* Aerobic exercise for adult patients with major depressive disorder in mental health services: a systematic review and meta-analysis. *Depression and Anxiety.* 2019;36(1):39–53.

[825] Al-Qahtani AM, Shaikh MAK, Shaikh IA. Exercise as a treatment modality for depression: a narrative review. *Alexandria Journal of Medicine.* 2018;54(4):429–435.

[826] Salehi I, Hosseini SM, Haghighi M, *et al.* Electroconvulsive therapy (ECT) and aerobic exercise training (AET) increased plasma BDNF and ameliorated depressive symptoms in patients suffering from major depressive disorder. *Journal of Psychiatric Research.* 2016;76:1–8.

[827] Doose M, Ziegenbein M, Hoos O, *et al.* Self-selected intensity exercise in the treatment of major depression: a pragmatic RCT. *International Journal of Psychiatry in Clinical Practice.* 2015;19(4):266–275.

[828] Martinsen EW, Medhus A, Sandvik L. Effects of aerobic exercise on depression: a controlled study. *British Medical Journal (Clinical Research Ed).* 1985;291(6488):109.

[829] Beck AT, Ward CH, Mendelson M, *et al.* An inventory for measuring depression. *Archives of General Psychiatry.* 1961;4(6):561–571.

[830] Legrand FD, Neff EM. Efficacy of exercise as an adjunct treatment for clinically depressed inpatients during the initial stages of antidepressant pharmacotherapy: an open randomized controlled trial. *Journal of Affective Disorders.* 2016;191:139–144.

[831]   Lovibond PF, Lovibond SH. The structure of negative emotional states: Comparison of the Depression Anxiety Stress Scales (DASS) with the Beck Depression and Anxiety Inventories. *Behaviour Research and Therapy*. 1995;33(3):335–343.

[832]   Carneiro LS, Fonseca AM, Vieira-Coelho MA, *et al.* Effects of structured exercise and pharmacotherapy vs. pharmacotherapy for adults with depressive symptoms: a randomized clinical trial. *Journal of Psychiatric Research*. 2015;71:48–55.

[833]   Åsberg ME, Perris CE, Schalling DE, *et al.* CPRS: development and applications of a psychiatric rating scale. *Acta Psychiatrica Scandinavica*. 1978;(Suppl 271):69.

[834]   Aas I. Global Assessment of Functioning (GAF): properties and frontier of current knowledge. *Annals of General Psychiatry*. 2010;9(1):1–11.

[835]   Kroenke K, Spitzer RL. The PHQ-9: a new depression diagnostic and severity measure. *Psychiatric Annals*. 2002;32(9):509–515.

[836]   Patten CA, Bronars CA, Vickers Douglas KS, *et al.* Supervised, vigorous intensity exercise intervention for depressed female smokers: a pilot study. *Nicotine & Tobacco Research*. 2016;19(1):77–86.

[837]   Guy W. *ECDEU Assessment Manual for Psychopharmacology*. US: US Department of Health, Education, and Welfare, Public Health Service; 1976.

[838]   Nolen-Hoeksema S, Davis CG. "Thanks for sharing that": ruminators and their social support networks. *Journal of Personality and Social Psychology*. 1999;77(4):801.

[839]   Olson RL, Brush CJ, Ehmann PJ, *et al.* A randomized trial of aerobic exercise on cognitive control in major depression. *Clinical Neurophysiology*. 2017;128(6):903–913.

[840]   Haghighi M, Salehi I, Erfani P, *et al.* Additional ECT increases BDNF-levels in patients suffering from major depressive disorders compared to patients treated with citalopram only. *Journal of Psychiatric Research*. 2013;47(7):908–915.

[841]   Brown RS, Ramirez DE, Taub JM. The prescription of exercise for depression. *The Physician and Sports Medicine*. 1978;6(12):34–45.

[842]   Kessler RC, Chiu WT, Demler O, *et al.* Prevalence, severity, and comorbidity of 12-month DSM-IV disorders in the National Comorbidity Survey Replication. *Archives of General Psychiatry*. 2005;62(6):617–627.

[843]   Forbes D, Creamer M, Phelps A, *et al.* Australian guidelines for the treatment of adults with acute stress disorder and post-traumatic stress disorder. *Australian & New Zealand Journal of Psychiatry*. 2007;41(8):637–648.

[844]   Hegberg NJ, Hayes JP, Hayes SM. Exercise intervention in PTSD: a narrative review and rationale for implementation. *Frontiers in Psychiatry*. 2019;10:133.

[845]   Mizzi AL, McKinnon MC, Becker S. The impact of aerobic exercise on mood symptoms in trauma-exposed young adults: a pilot study. *Frontiers in Behavioral Neuroscience*. 2022;16.

[846]   Goldstein LA, Mehling WE, Metzler TJ, *et al.* Veterans group exercise: a randomized pilot trial of an integrative exercise program for veterans

with posttraumatic stress. *Journal of Affective Disorders*. 2018;227: 345–352.

[847]   Rosenbaum S, Sherrington C, Tiedemann A. Exercise augmentation compared with usual care for post-traumatic stress disorder: a randomized controlled trial. *Acta Psychiatrica Scandinavica*. 2015;131(5):350–359.

[848]   Babson KA, Heinz AJ, Ramirez G, *et al.* The interactive role of exercise and sleep on veteran recovery from symptoms of PTSD. *Mental Health and Physical Activity*. 2015;8:15–20.

[849]   Fetzner MG, Asmundson GJ. Aerobic exercise reduces symptoms of posttraumatic stress disorder: a randomized controlled trial. *Cognitive Behaviour Therapy*. 2015;44(4):301–313.

[850]   LeBouthillier DM, Fetzner MG, Asmundson GJ. Lower cardiorespiratory fitness is associated with greater reduction in PTSD symptoms and anxiety sensitivity following aerobic exercise. *Mental Health and Physical Activity*. 2016;10:33–39.

[851]   Diaz AB, Motta R. The effects of an aerobic exercise program on post-traumatic stress disorder symptom severity in adolescents. *International Journal of Emergency Mental Health*. 2008;39(2):179–187.

[852]   Newman CL, Motta RW. The effects of aerobic exercise on childhood PTSD, anxiety, and depression. *International Journal of Emergency Mental Health*. 2007;9(2):133–158.

[853]   Manger TA, Motta RW. The impact of an exercise program on posttraumatic stress disorder, anxiety, and depression. *International Journal of Emergency Mental Health*. 2005;7(1):49–57.

Index